ELECTRICAL CHARACTERIZATION OF SILICON-ON-INSULATOR MATERIALS AND DEVICES

The Kluwer International Series in Engineering and Computer Science

ELECTRONIC MATERIALS: SCIENCE AND TECHNOLOGY

Series Editor
Harry L. Tuller
Massachusetts Institute of Technology

Other books in the series:

Diamond: Electronic Properties and Applications, Lawrence S. Pan and Don R. Kania, editors
Ion Implantation: Basics to Device Fabrication, Emanuele Rimini, author
Solid State Batteries: Materials Design and Optimization, C. Julien and G. Nazri, authors
Sol-Gel Optics: Processing and Applications, L.C. Klein, editor

---

The Series **ELECTRONIC MATERIALS: Science and Technology** will address the following goals

* Bridge the gap between theory and application.

* Foster and facilitate communication among the materials scientists, electrical engineers, physicists and chemists.

* Provide publication with an interdisciplinary approach in the following topic areas:

* Sensors and Actuators
* Electrically Active Ceramics and Polymers
* Structure-Property-Processing -Performance Correlations in Electronic Materials
* Electrically Active Interfaces
* High T_c Superconducting Materials
* Optoelectronic Materials
* Composite Materials
* Defect Engineering
* Solid State Ionics
* Electronic Materials in Energy Conversion-Solar Cells, High Energy Density Microbatteries, Solid State Fuel Cells, etc.

With the dynamic growth evident in this field and the need to communicate findings between disciplines, this book series will provide a forum for information targeted toward

• Materials Scientists • Electrical Engineers • Physicists • Chemists

ELECTRICAL CHARACTERIZATION OF SILICON-ON-INSULATOR MATERIALS AND DEVICES

by

Sorin Cristoloveanu

Polytechnic Institute of Grenoble

and

Sheng S. Li

University of Florida

KLUWER ACADEMIC PUBLISHERS
Boston / Dordrecht / London

Distributors for North America:
Kluwer Academic Publishers
101 Philip Drive
Assinippi Park
Norwell, Massachusetts 02061 USA

Distributors for all other countries:
Kluwer Academic Publishers Group
Distribution Centre
Post Office Box 322
3300 AH Dordrecht, THE NETHERLANDS

Library of Congress Cataloging-in-Publication Data

Cristoloveanu, Sorin.
Electrical characterization of silicon-on-insulator materials and devices / by Sorin Christoloveanu and Sheng S. Li.
p. cm. -- (The Kluwer international series in engineering and computer science ; 305)
Includes bibliographical references and index.
ISBN 0-7923-9548-4
1. Semiconductors--Design and construction. 2. Silicon-on -insulator technology. 3. Semiconductors--Electric properties.
I. Li, Cheng S., 1938- . II. Title. III. Series: Kluwer international series in engineering and computer science ; SECS 305.
TK7871.85.C75 1995
621.3815'2--dc20 94-43971
CIP

Printed on acid-free paper.

Printed in the United States of America

To our SOI and non–SOI families

Contents

Preface

Silicon on Insulator is more than a technology, more than a job, and more than a venture in microelectronics; it is something different and refreshing in device physics. This book recalls the activity and enthusiasm of our SOI groups. Many contributing students have since then disappeared from the SOI horizon. Some of them believed that SOI was the great love of their scientific lives; others just considered SOI as a fantastic LEGO game for adults. We thank them all for kindly letting us imagine that we were guiding them.

This book was very necessary to many people. SOI engineers will certainly be happy: indeed, if the performance of their SOI components is not always outstanding, they can now safely incriminate the relations given in the book rather than their process. Martine, Günter, and Y.S. Chang can contemplate at last the amount of work they did with the figures. Our SOI accomplices already know how much we borrowed from their expertise and would find it indecent to have their detailed contributions listed. Jean-Pierre and Dimitris incited the book, while sharing their experience in the reliability of floating bodies. Our families and friends now realize the SOI capability of dielectrically isolating us for about two years in a BOX. Our kids encouraged us to start writing. Our wives definitely gave us the courage to stop writing. They had a hard time fighting the symptoms of a rapidly developing SOI allergy.

There is something good about writing a book. At the least, one has an alibi for missing bad TV sessions, escaping gardening, and avoiding losing tennis games. Among the various languages we speak, the least worst common denominator was English. Since each of us has mastered about 2×10^3 words/m^3eV^{-1}, we secretly hoped to double this capital and to achieve something readable by cooperating, speeding on information super-highways, and e-mailing SOI dreams. By chance, Kluwer was not convinced at all by our optimism and generated a lot of tough editing work. To speed up the process, it was decided that words such as *voltage, interface*, and *transistor* will be maintained in the original version without being translated from French to English. A number of mistakes remain. We would be happy if they were to be considered as intentionally inserted in order to test the degree of somnolence of the reader.

Good luck and Bonne chance!

Sorin (Grenoble, France)[1]
Sheng (Gainesville, Florida)[2]

---

[1]Sorin Cristoloveanu is a Director of Research CNRS, currently acting as the Head of the Laboratoire de Physique des Composants à Semiconducteurs, associated with the Institut National Polytechnique de Grenoble and the Centre National de la Recherche Scientifique—LPCS, ENSERG, BP 257, 38016 Grenoble Cedex 1, France.

[2]Sheng S. Li is a Professor at the Department of Electrical Engineering, University of Florida. Gainesville—227 Benton Hall, University of Florida, Gainesville, FL 32611.

ELECTRICAL CHARACTERIZATION OF SILICON-ON-INSULATOR MATERIALS AND DEVICES

Chapter 1

Introduction

Silicon On Insulator (SOI) is a fascinating technology for ultradense integrated circuits. SOI structures were first introduced about 30 years ago for quite a different purpose, which was related to radiation hardness. At that time, it was difficult to imagine that bulk silicon could one day become rather obsolete with respect to ultra-large scale integration (ULSI) requirements. Possibly a victim of its unprecedented rapid development, bulk-silicon technology is now facing very fundamental limitations. At this point, the potential of SOI technologies has become very attractive. SOI devices are more tolerant to the constraints of scaling-down rules, and they perform superbly.

1.1 Why SOI?

In an MOS transistor, only the very top region (0.1–0.2 μm thick) of the silicon wafer is actually useful for electron transport. The inactive volume, more than 99.9% of the wafer, induces only undesirable, parasitic effects. SOI structures emerged from the idea of separating or isolating the active device overlay from the detrimental influence of the silicon substrate. The thickness of SOI films can be precisely adjusted to meet the capriciousness of any performing devices. Figures 1.1a and 1.1b reproduce the legendary picture that was unsparingly used to advertise the merits of SOI.

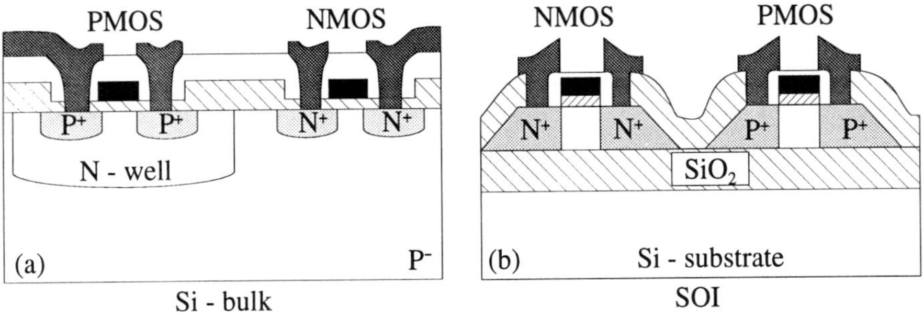

Figure 1.1 Schematic configuration of CMOS transistors in bulk-silicon and SOI wafers.

Dielectric isolation. SOI circuits consist of single-device islands, dielectrically isolated from each other and from the underlying substrate (Fig. 1.1b). The interdevice distance is far more shrinkable in SOI. The critical limitation of bulk silicon for VLSI circuits comes from the inherent proximity of diffused regions that belong to adjacent components. Latch-up means unintentional activation of parasitic devices. In Figure 1.1a, a NPN bipolar transistor is formed by the N^+ drain of the n-channel MOSFET, the p-type substrate, and the n-well. Similarly, a PNP transistor subsists between the substrate, the well, and the source of the p-channel MOSFET. The superposition of PNP and NPN transistors, which share the same diffused regions, is nothing but a thyristor whose turn-on causes uncontrollable high currents and circuit failure. Sophisticated schemes of trench isolation are necessary in bulk silicon. By contrast, the word *latch-up* is simply missing from the SOI dictionary.

Vertical junctions. In regular SOI films, the source and drain regions extend to the insulator, and only their lateral sides serve as junctions. The surface of such a vertical junction is much smaller than in bulk silicon. This smaller surface yields a substantial reduction in parasitic capacitances, and hence in commutation delay and dynamic power dissipation. In other words, for a predefined power consumption, much denser and faster circuits can be integrated on SOI wafers.

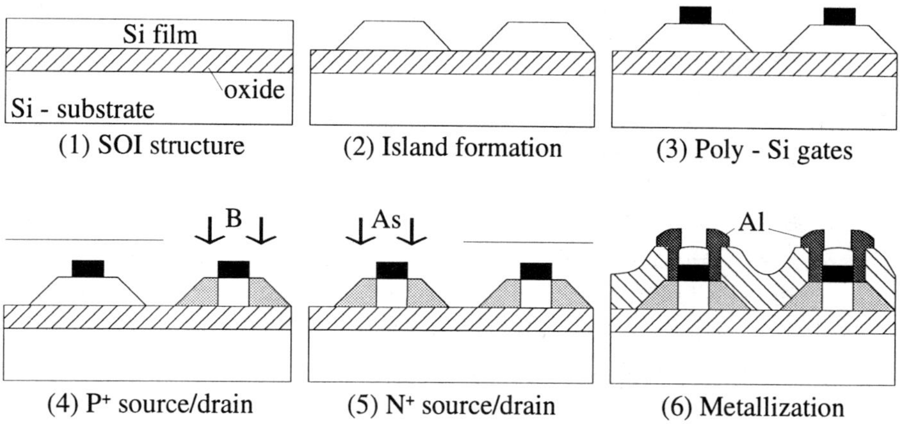

Figure 1.2 The generic sequence for processing CMOS/SOI integrated circuits.

Circuit design and processing. According to the sequence shown in Figure 1.2, the fabrication of CMOS circuits on SOI is simpler than in bulk silicon. The absence of wells and interdevice trenches reduces the number of processing steps and offers additional design flexibility. The location of individual device islands does not have to obey any isolation constraint and is only aimed at minimizing packing density and interconnections. Moreover, the particular configuration of SOI structures enables the fabrication of many innovative devices. The basic advantages of SOI—less processing, enhanced performance, more chips on the wafer—are key sources of profit. Economical evaluations are very positive about changing the mass production of ULSI circuits from bulk silicon to SOI.

Short-channel effects. In small-geometry MOS transistors, the depletion zones induced by the source and drain junctions become relatively significant and impede the gate control over the whole space-charge region. Annoying short-channel effects originate from "charge sharing" between gate and junctions: threshold voltage roll-off, degradation of the subthreshold slope, drain-induced barrier lowering, punch-through, bipolar transistor action, etc. SOI devices are more immune to short-channel effects. Basically, the extension of source/drain depletion regions is restricted by the junction size and by the dual-gate control (via front-gate and silicon substrate) of the surface potential.

Reliability. The primary motivation for developing SOI technologies was their excellent tolerance of *transient* radiation effects. Incoming particles generate electron-hole pairs in proportion to the device volume. Resulting photocurrents act as leakage currents, causing charge collection and soft errors. Logic upsets are dramatically reduced in SOI, since the volume exposed to carrier generation is 2–3 orders of magnitude smaller than in bulk Si. Moreover, the problem of radiation-activated latch-up paths makes no sense in SOI. The *permanent* damage induced by the cumulative dose of ionizing radiations is a more complex aspect discussed in Chapter 10. Another reliability issue, addressed in the last chapter of this book, is the vulnerability of submicron devices to hot-carrier injection. The basic argument, which is regarded as a promise for enhanced immunity to hot carrier damage, is the reduction of the electric-field peak in SOI.

The poet would conclude that there is something about SOI that has the aura of inevitability.

1.2 Why Not Yet SOI?

The enthusiastic enumeration of SOI merits and potential advantages over bulk silicon has not been enough to stop the fabulous progress of silicon-based microelectronics. For each new generation of components, bulk-silicon technology was able to overcome its limitations and to progress beyond the most optimistic expectations. It is primarily this success that has delayed the entry of SOI into the commercial arena. Anticipated initially for 1 μm technology, SOI was postponed to half-micron, and more recently to subquarter-micron devices. It is fair to point out that bulk-silicon advocates used to claim ironically that SOI would be more and more suitable as devices became smaller and smaller, and that SOI would certainly reach the stage of a perfect technology when nothing was left. Nevertheless, most of the major semiconductor companies have asked their specialists to make the pilgrimage to the SOI world and to take a careful look at the future.

Several niches are envisaged for SOI: radiation-hard circuits, 3–dimensional integration, sensors, power, high temperature, quantum, and exotic devices. However, the credibility of SOI depends on the extent of applications, and in this respect, ultrasmall, low-power CMOS circuits are the best prospect—although the potential of bulk silicon is far from being exhausted.

Besides the health of bulk-silicon technology, other factors have thus far inhibited the blooming of SOI. SOI substrates are not fully optimized in terms of uniformity, cost, and availability. The SOI version of the chicken/egg dilemma is that mass production cannot start without massive wafer supply, and nobody can afford huge investment in wafer fabrication without the security of mass production.

An unexpected drawback of the impressive SOI achievements is that parasitic effects, which were in the past obscured by the insufficient crystal quality, become more severe. The control of specific effects in SOI transistors, such as floating body, dynamic transients, and interface coupling, is still a challenging issue. One type of floating-body mechanism, quite detrimental for the progress of the SOI technology, was Gorbatchov's effect. Slowing down the defense involvement in SOI has reopened many questions about the SOI myth.

After a long period of incubation, SOI technology now is in a period of rapid and successful development. Impressive devices with superior performance have been demonstrated. However, SOI has yet to overcome the reputation of being an expensive material for confidential and sophisticated circuits. This is why the SOI community watches impatiently for signs of weakness in the bulk-silicon giant.

1.3 Why an SOI Book?

Attainment of the threshold of SOI technology for commercial applications appears to be imminent. A major condition is the degree of understanding of SOI material properties, fabrication techniques, and device operation. It seems to be the right time to have a detailed book that describes various characterization techniques as well as the state of the art in SOI physics and technology.

Adequate characterization techniques, enabling us to overcome the difficulties induced by the thinness of the film and by the stacked interfaces, are a prerequisite for the further optimization of SOI technologies.

Since the quality of SOI structures has great impact on the performance of integrated circuits, reciprocal device-based characterization is very suitable for exploring the properties of the starting material as well as the process-induced modifications. A key monitor for SOI technologies is the concentration of defects in the film and at the front/back interfaces.

As shown in Chapter 2, it is the pedigree of each SOI material that defines the nature and the amount of the defects. The purpose of Chapter 3 is to give a flavor of the variety of SOI components. The discussion of characterization techniques will proceed systematically from the wafer to the device. The methods for inspecting the quality of as-grown wafers will be presented in Chapter 4, whereas Chapter 5 will review the more or less conventional transport measurements in unprocessed SOI samples.

The MOS capacitance and conductance techniques are part of the history of bulk silicon. Their adaptation to SOI will be discussed in Chapter 6. Nevertheless, these methods are far more difficult and fragile in SOI, where both the measurement and the parameter extraction are complicated by the contributions of several interfaces. Current-based techniques are more appropriate for thin films: the current is easily measurable even in small area devices, the experiment can be conducted separately from the front or the back interface, and simple single-interface models often are sufficient for data analysis. The use of diodes and transistors for characterization purposes will be addressed in Chapters 7, 8, and 9. In the last chapter, the characterization techniques will serve to analyze the reliability of SOI MOSFETs in terms of vulnerability of the front-gate oxide and synthesized buried oxide to radiation effects and hot-carrier injection.

Finally, it is the purpose of this book to provide SOI colleagues with clear technical information that can be readily utilized in their laboratories. In the meantime, we hope that non-SOI readers will have their curiosity satisfied by a comprehensive overview of our world. They will find that SOI is more than a refreshing hobby for physicists who became tired with bulk silicon, thermal oxides, and single-gate MOS transistors. For the purpose of clarity, the selection of references and mathematical manipulations was very conservative. Rather than making an SOI telephone book, we found it preferable to miss a number of valuable citations.

Chapter 2

Methods of Forming SOI Wafers

SOI technologies can in general be divided into two groups. In the first, the silicon film is deposited directly onto an insulating substrate, as is the case for silicon on sapphire (SOS) and silicon on zirconia (SOZ). In the second group, a thin insulating layer is used to separate the active semiconductor layer from the semiconductor substrate (i.e., the SIS structure). Methods of forming SIS structures include Separation by IMplantation of OXygen (SIMOX), Zone-Melting Recrystallization (ZMR) of polysilicon, Epitaxial Lateral Overgrowth (ELO), Full Isolation by Porous Oxidized Silicon (FIPOS), and Wafer Bonding (WB).

In this chapter, the SOI technologies will be reviewed in terms of material synthesis and related mechanisms, processing variants, typical advantages, and drawbacks. Without entering into tedious technical details, attention will be given to the processing parameters that impact on the quality and properties of SOI structures. The electrical image of each SOI material will be revealed by putting together the fragments collected from the various characterization techniques that are the subject of the forthcoming chapters.

2.1 SIMOX

Among the SOI technologies, SIMOX is considered to be the most advanced and promising for high-density CMOS circuits. What just a few years ago was regarded as a crude and exotic method of synthesis has

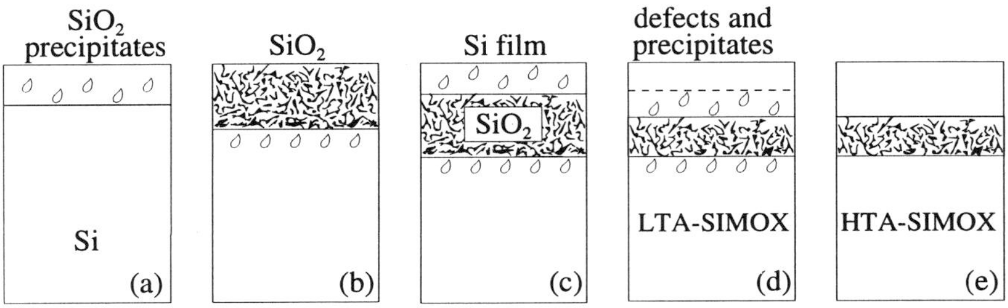

Figure 2.1 Curriculum vitae of SIMOX: (a) oxygen precipitation, (b) formation of superficial oxide, (c) buried-oxide synthesis, (d) "low"-temperature annealing, (e) "high"-temperature annealing.

rapidly evolved into an exciting and rather mature process.

2.1.1 SIMOX Synthesis

The buried oxide is synthesized by internal oxidation during the deep implantation of oxygen ions into silicon. A postimplantation annealing at high temperature is necessary to recover the crystalline quality of the Si overlay. Figure 2.1 illustrates the milestones in the development of SIMOX structures:

- Shallow implantation of oxygen into Si wafers leads to oxygen precipitation (Fig. 2.1a).
- High-dose implantation ($\sim 10^{18}\,O^+cm^{-2}$) forms a stoechiometric SiO_2 layer at the wafer surface (Fig. 2.2b), which demonstrates the feasibility of SIMOX.[1]
- Deep implantation at high energy ($\geq 150keV$) allows the synthesis of a buried oxide 0.2–0.4 μm thick underneath a thin Si film (Fig. 2.1c). The first CMOS circuits were successfully fabricated by Izumi et al. in 1978.[2]
- Annealing at relatively low temperature (LTA: 1150–1200 °C) results in an inhomogeneous Si film (Fig. 2.1d) in which only the upper 0.1 μm-thick region stands as a good-quality crystal. The bottom interfacial region is still greatly disordered and contains many defects. Its composition (mono- and polycrystalline, oxide

precipitates, etc.) depends upon the implantation temperature and dose.[3]

- High-temperature annealing (HTA: 1250–1405 °C) improves and simplifies the vertical SOI structure (Fig. 2.1e): the high-quality device-grade region extends over the whole film, and the interfaces become sharp.[4]

The rapid optimization of SIMOX in the mid-1980s was made possible by the enthusiastic involvement of several teams. In particular, those led by Hemment, Stoemenos, and Bruel developed the understanding of the fundamental mechanisms of implantation and annealing.

During implantation, small oxide precipitates form and tend to coalesce. The increase in the molar volume needed for conversion of silicon into SiO_2 is accommodated by the emission of interstitial Si atoms. The energy needed to break the silicon bonds (5 eV) is naturally provided by the ion beam. This is the major difference from thermal oxides where such an energy is not available and the oxidation proceeds from the surface to the bulk of the wafer. The Si interstitials migrate and reconstruct the surface, which alleviates the erosion effect of incoming oxygen ions.

In principle, the emission and migration of silicon interstitials is an ideal condition for strain-free oxidation. Unfortunately, the path for migration to the surface becomes more difficult as blocking SiO_2 precipitates form. Silicon supersaturation occurs, which in turn slows down the oxidation rate, increases the stress and eventually results in oxide densification. The oxidation moves towards the surface where the migration is easier, leaving behind oxygen deficiency and unoxidized silicon, which materialize in Si islands at the bottom of the buried oxide (BOX). In the meantime, small dislocations and overlapping stacking faults develop in the film close to the interface. It is clear that the use of thick capping layers for implantation or annealing inhibits the migration of Si interstitials and therefore enhances the defect generation.[5]

During annealing, a cascade-like thermodynamic process takes place: the smaller oxide precipitates subsisting in the Si film dissolve while larger precipitates are being formed.[6] This gradually leads to the pumping of oxygen from the film towards the buried oxide, which acts as an attracting precipitate of infinite radius. As a final consequence, the residual concentration of oxygen in the overlay drops below $10^{18}\,cm^{-3}$, the interfaces become sharper, and most of the film defects are eliminated (they are pinned by the growing SiO_2 precipitates and annihilated at the Si–BOX interface).

The implant parameters, mainly the energy, dose, and temperature, can be used to improve and also to *engineer* the SIMOX structure and interface properties. The lower the dose, the thinner the buried oxide (Fig. 2.2a), the thicker the film, and the smaller the amount of implantation-induced damage. Buried oxides as thin as 90 nm are obtained with 4×0^{17} O^+/cm^2 dose.[7, 8] Decreasing the energy makes the silicon overlay thinner (Fig. 2.2b). Alternatively, thin films can be produced by sacrificial oxidation or by implantation through a capping oxide. Increasing the implantation temperature from 77 K to 650 °C changes the morphology of the film from amorphous to polycrystalline and improves the back interface quality. The wafer temperature is adjusted by beam heating or lamp irradiation. Temperatures in excess of 650 °C are unsuitable due to the overwhelming effect of surface sputtering.

High-current implanters (150–200 mA) have been specially designed to produce large-diameter SIMOX wafers with excellent purity and homogeneity in a reasonable short time (a few hours per batch). The typical implantation conditions of SIMOX wafers, commercially available from IBIS and SOITEC, are 1.8×10^{18} O^+/cm^2, 190 keV, and 650 °C. The thicknesses of the silicon overlay and buried oxide are 210 nm and 380 nm, respectively.

The increase in the annealing temperature and the resulting gain in quality represent the actual threshold of SIMOX technology.[4] High-temperature annealing, up to 1360 °C, is performed in polysilicon furnaces. A maximum temperature of 1405 °C was reached by focusing lamp irradiation on the back of the wafer to provoke the partial melting of the rear surface (1412 °C).[9] The recrystallization is a combined and nonlinear effect of annealing temperature, duration, and temperature ramping. Although the ideal conditions are not fully specified yet, it is admitted that very good quality SIMOX needs annealing at about 1320 °C, for six hours, in argon ambient containing 1% of oxygen. Annealing in nitrogen has been abandoned because background contamination arises.

A number of SIMOX variants have been explored. The implantation of nitrogen (SIMON) instead of oxygen results in sharper interfaces after LTA, but the level of donor autodoping is unacceptable. Mixed implantations of oxygen and nitrogen or oxygen and carbon were tested with the aims of hardening the BOX and attenuating the electrical activity of oxygen donors. The best-quality SIMOX material so far is prepared

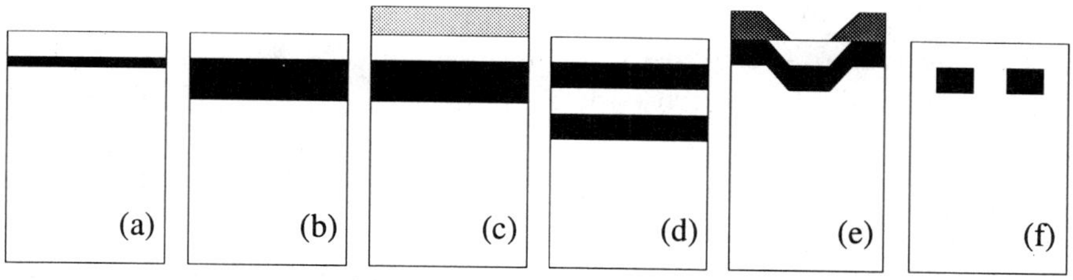

Figure 2.2 Alternative SIMOX structures: (a) thin BOX (lower dose), (b) thin silicon film (lower energy), (c) thick film completed by epitaxy, (d) double SIMOX, (e) isolated single-transistor islands, (f) interrupted BOX.

by a multistep process composed of sequential implants of low doses (0.3–$0.6 \times 10^{18}\, O^+/cm^2$) and intermittent HTA anneals.

A recent development consists of gradually reducing the energy of the successive implantation steps in order to fully oxidize the oxygen-rich transition region located above the BOX.[10] A multiple SIMOX process is effective in reducing the density of dislocations and precipitates in the film and of Si islands in the BOX.

Further alternatives for SIMOX structures are shown in Figure 2.2. Thick films (1–2 μm), requested for both bipolar and bulk compatible CMOS processes, are formed by epitaxial overgrowth (Fig. 2.2c). Two sequential oxygen implants at high and low energies are used to fabricate a double SIMOX (Fig. 2.2d).[11] The Si layer sandwiched between the two distinct buried oxides may serve for electrical and optical interconnections or for innovating devices.

Totally isolated Si islands (single-transistor islands, Fig. 2.2e) were processed by masked implantation through a patterned capping oxide: underneath the cap, the BOX reaches the surface, whereas in the window region it is situated at normal depth.[12] Similarly, interrupted BOX (Fig. 2.2f) was processed to preserve some of the SOI advantages while providing a substrate contact to the film.

Emphasis is given to the fabrication of large SIMOX wafers, with ultrathin Si film and BOX, for ULSI CMOS circuits. Heat dissipation and short-channel effects are improved with thin oxides. The lower oxygen dose needed for these structures implies a considerable gain in processing time and wafer cost.

2.1.2 Typical Defects and Electrical Properties

Most of the available diagnostic techniques, including microscopy (TEM, XTEM, AFM, etc.), spectroscopy and spectrometry (SIMS, RBS, XPS, ESR, etc.), and optical, physical, chemical, and electrical analyses, have been applied to characterize SIMOX structures.[13] We will briefly review the properties of the film, oxide, and interfaces.

In typical SIMOX, the *silicon film* is stress-free. Raman scattering measurements reveal that the compressive stress, developed during implantation via oxide precipitation, decreases drastically with increasing anneal temperature.[14] The dominant defects observed in SIMOX are the threading dislocations. They are introduced during high-temperature annealing when three-dimensional matrix reconstruction occurs. The dislocations accommodate the misorientation of neighboring Si sites.[5] Their density, as large as 10^8–$10^{10}\,cm^{-2}$ in LTA SIMOX, is reduced to 10^5–$10^6\,cm^{-2}$ in standard HTA SIMOX (annealed at 1320 °C) and even further down to 10^3–$10^4\,cm^{-2}$ in multi-implanted material.

Small stacking faults with a variety of shapes are frequently observed at the bottom of the Si overlay. They appear during BOX formation (stress effect) and/or annealing (Si redistribution effect). Additional stacking fault generation is the consequence of the sacrificial oxidation, which is primarily used to adjust the film thickness to CMOS process constraints.[15, 16] Since the film is packed between two oxides, the Si interstitials cannot escape, so the density of stacking faults exceeds $10^6\,cm^{-2}$.

Hole traps with activation energy of 0.44 eV and 0.63 eV above the valence band have been identified by DLTS in LTA SIMOX.[17] Film contamination from implanter, furnace, or capping layers—a primary concern in early days—has been drastically alleviated (less than $10^{15}\,cm^{-3}$ for metals and $10^{18}\,cm^{-3}$ for carbon).

An intrinsic source of contamination of SOI layers, especially pronounced in SIMOX, comes from subsisting oxygen atoms. *Thermal donors* (TDs) and *new donors* (NDs) are activated by short anneals at 450–550 °C and 750 °C, respectively.[18] The origin, kinetics, and activation energies of oxygen donors are still under consideration in bulk Si. Even more complex is the case of SIMOX in which the donor signature may vary with the conditions of synthesis (see also Fig. 5.6).[19] Thermal donors are double donors resulting from the formation of metastable oxygen–silicon aggregates. During annealing, the size of the cluster increases, and the energy levels become shallower. This explains the large variety of thermal donors—more than 10 species have been identified.

New donors are presumed to originate from positively charged states that cause band bending at the interface between small SiO_x precipitates and the silicon matrix. This scenario implies that small accumulation regions surround the precipitates and that electron transport, dominated by interface scattering, takes place by percolation.[19] Most new donors are situated in a narrow region close to the BOX. Above 750 °C, the precipitates dissolve, causing the extinction of new donors. The guideline for limiting the donor activation is to reduce the concentration of oxygen N_{ox}: the initial generation rate of TDs is indeed proportional to N_{ox}^4, and the total number of TDs is proportional to N_{ox}^3.

Although the starting silicon wafer is p-type ($\sim$ 20 Ωcm), SIMOX usually exhibits n-type conduction. This doping conversion is probably related to the formation of oxygen donors. Our Hall effect measurements indicate that the minimum residual doping level (also called background doping or autodoping) of commercially available SIMOX wafers is low enough ($N_D \simeq 2$–5×10^{15} cm^{-3}).[1]

Typical Hall mobilities, rated from 800 to 1200 cm^{2}V^{-1}s^{-1} for electrons, are among the best ever measured on SOI and are not far from the ideal value in bulk silicon (1400 cm^{2}V^{-1}s^{-1}). The carrier scattering mechanisms are also comparable, whatever the temperature, with the case of bulk Si.[20] This demonstrates the overall excellent quality of SIMOX crystal as well as its in-depth homogeneity. By contrast, early LTA–SIMOX showed a distinctive two-layer behavior: the bottom region of the film had a very poor carrier mobility dominated by crystalline defects.[21]

The minority carrier lifetime is more sensitive to processing conditions than the mobility. It is very poor (10–100 ns) in LTA SIMOX, but increases by three orders of magnitude (10–150 μs) after annealing above 1300 °C.[22] Multi-implanted SIMOX always provides the best set of electrical parameters.

No obvious electrical activation of dislocations is visible for concentrations below 10^6 cm^{-2}. Only above 5×10^6 cm^{-2} are the carrier mobility and lifetime slightly degraded. The presence of dislocations has been incriminated to explain more subtle effects:[19] some untypical low-

[1]Notice that the manufacturers' specifications ($\simeq 10^{15}$ cm^{-3}, p-type) refer only to the starting Si wafer.

frequency noise behavior, decoration with metal impurities leading to a shortage in bipolar transistors, some weaknesses in tolerating radiation effects, sporadic failure of gate oxides to breakdown tests, etc.

An accelerated diffusion of dopant impurities (in particular, arsenic) has been noted in thin films. The pileup of inactive dopants near the upper interface of the film, which was reported for early SIMOX, is no longer a problem.[23] The thermal activation of dopant impurities follows bulk-silicon laws.

Due to its particular method of synthesis, namely, the implantation-induced emission of silicon interstitials, the *buried oxide* attracts much interest. In spite of the fact that ellipsometry does not reveal unusual refractive index or optical properties, the only certitude so far is that the magic BOX is definitely not a thermal oxide. It is Si-rich, which would explain the high density of electron traps (strained Si–Si bonds) and E' centers (acting as hole traps).[5, 24]

Slow traps have been equally observed. In bulk silicon, the traps are easily eliminated by a brief anneal at 1000 °C in oxygen ambient. The analogous procedure in SIMOX consists in overcoming the screening effect of the Si film by implanting an additional low dose of oxygen and annealing at 1000 °C for one hour in argon ambient.[10] The densified nature of the buried oxide is proved directly by densimetry and indirectly by a 20% lower etch rate than in thermal oxides.[5]

The breakdown voltage in good-quality buried oxides exceeds 300 V or 8 MV/cm. Localized defect conduction may be superimposed on the normal bulk-oxide conduction.[25] Silicon pipes in the BOX lead to a high-current ohmic conduction. For the case of nontraversing defects (Si–Si bonds in silicon-rich regions), the current is smaller, but a superlinear I–V characteristic develops before breakdown. Crystalline silicon inclusions and islands may be encountered at the bottom of the oxide, reducing its effective thickness and hardness.

Copper and several other metal impurities (Fe, Ti, Cr, etc.) were detected during the formation of SIMOX structures.[26] The metal contamination that results from the interaction of high-flux oxygen beam with the implanter was dramatically reduced by coating the implanter parts with silicon.

Postimplantation techniques have been developed for gettering subsisting metal impurities. Annealing causes the contaminants to diffuse from the surface of the SIMOX film, through the buried oxide, into the Si substrate. The implantation-induced lattice defects, located just underneath the buried oxide, serve as effective gettering sinks for Cu and Ni.[27]

Lateral fluctuations of the BOX thickness are below $\pm 10\,\text{nm}$. Other interesting aspects were reported: mobile charges due to alkali ions, electron trapping into Si islands, oxide polarization effects,[28] etc. Oxide pipes that cause leakage paths are accidentally formed due to the masking effect of particles during implantation. Transition regions with inhomogeneous and higher refractive index may exist at each side of the BOX.[29] Fortunately, these transition regions shrink with increasing anneal temperature.

In regular-quality SIMOX wafers, the buried-oxide *interfaces* are sharp and uniform. The density of traps and fixed charges at the upper interface of the buried oxide is typically in the range of $D_{it_2} \simeq 0.5$–$2\times 10^{11}\,\text{cm}^{-2}\text{eV}^{-1}$. It is definitely larger than at the interface between film and gate oxide ($D_{it_1} \simeq 10^{10}\,\text{cm}^{-2}\text{eV}^{-1}$, as in bulk silicon), but small enough not to adversely affect the performance of integrated circuits. As a matter of fact, the carrier mobility at this interface reaches 80%–95% of the front-channel mobility.

In order to fully realize the technology achievements, it is worth remembering that, in LTA SIMOX, densities of upper BOX interface traps in excess of 10^{12}–$10^{13}\,\text{cm}^{-2}\text{eV}^{-1}$ and back-channel mobilities as low as 1–$10\,\text{cm}^2\text{V}^{-1}\text{s}^{-1}$ have been reported.[22] At the opposite side of the buried oxide, the implantation-induced damage of the silicon crystal is attenuated and, therefore, the defect density is always smaller than at the upper interface.

In conclusion, although the SIMOX synthesis may appear to be a seemingly destructive process, it does result in device-grade SOI structures. The thin silicon layer is wafer-scale monocrystal with high quality and excellent electrical properties. The interface sharpness, the homogeneity of the buried oxide, and the overall flexibility of engineering the structure represent sound arguments that promote SIMOX as the most suitable technology for thin-film applications.

Most SOI devices have first been demonstrated on SIMOX and will therefore be revisited in Chapter 3. As a matter of fact, it is the successful and rapid optimization of SIMOX that has inhibited the development of other SOI contenders (ZMR, FIPOS, SOZ, etc.), except for wafer bonding.

2.2 Wafer Bonding

The wafer-bonding (WB) technique provides undamaged crystal quality and more flexibility than SIMOX for both the SOI film and the buried-oxide layer. There are three basic steps required for the WB process:[30] (1) mating two silicon wafers together at room temperature, (2) annealing the bonded wafers at temperatures above 800 °C for several hours to increase bonding strength, and (3) thinning down one of the two wafers to a proper thickness by grinding and polishing and/or etching. Figure 2.3 shows the processing of a 2 μm or thicker SOI bonded wafer.[31] In the first two steps, bonding interfaces have to be homogeneous without unbonded areas or bubbles, and the bonding strength needs to be high enough to maintain well-bonded interfaces. In the third step, individual thinning processes have to be developed in response to the needs of each specific SOI application.

In the conventional polishing technology, the thinning of SOI layers down to $2 \pm 0.5\,\mu$m can be easily realized, which satisfies the requirement of bipolar devices and power MOS devices. For dielectric isolation wafers, protruded oxide layers stop polishing selectively, and the desired dielectrically isolated areas of high-quality silicon can be obtained.

The main drawback of the wafer bonding technique is its difficulty in producing extremely thin, uniform Si film via the conventional polishing technique. Therefore, the third step is becoming more important for manufacturing ultrathin Si films with good uniformity. Recently, a plasma-assisted chemical etching (PACE) method for producing extremely thin, uniform SOI films has been developed by Mumola et al.[32] The process incorporates a high-speed, high-density measurement and mapping of film thickness, which is used to guide the localized plasma removal of silicon. Using this computer-controlled process, silicon film thicknesses less than 100 nm have been achieved with uniformity better than 10% across the wafer.

Another approach to attain ultrathin SOI layers (i.e., film thickness less than 0.1 μm) on bonded wafers is by combining the SIMOX and bonded-wafer technologies. The complementary nature of the two SOI technologies suggests the possibility of transferring the SIMOX layer on a thermally oxidized Si wafer by using the SIMOX-wafer-bonding (SWB) method.[33]

The process steps for SWB are similar to those shown in Figure 2.3. In this case, a SIMOX wafer and an oxidized Si wafer are placed face-to-

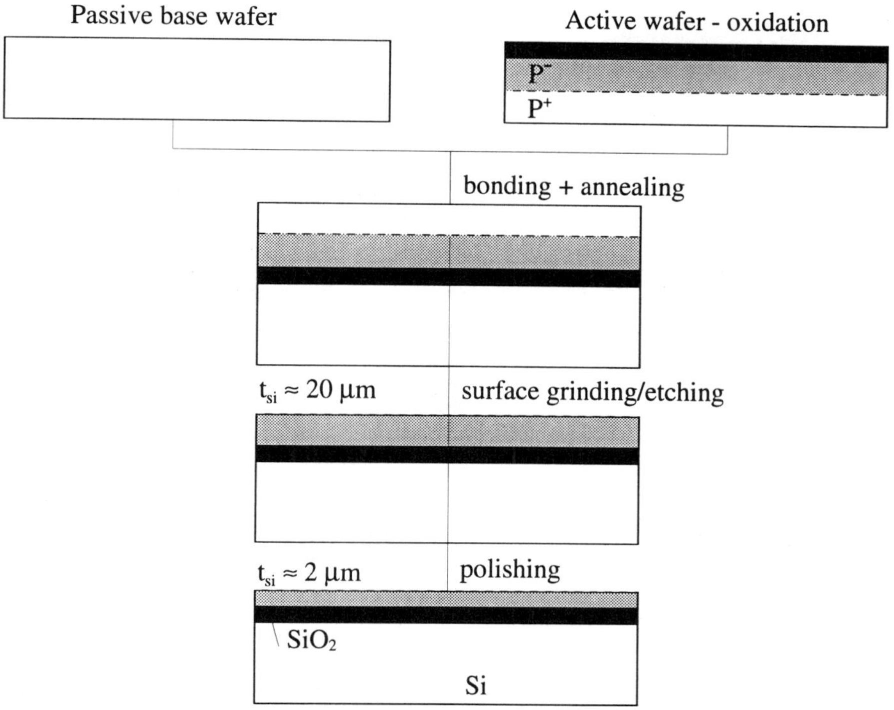

Figure 2.3 The process steps for a $2\,\mu m$ or thicker SOI bonded wafer fabri-. cation (after Abe et al.[31])

face with a gap of about 1.5 mm. After cleaning and drying the surfaces, the two wafers are brought into contact at room temperature, followed by annealing at temperatures above 960 °C. The annealed wafer pair is then thinned in KOH solution to etch the Si substrate of the SIMOX wafer. Finally, the buried oxide, which served as a natural and ideal etch-stop layer, is removed by a buffer oxide etching solution. This leaves the silicon film of the original SIMOX wafer on the top of the thermal oxide of the oxidized wafer.

Although rather complicated processes are involved, SWB retains the advantages of the two key SOI technologies, namely, the uniformity of ultrathin silicon films (SIMOX) and the thermal buried oxide of variable thickness (WB). However, the costs of WB and SIMOX processes are also combined.

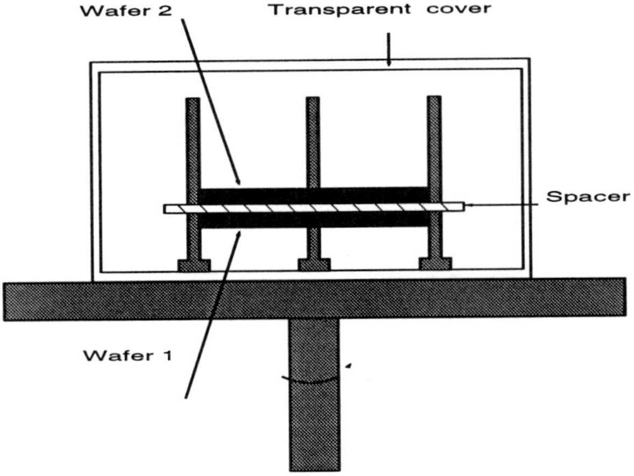

Figure 2.4 The front view of a microcleanroom setup for achieving particle-free bonded interfaces (after Mitani et al.[35]).

2.2.1 Bonding Mechanisms

In order to homogeneously mate two wafers together at room temperature, there are four requirements regarding the wafer and process conditions. These include (1) hydrophilic silicon surfaces, (2) sufficiently flat silicon surfaces, (3) particle-free surfaces and environments, and (4) no air trapping during bonding. Otherwise, wafers are sometimes unable to stick together or, even if they are bonded partially, unbonded areas or bubbles appear at the bonding interface. These bubbles, referred to as *voids*, are actually not empty but contain gas under high pressure.

Hydrophilic surfaces are thought to allow the initial bonding (at room temperature without external forces) due to van der Waal forces between adsorbed monolayers of water molecules. Thus, in combination with the realization of particle-free surfaces, a procedure of wet chemical cleaning prior to bonding has to be carefully considered. For effective particle prevention, the initial bonding has to be performed in a clean room of class 1 or better, right after chemical cleaning. Trapping of air can be prevented by propagating the bonding area radially from just one chosen contact point.

Among the above four requirements, the one that causes the most problems is that of particle-free bonding. To cope with this difficulty, a microcleanroom setup has been developed recently by Mitani et al.[35] as shown in Figure 2.4. This setup allows wafer bonding without bubbles in a normal noncleanroom laboratory environment by flushing deionized water to remove the local particles between the two silicon wafers that are 1 mm apart. Particle-free bonded wafers with a diameter up to 8 inches have successfully been fabricated.

The bonding strength is usually low ($\sim$ 70 erg/cm^2) after initial room-temperature bonding. In order to increase the bonding strength and to keep the wafers tightly bonded during thinning down and device processing, the bonded wafers are annealed at 800 °C or a higher temperature. The bonding strength increases monotonically with temperature and becomes more pronounced above 600 °C. It is not modified by the surrounding atmosphere such as N_2 or O_2 during annealing, nor does it depend on annealing time. In general, the bonding strength of the Si/SiO_2 interface is found to be much lower than that of the Si/Si interface, but higher than that of SiO_2/SiO_2 interfaces. For example, the bonding strength of the Si/Si interface annealed at 700 °C is as strong as the bonding strength of Si/SiO_2 interfaces at 1100 °C annealing.[31] The quartz cohesion (2200 erg/cm^2) was achieved after annealing above 1250 °C.[34]

Another important issue that needs to be discussed here is the formation of bubbles at the bonded-wafer interfaces. In contrast to the conventional opinion that temperature-dependent bubbles are directly caused by water, Mitani et al.[35] have recently demonstrated that the amount of bubble generation corresponds to the contamination level of hydrocarbons. They suggest that the desorption of the hydrocarbons contaminating the surfaces basically opens the bonded interface and forms bubbles, and that the bubbles disappear when the hydrocarbons dissolve into silicon bulk during high-temperature annealing. The observation of Bengtsson et al.[36] that temperature-dependent bubbles also form if wafers with hydrophobic surfaces are bonded supports the concept that interface water is not the primary cause of bubbles. In general, bubble formation can be prevented by prebonding annealing at 600–800 °C.

2.2.2 Thinning of Bonded Wafers

We next discuss the procedures of thinning bonded wafers. Two basic steps are required to obtain useful bonded SOI substrates for device fabrication.

Polish-Back. The conventional grinding and polishing techniques can be applied to achieve SOI layers of 2 μm or thicker, which satisfy the requirement of bipolar devices and power MOS transistors. The thickness variations and their control are mostly determined by the parallelism and flatness of the base wafer. The drawback of this technique is that it cannot be extended to ultrathin SOI production in which film thickness less than 0.1 μm is suitable. Local polishing techniques may improve the thickness uniformity when an in situ thickness measurement is introduced. However, a method that makes the bonding interfaces a reference plane for thinning processes is necessary to realize SOI layers for ULSI MOS devices. For a modified polish-back technology, oxide islands are fabricated in the substrate. When the oxide is reached, the polishing process is stopped, which produces silicon islands of 0.1 μm thickness.

Etch-Back. The etch-back process, leading to the acronym of BESOI (bonded and etched-back SOI), is applied to SOI layers after polishing down to a few microns of thickness. The etching thins the SOI film to a desired thickness that is determined by an etch-stop layer. The process makes use of the etching selectivity between the etch-stop layers and the normal silicon layers. High boron doping is usually used in the etch-stop layers. Figure 2.5a shows a simple structure in which a single etch-stop is fabricated by silicon epitaxy on P^+ substrate. After bonding the Si wafer on an oxidized wafer, the P^+ substrate is removed by etchant consisting a mixture of HF, HNO_3, and CH_3COOH in a ratio of 1:3:8. The etching is stopped at the epilayer. However, since the etching selectivity is not excellent and the P^+ wafer doping is not spatially uniform but shows striations, the attainable thickness uniformity of the Si film is moderate. An improved double etch-stop structure is shown in Figure 2.5b. It can be realized on low-doped substrates by using (1) a high-energy ($\geq$ 1 MeV) boron implant, (2) a double epilayer consisting of a P^+ layer followed by a low-doped layer, or (3) a boron implant followed by a low-doped layer.[30] Solutions such as EDP and KOH show excellent etching selectivity between the highly boron-doped layer and low-doped layer. Very reasonable SOI film uniformity has been achieved by using this double etch-stop method.

Totally isolated islands can also be fabricated by wafer bonding. Grooves are etched with standard lithography on the active wafer, which is then thermally oxidized. A polish-stop layer (Si_3N_4 or densified SiO_2) is deposited, and the wafer is bonded to an oxidized passive wafer. Me-

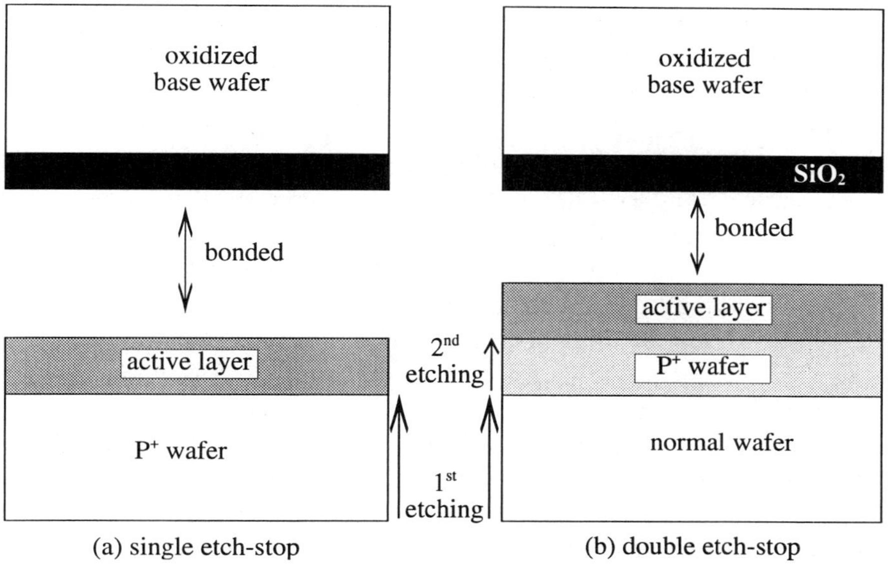

Figure 2.5 Schematics of etch-back: (a) with a single etch-stop and (b) with a double etch-stop.[30]

chanical and electrochemical thinning is done until the polish-stop layer is met. This results in silicon islands with selected size in an ocean of SiO_2.[37]

2.2.3 Properties and Applications of Bonded Wafers

The behavior of native oxide layers, present at the Si/Si bonding interface at room temperature, as a function of annealing temperature, time, and rotational misorientation of the two wafers has been studied by Ahne et al.[38] For bonded float-zone (FZ) wafers, the native oxide layer dissolves at high temperatures due to the diffusion of oxygen into the FZ silicon, which contains a very low oxygen concentration.

By contrast, for the bonded CZ wafers, the oxide layer at the interface increases due to the oxygen diffusion from the oxygen-rich CZ wafers to the interface. The results have been verified by TEM and SIMS analysis, which showed the presence of H, C, N, F, and Cl impurities at the bonding interfaces.[31] In addition, boron has also been detected at the bonding interface as contamination from either air or chemical cleaning before bonding.[39]

Capacitance–voltage (C–V) measurements on both the SIS and MOS structures formed on bonded wafers have been performed. The density of interface traps is typical for the system silicon–thermal oxide. Negative charges with density of around 10^{11} cm^{-2} at the Si/Si bonding interface have been reported by Bengtsson et al.[36] and at the Si/SiO_2 interface by Maszara et al.[40] The negative charges may be caused by excess water at the bonding interfaces.

The electric fields for oxide breakdown in bonded wafers were found to be lower than in thermal oxidized wafers. The degradation may be due to a relatively low bonding strength. Also, the charge trapping induced by Fowler–Nordheim electron injection preferentially takes place at the bonded SiO_2/SiO_2 interface.[41]

However, reports on carrier lifetimes and mobilities in bonded SOI wafers show values comparable to bulk-silicon materials, and no influence of bonding processes on the electrical properties of the active silicon layers has been detected.[42, 43] The background (unintentional) doping of the silicon film, deduced from Hall effect measurements, is p-type and may exceed 10^{16} cm^{-3}. Such a contamination is related to the use of boron in the etch-stop layer.

The immediate applications of wafer bonding are in the areas of bipolar and power MOS devices with film thickness exceeding 2 μm.[44] The WB technology is also a potential asset for the integration of low-voltage control circuits for high-voltage components in power ICs. A linear power supply has been built in high-voltage bonded wafer technology.[44] In micromechanics, because the WB technology has made it possible to fabricate thin diaphragms and cantilevers more easily, pressure sensors and accelerometers have been developed that require smaller chip sizes than conventionally processed devices.[45, 46]

More complicated structures such as microvalves have also been realized recently.[45] In electronic device applications, WB technology can replace the diffusion process or epitaxy process for fabricating the p–n junctions.[47] Wafer bonding allows the production of abrupt junctions with a low density of process-induced defects.

For VLSI circuit applications, a fully functional 16 K bipolar PROM has successfully been fabricated on early SOI bonded wafers using lateral trench isolation,[48] and high-performance fully depleted p- and n-channel CMOS/SOI circuits have been demonstrated on 0.1 μm plasma-thinned (PACE) bonded silicon wafers.[49]

2.3 Zone-Melting Recrystallization

Zone-Melting Recrystallization (ZMR) technology produces SOI structures by recrystallization of polysilicon films, deposited on oxidized silicon wafers. In the ZMR process, a thermal oxide (1–2 μm thick) is first grown on a bulk-silicon substrate, followed by deposition of LPCVD amorphous or polycrystalline silicon film (0.5–1.0 μm thick) on the thermal oxide. The whole structure is capped with a 2 μm thick layer of deposited thermal oxide covered by a thin nitride (Si_3N_4) layer. A molting zone, created by lamps,[50] a graphite strip heater,[51] an electron beam,[52] or a laser,[53] is scanned across the entire silicon wafer. As a result, full liquid-phase recrystallization of a silicon wafer can be carried out in a single pass.

The ZMR process of SOI wafers using a graphite strip heater is shown in Figure 2.6a. A resistively heated graphite susceptor raises the temperature of the entire silicon wafer up to a few hundreds degrees below the melting temperature of silicon. Additional heating is locally produced at the surface of the wafer through the use of a heated graphite strip, located a few millimeters above the sample and scanned across it. The graphite susceptor and the graphite strip heater can be replaced by lamps. As shown in Figure 2.6b, a lamp recrystallization system consists of (i) a bank of halogen lamps to heat the silicon wafer from the back side to 1100 °C or higher temperature, and (ii) a bank of halogen or mercury lamps with an elliptical reflector to focus the beam on the top side of the wafer. To homogenize the energy deposition at the back of the silicon wafer, an unpolished quartz plate may be inserted between the lamp bank and the wafer. As in the case of strip-heater recrystallization, a narrow, wafer-long molten zone is created and scanned across the wafer at a speed of 0.1 to 1 mm/s. In general, the ZMR process is relatively independent of the type of radiation source (i.e., graphite strip heater or lamps), but it is quite different from that of laser recrystallization. For example, the dwell time for the CW laser processing is on the order of 1 ms or less, while it is on the order of seconds for halogen lamps.

The ZMR process can be seeded or unseeded. For seeded ZMR, the oxidized Si wafer is patterned with continuous seeding lines or discrete seeding windows ($1 \times 1\,\mu m^2$). Seeding is responsible for thermal and topological discontinuities, which may induce mass transport. Surface uniformity can be recovered by planarization and thinning. The rapid cooling of the molted zone assists the defect generation near the seeds.

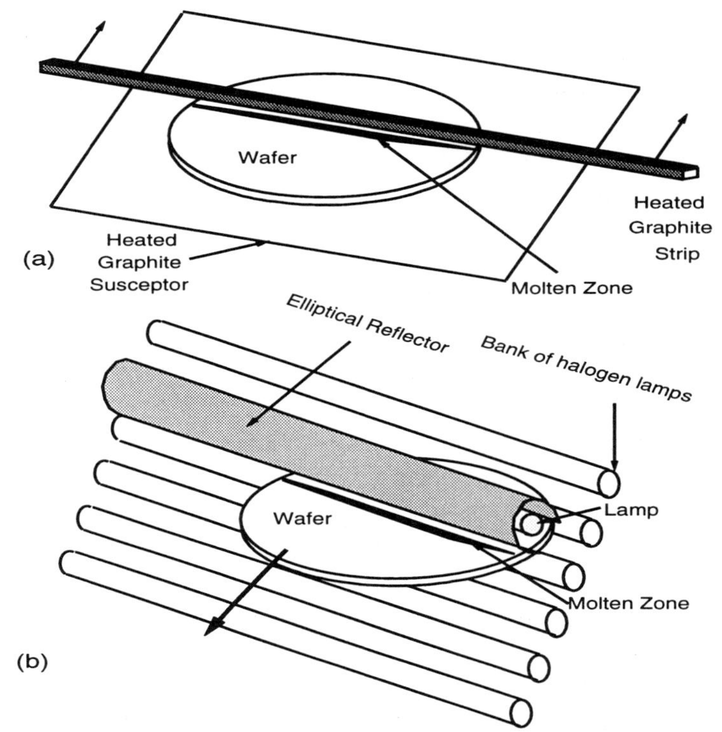

Figure 2.6 Zone-Melting Recrystallization (ZMR) of an SOI wafer using (a) a graphite strip heater and (b) lamps.

SOI devices must be located, above the oxide, in "defect-free" islands.

Optimization parameters are the thicknesses of the oxide, Si film and capping layer, scanning speed and direction, seed interdistance, and heating power. Several stages are identified by increasing the heating: (1) silicon melting over oxide, (2) melting of silicon film at the top of the seed, (3) substrate melting underneath the seed, and (4) substrate overheating below oxide. The crystal quality reached after stages 1–2 is not good enough, whereas stage 4 is prohibited for three-dimensional integration.

Unseeded ZMR–SOI films are basically composed of large polycrystalline grains. The location of grain boundaries can be controlled by depositing a patterned antireflective Si_3N_4 layer over the silicon film.

The grain boundaries are confined underneath the patterns as a result of the discontinuity in thermal gradient. Pseudo-unseeded ZMR is obtained by locating the seeds on the periphery of the wafer. After the ZMR process is completed, the seeded region is removed and the SOI film looks like a wafer-scale unseeded monocrystal.

The predominant defects that hamper the wide application of ZMR SOI materials are grain subboundaries. These are usually parallel to the scanning direction and are spaced 20 to 50 μm apart. The misorientation of the subboundaries is near 1°, and the resulting stress causes the formation of dislocation arrays. In films over 1 μm thick, subboundaries are essentially eliminated, but in very thin films, which are of primary interest for CMOS devices, subboundaries are still a significant problem to overcome. Between subboundaries, ZMR material is very close to bulk silicon in quality. The subboundary spacing is directly related to the thermal gradient and scan speed. The higher the thermal gradient, the smaller the spacing between the subboundaries. The thermal gradient is controlled by the profile of beam energy deposited on the wafer by the incoherent light source.

The misalignment between adjacent crystals has been found to decrease with subboundary spacing.[54] The subboundaries in ZMR materials do not affect majority carrier transport, but do significantly lower the minority carrier lifetimes. They can also cause problems by enhancing diffusion of dopant impurities.

Actual ZMR processing consists of both faceted and cellular growth. The faceted growth seems to be dominant near the melting isotherm, while the cellular-dendritic growth is more important near the solidification isotherm.[55, 56] The establishment of an equilibrium between cooling of the high-reflectivity liquid silicon and heating of the lower-reflectivity solidified silicon explains the formation of a stable solidification interface that produces straight dislocation trails as the only observable defect in ZMR material.[57] In fact, if optimal ZMR processing conditions such as high-purity silane for polysilicon deposition, uniform backside heating, low thermal gradients, and controlled scan speed are used, a stable solidification front and the "elimination" of subboundaries can be achieved simultaneously. As a result, only parallel trails of isolated dislocations are expected to be observed.[58] Average dislocation densities lower than $5 \times 10^4\,\mathrm{cm}^{-2}$, residual dopant density of $2 \times 10^{15}\,\mathrm{cm}^{-3}$, junction leakage below $10^{-6}\,\mathrm{A/cm}^2$, high carrier mobilities, and minority carrier lifetime greater than 30 μs have been obtained in the optimized ZMR material.[57]

ZMR SOI has been used in the fabrication of a wide variety of electronic devices and VLSI circuits. Among them, ring oscillators with 0.8 μm gate length and propagation delay times of 95 ps at 5 V have been demonstrated; these were 30% faster than bulk-silicon counterparts. CMOS VLSI circuits including gate arrays and SRAMs have also been built on ZMR materials. The outstanding merit of the ZMR process is connected with the production of three-dimensional (3–D) integrated circuits. Potential advantages of 3–D integration include multifunctional operation, facilitation of parallel processing, optical sensing functions, and high speed. VLSI circuits with two and three active layers, fabricated on ZMR SOI wafers, will be reviewed in Section 3.6. One straightforward application of 3–D circuits is to use CMOS devices fabricated in the SOI film to control the high-voltage power devices built in the SOI substrate.

2.4 Epitaxial Lateral Overgrowth

The Epitaxial Lateral Overgrowth (ELO) technique refers to the epitaxial growth of silicon from seed windows that are opened on an oxidized silicon substrate. Epitaxial growth is initiated selectively in the seed windows and progresses vertically until it reaches the level of oxide. If the process is allowed to continue, the epitaxy will grow laterally over the insulating film, producing a single crystalline silicon SOI layer, as shown in Figure 2.7a.[59]

Device-quality ELO layers can be obtained at temperatures as low as 750 °C in an ultrahigh-vacuum epitaxy system or in commercially available reactor systems (with pressure ranging from 50 to 150 Torr) at temperatures of around 850 °C.[60] If the process is done properly, no polycrystalline silicon is nucleated on the oxide surfaces, and stacking faults on the edges are minimal. Typical sample preparation for the ELO process involves the development of patterns for seed windows in an oxide layer grown on a (100) silicon wafer. The edges of the windows are oriented along the ⟨010⟩ direction. After cleaning, the wafers are loaded into a LPCVD reactor and undergo a high-temperature (860–1000 °C) hydrogen bake to remove the native oxide from the seed windows. Epitaxial growth is performed using a $SiH_2Cl_2 + H_2 + HCl$ gas mixture. The HCl–to–SiH_2Cl_2 gas ratio can be adjusted to give growth rates from

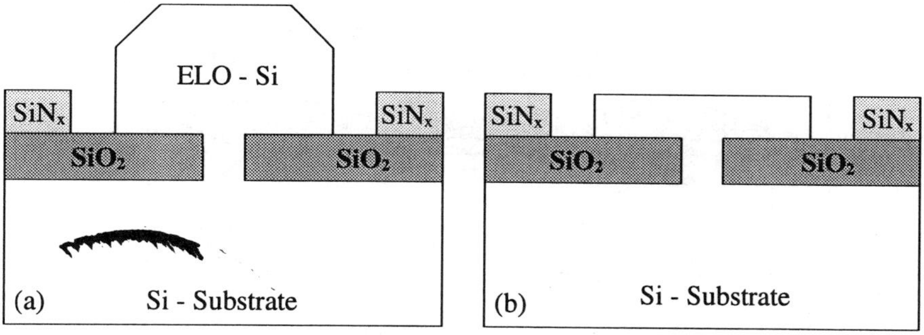

Figure 2.7 (a) The growth of ELO SOI material with aspect ratio close to 1, as measured from the seed hole. (b) Chemical–mechanical polishing can planarize ELO to the thickness determined by the silicon nitride etch-stops (after Neudeck[61]).

less than 0.1 to 0.25 μm/min with no nucleation on the oxide. Phosphine gas can be added as the n-type dopant during growth, and diborane as p-type dopant. If needed, the seeding regions can be removed by localized postepitaxy oxidation.

The main drawback of the ELO process is its growth aspect ratio (i.e., the lateral-to-vertical growth ratio), which is slightly less than 1 for device-quality material. This creates a large step topology on the surface of the ELO material that is not compatible with the silicon planar technology for VLSI circuits. To overcome this problem, one needs to confine the vertical growth or to use chemical–mechanical polishing to planarize the ELO material. A self-limiting chemical–mechanical planarization yields good uniformity across the ELO wafer and excellent control of the SOI film thickness, as illustrated in Figure 2.7b.[61] The technique makes use of the difference in the removal rates of silicon, SiO_2, and Si_3N_4. The polishing rates are dependent on the balance between chemical and mechanical attack on the exposed ELO silicon, SiO_2, and Si_3N_4 surfaces. Either silicon dioxide or silicon nitride can be used to control the final thickness of the ELO layers.

To eliminate the planarization problem of the ELO process, it is possible to grow ELO inside a confined cavity so that vertical growth is restricted while maintaining the lateral growth. The technique, known as *confined lateral selective epitaxial growth* (CLSEG) or *tunnel epitaxy*, consists of filling a lateral cavity whose walls are made of dielectric mate-

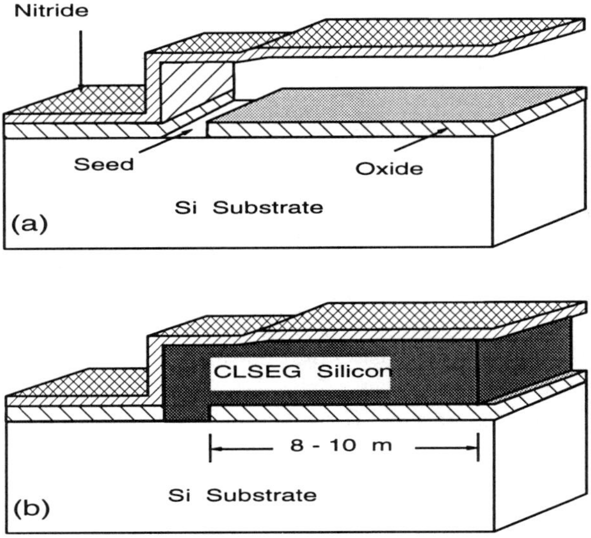

Figure 2.8 (a–b) The growth of ELO SOI material by CLSEG or tunnel epitaxy (after Neudeck[61]).

rial with selective silicon epitaxy.[62] As shown in Figure 2.8, the CLSEG process starts with a thick layer of thermal oxide grown on a (100) Si substrate and a seed hole that is patterned along the ⟨010⟩ direction. A sacrificial layer of amorphous silicon is deposited and patterned to cover the seed hole and to extend some lateral distance beyond it. The thickness and shape of the sacrificial layer will determine the cavity dimensions. The amorphous silicon is converted to polysilicon and then oxidized.

The entire structure is covered with a thin layer of silicon nitride for mechanical support. A via-hole is then etched through the nitride and oxide top layers, which is followed by a selective wet etch that removes the entire sacrificial layer. Once the protective thin oxide layer is removed from the seed area, the selective epitaxy growth (SEG) and ELO processes can be performed on this wafer. Using CLSEG method, an aspect ratio of 7:1 has been achieved in the ELO SOI materials.[63] Three-dimensional devices and stacked inverters have been realized by ELO of silicon over MOS devices.[61]

Aspect ratios as large as 40:1 were obtained by associating liquid-phase epitaxy with ELO process. Seeding windows of size $20 \times 500\,\mu m^2$ were opened in the $0.1\,\mu$m-thick capping oxide. The epitaxy was conducted at 800–950 °C in a tipping-boat system. The epitaxial growth proceeds selectively in the seeding window and then spreads laterally until the Si film thickness reaches 2–3 μm. The single-crystal device regions can extend over 120 μm and present a very low density of defects.[64]

The obvious advantage of ELO for 3–D integration is that it does not require any high-temperature postgrowth recrystallization. In other words, devices processed earlier are not damaged at all during the stacking of a new ELO layer. However, this SOI technology is limited severely by the small size of monocrystalline silicon islands and by the overall complexity of the process.

2.5 Full Isolation by Porous Oxidized Silicon

Another SOI technology, known as FIPOS (Full Isolation by Porous Oxidized Silicon), involves the use of oxidized porous silicon.[65, 66] This technique is more complicated than the SIMOX and ZMR processes depicted in the previous sections but offers the potential for essentially defect-free active silicon layers. The process is intrinsically clean. Oxidation of the porous silicon layer leads to standard thermal oxide and does not disturb the high quality of the original silicon film.

In a typical FIPOS sequence, as shown in Figure 2.9a, phosphorus islands are implanted in a p-type Si wafer. Anodic reaction in a concentrated hydrofluoric acid (HF) solution is used to convert the superficial p-type regions into porous silicon. The implanted n-type islands are not affected. In a standard electrochemical cell, the anode is the silicon wafer and the cathode is an inert metal like platinum. Since porous Si has a very large surface-to-volume ratio (200–1000 m^2 per cm^3!), it is selectively oxidized at 750–800 °C. The final structure is obtained after annealing at 1100 °C for oxide densification.

The key aspect in FIPOS technique is the formation of porous silicon, which is controlled by the type and concentration of the doping as well as by the current density and HF solution. The pore diameter varies from 1 to 30 nm. The ideal porosity is 56% in order to accommodate the change in volume from silicon to oxide and thus to attenuate the strain. The porosity increases for higher anodic reaction current (10–100 mA) or for lower HF concentration.[66]

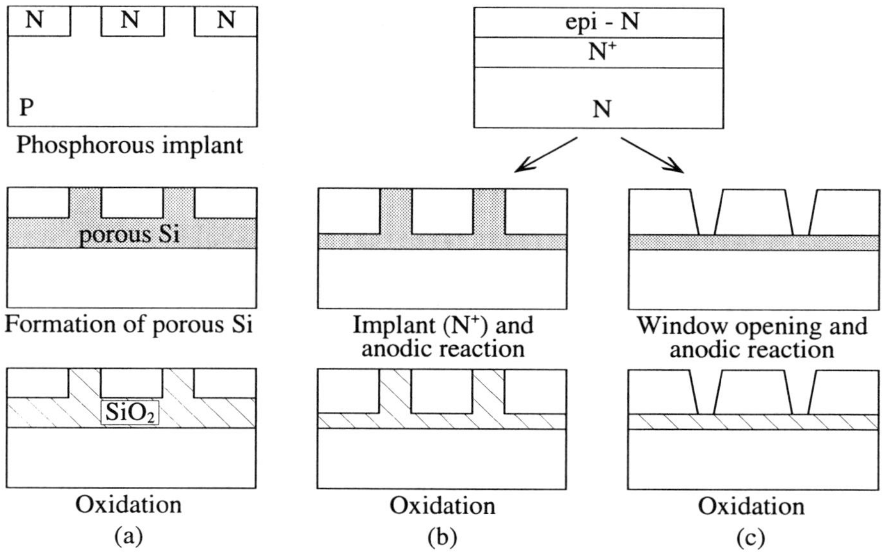

Figure 2.9 (a) Conventional processing sequence of FIPOS and alternatives for isolation (b) by a lateral oxide or (c) by mesa etching (after Cristoloveanu[69]).

Selective formation of porous Si may be obtained by changing not only the type but also the concentration of the dopant. Abrupt NN^+N structures, shown in Figure 2.9b, are prepared by homogeneous phosphorus implantation ($10^{19}\,cm^{-3}$) and epitaxial growth ($10^{15}\,cm^{-3}$). Contacts to the heavily implanted N^+ layer are made by local implantation. The N^+ regions are transformed in porous Si, and then oxidized. In Figure 2.9c, the FIPOS process is achieved by first depositing a silicon epilayer on silicon substrate with a heavily doped top layer. Openings are etched through the epilayer to the substrate, and anodization converts the doped layer to porous silicon. The porous silicon is oxidized readily, leaving the original silicon layer on top. According to the options of Figures 2.9b and 2.9c, the silicon islands are either separated by a lateral oxide or mesa isolated.[67]

Although FIPOS stands as a relatively inexpensive material that offers variable thicknesses of the SiO_2 insulator and film, as well as good crystal quality, it has not been fully optimized. Its inconveniences are mainly associated with the wet process, relatively large thickness of the

Si film (0.5 μm), residual stress, and contamination induced by the chemical process. An interesting alternative is to form, at the surface of the Si wafer, a porous overlay that serves as a substrate for the epitaxial growth of a thin and low-doped Si film. Windows are opened to allow the oxidation of the porous buried layer.

The crystalline quality of the Si epilayer on FIPOS material made it a favorable contender for advanced SOI bipolar applications. High-speed 1.5 μm 64 K CMOS SRAMs have been fabricated by NTT using the FIPOS process.[68]

However, a few years ago, FIPOS was asphyxiated by the successful development of SIMOX.[2] Ironically, the research on FIPOS stopped just prior to an enormous increase of interest in porous silicon in the context of luminescence and electroluminescence. FIPOS pioneers returned to prepare more material. If the electroluminescence of porous silicon can be harnessed, it is clear that FIPOS will be offered a second chance. It would be the only silicon-based material that would combine light emission and electronic functions on the same chip.

2.6 Silicon on Sapphire

2.6.1 SOS Synthesis

Silicon on Sapphire (SOS) is the founding member of the SOI club, still in reasonable good shape and with a healthy business activity. It has been speculated that SOS was the first distress signal launched, more than 30 years ago,[70] by the conventional silicon technology. At that time, the choice of sapphire (Al_2O_3) rather than spinel (MgO–Al_2O_3) as an insulating substrate was a hopeful compromise between its relative availability and its acceptable properties; in particular, the mismatch of its thermal and crystallographic parameters with those of epitaxial Si film is not totally unresolvable. An impressive amount of research, primarily in the 1970s, was dedicated to the numerous fundamental and technological problems. The history of SOS development can be traced from several well-documented reviews.[71, 72]

Due to the high melting temperature of sapphire, special skills have been developed or inspired from the jewelry industry for the manufacture of sapphire crystals.[71] Czockralsky growth and edge-defined film-fed

[2]SIMOX consumed most of the oxygen available on the SOI planet.

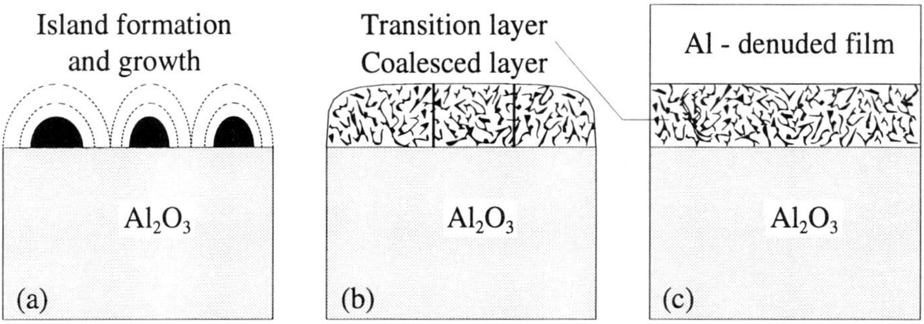

Figure 2.10 (a–c) Gradual formation of the Si–Al_2O_3 interface, transition layer, and silicon film.

growth are the most frequent methods to fabricate device-grade substrates. Other techniques include the flame fusion or Verneuil method and the boat exchanger or gradient furnace method. A careful surface finishing of sapphire wafers, which consists in multistep polishing and annealing, is necessary before they receive the Si film.

The sapphire crystal is rhombohedral and, therefore, different from the face-centered cubic Si crystal. Among several compatible crystallographic orientations for epitaxial growth, the standard combination is Si(100)/$Al_2O_3(1\bar{1}02)$. Silane is used as the gaseous source of chemical vapor deposition (CVD) of silicon. The heteroepitaxial growth is conducted at a temperature of 900–1000 °C in order to minimize contamination from the reactor, intrinsic autodoping with aluminum by sapphire reaction, and unwanted crystal orientations.[73]

The growth starts with the formation of small, hemispherical silicon islands whose size is gradually enlarged (Fig. 2.10). When the islands have completely coalesced, the film thickness is about 20 nm. This transition region is characterized by lateral stress (strong silicon–sapphire bonds), many crystallographic defects due to the lattice mismatch, and aluminum contamination. The erosion of the sapphire substrate during epitaxy leads to the incorporation of Al atoms into silicon aggregates, which explains the autodoping of the transition layer. Since the contaminating reaction takes place prior to coalescence, the initial growth rate (2 μm/min) must ensure a rapid coverage of the substrate. The subsequent growth is conducted at a lower rate, which avoids defect

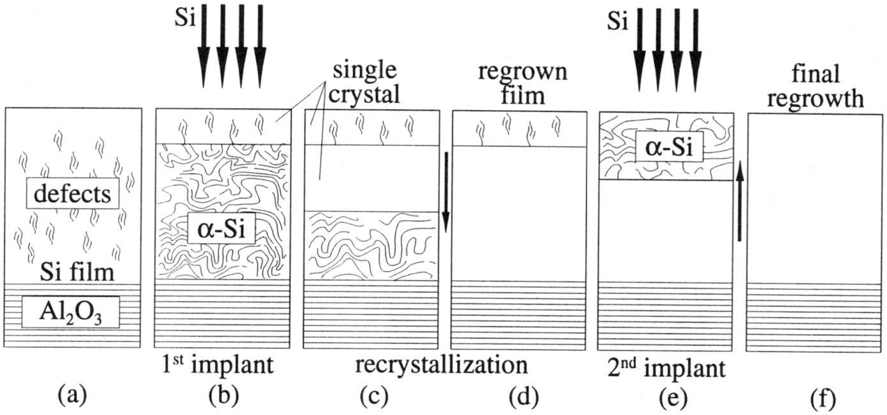

Figure 2.11 Single and double sequence of recrystallization of SOS by ion implantation and solid-phase regrowth: (a) as-grown film, (b) first implant, (c) film recrystallization, (d) regrown film, (e) second implant and film regrowth, (f) "defect-free" regrown film (after Cristoloveanu[72]).

propagation and improves the film quality.[71]

More damaging than the lattice mismatch (10%) is the difference in thermal expansion coefficients between Si (3.8×10^{-8} °C^{-1}) and Al_2O_3 (9.2×10^{-8} °C^{-1}). During the postgrowth cooling phase, differential contraction occurs and causes strong compression (10^{10} dyn/cm^2) of the Si film.[74] Only part of the contaminant species and crystallographic defects (dislocations, microtwins, and stacking faults) are eliminated by annealing at about 1100 °C.

Solid-phase epitaxial regrowth (SPER, Fig. 2.11) remarkably improves the standard SOS material.[75] Silicon ions are implanted to render the film amorphous, except for a thin superficial layer. This complete disordering erases the memory of the silicon lattice and the damaged interface. Annealing at about 600 °C allows the epitaxial regrowth of the film, starting from the less defective surface region, which plays the role of a seed, towards the Si–Al_2O_3 interface. The benefits are the substantial reduction of the compressive strain, suppression of the transition layer, and reduction of the density of microtwins, stacking faults, and interface traps. This improvement is visible in carrier mobility and lifetime, but also affects the cost of already expensive SOS wafers.

The front-surface crystallinity can be enhanced as well by performing an additional, shallower Si implant.[76] The epitaxial regrowth is now directed from the already recrystallized bottom of the film towards the amorphous surface region (Fig. 2.11e,f). This second implant/regrowth phase can be avoided by oxidizing and etching off the nonrecrystallized top-surface region; the film thickness is then completed by epitaxy. Other variants of the SPER process include (i) implantation at 77 K to facilitate the amorphization,[77] (ii) oxygen rather than Si implants,[78] and (iii) higher temperature rapid annealing.[79]

There are different approaches that have been explored for improving SOS material:

- Liquid-phase epitaxial regrowth (LPER)—the top region of the film is made molten by laser irradiation while the interface region serves as a seed.[80]
- α-Si buffer layer—a thin amorphous Si layer 1–4 nm thick is deposited by sputtering on the sapphire surface prior to the epitaxy.
- Slightly misoriented sapphire substrate.[81]
- Film hydrogenation.[82]

2.6.2 Electrical Properties

The electrical properties of SOS structures are governed by three key aspects: (1) lateral stress, (2) in-depth inhomogeneity of the film, and (3) transition layer.

The stress modifies the configuration of silicon energy bands.[72] Two of the six isoenergy valleys of the conduction band are split by 75 meV. This is large enough to transfer 95% of electrons from these valleys into the four valleys that remain electrically active. As far as the valence band is concerned, the energy levels of heavy and light holes are split by 15 meV. The metamorphosis of the conduction and valence bands implies that qualitatively distinct carrier transport features must be expected in SOS and in bulk Si. Even in the ideal case of defect-free SOS, the stress would be responsible for increased hole mobility, degraded drift mobility (by 25%) and Hall mobility (by 50%) of electrons, modified anisotropy, strong longitudinal magnetoresistance, etc. A special transport model has been developed for strained SOS and experimentally verified by the Hall effect and magnetoresistance.[83]

The measurements also show that the electrical parameters (relaxation time, mobility, lifetime) are thickness dependent and degrade in thinner films (see Fig. 5.5). This finding reflects a strong inhomogeneity along the vertical direction. Profiling techniques are based on gradual thinning of the film, either effective by anodic oxidation and etching, or virtual using depletion-controlled MOS structures. Experimental profiles of carrier mobility, concentration and lifetime, magnetoresistance, and scattering process will be discussed in Section 9.1. They clearly evoke the vertical distribution of crystalline defects. The mobility degradation rate is in the range of 5%–30% per 100 nm depth. The minority carrier lifetime—determined with gate-controlled diodes, MOS transistors, and magnetodiodes—is typically 50 ns near the top interface and decreases by one order of magnitude at the bottom of the film.[72]

The net degradation of the transport parameters near the Si–Al_2O_3 interface is associated with the presence of the transition layer. Besides the poor crystal quality, the autodoping is of great concern since aluminum contaminant is electrically active. Elemental Al could be substitutional in the silicon lattice involving a p-type doping, whereas Al–O complexes are responsible for donor-like doping. Both types of overdoping of the transition layer have been reported.[71]

The properties of the silicon–sapphire interface cannot be dissociated from those of the transition layer. The interface is incoherent but planar. The electrical parameters were mostly inferred from the characteristics of front-channel devices. The subthreshold leakage current of MOS transistors, much higher than in bulk-Si MOSFETs, is attributed to generation–recombination centers located at the Si–Al_2O_3 interface. Their presence is confirmed by low-frequency noise measurements. The usual $1/f$ spectrum transforms into a large generation–recombination $1/f^2$ noise as soon as the transistor body is biased such as the depletion region extends to the back interface.[84]

It was found that, surprisingly, the dispersion in the interface properties is suppressed in fully depleted MOSFETs, thanks to the poor properties of the interface! The Fermi level and back-threshold voltage are clamped; thereby, the defects are screened.[85]

Direct investigations of the Si–Al_2O_3 interface were conducted by depositing a metal gate on the back of the wafer and operating MOS capacitors and transistors.[86] To alleviate the inconvenience of very high gate voltages (several kV), the sapphire can be thinned locally, with ultrasonic method, from 200–500 μm down to 30–40 μm. Very poor electron mobility (10–40 $cm^2V^{-1}s^{-1}$) and extremely large concentrations of interface traps (8×10^{12} $cm^{-2}eV^{-1}$) and fixed charges (10^{12} cm^{-2}) have been reported.[87] These findings correspond to standard SOS wafers and must be reconsidered in the context of SOS improved by recrystallization. Unfortunately, less information is available on SPER because most of the characterization effort has been slowed down, and channeled towards the new, emerging SOI technologies (ZMR, SIMOX).

By comparison, the properties of the top Si–SiO_2 interface are much better and tend to equate with the case of bulk silicon, except that the carrier mobility is constantly smaller by 10%–40%.[71, 75] However, this difference is less of a problem in deep-submicron MOSFETs, since the saturation velocities are nearly identical.[88] The poor carrier lifetime in SOS inhibits the fabrication of bipolar transistors, so most SOS circuits—including memories, gate arrays, and microprocessors—are MOS-type. The problem of the transition layer is partially solved by overimplanting the bottom of the film or by processing fully depleted devices with thinner layers (0.1–0.3 μm instead of 0.5 μm).[89]

As an interesting exponent coincidence, it must be noted that 0.5×10^9 s time was necessary to achieve components able to switch in 0.5×10^{-9} s. But rather than competing in the arena of ultrafast microelectronics, SOS is dedicated to the market of radiation-hard applications where it is still successful.

2.7 Silicon on Zirconia

Yttria-stabilized zirconia (YSZ) has been proposed as a substitute for sapphire because of its reduced lattice mismatch and difference in thermal expansion coefficients. The heteroepitaxy of silicon proceeds very similarly in silicon on zirconia (SOZ) and in SOS.

The major point of interest is that the YSZ substrate becomes an ionic conductor above 1200 °C. Oxygen ions penetrate through the ZrO_2 substrate and oxidize the bottom of the film. This internal oxidation is effective in suppressing/consuming the transition layer and related defects. In the meantime, the interface is converted from Si–ZrO_2 to traditional Si–SiO_2 culture. Microscopic, physico–chemical (RBS, SIMS, X-ray diffraction), and electrical investigations have concluded that the quality of the silicon crystal is acceptable.[90]

The development of SOZ has suffered from the difficulty of fabricating

YSZ substrates with large enough diameter. An interesting solution may come from *silicon-on-zirconia-on-silicon* (SOZOS) structures. A thin layer of zirconia is epitaxially grown on a silicon wafer and then serves as a support to receive the epitaxy of the silicon film. There is, however, little chance for SOZ to become a major SOI competitor. SOZ came out too late after SOS, and it would demand too much energy for optimization. Such efforts would hit the marks of the recent SOI approaches that are far more economical and attractive.

The bottom line of this chapter could be the reasonable conclusion that SIMOX is the most mature and promising SOI technology. However, the comparison of SOI materials is not a simple exercise. One has to regard not only the structure properties but also the way they fit a practical application. For instance, ZMR is unrivaled for three-dimensional integration, while FIPOS may bring some light in microelectronics. SOS is still the leader in the field of radiation hard-devices and is expected to continue as such. SOI layers produced by wafer bonding and subsequent thinning show high quality because the process does not induce crystal damage. Another argument concerns flexibility with reference to the thicknesses of the buried oxide and top silicon film. While wafer bonding is advantageous for bipolar and power devices, its drawback is the lack of simple and reliable thinning procedure for films thinner than 0.1–0.2 μm. The domain of ultrathin films, which possess enormous potential for high-speed, low-voltage CMOS devices, by no means belongs to SIMOX, in spite of the ion implantation damage of the film and BOX.

It is easy to predict that not all SOI materials will be able to survive their fratricidal competition. The reduction in numbers has already begun and will be catalyzed by the efforts made to improve the material and device quality and bring down the cost of SOI wafers.

Bibliography

[1] M. Watanabe and A. Tooi, "Formation of SiO_2 films by oxygen-ion bombardment," *Jpn. J. Appl. Phys.*, vol. 5, p. 737, 1966.

[2] K. Izumi, M. Doken, and H. Ariyoshi, "CMOS devices fabricated on buried SiO_2 layers formed by oxygen implantation into silicon," *Electron. Lett.*, vol. 14, p. 593, 1978.

[3] C.G. Tuppen, M.R. Taylor, P.L.F. Hemment, and R.P. Arrowsmith, "The effects of implantation temperature on the properties of buried oxide layers in silicon formed by oxygen ion implantation," *Thin Solid Films*, vol. 131, p. 233, 1985.

[4] J. Margail, I. Stoemenos, C. Jaussaud, M. Dupuy, P. Martin, B. Blanchard, and M. Bruel, "Structural characterization of SIMOX structures," in *Energy Beam Solid Interactions and Transient Thermal Processing,* V.T. Nguyen and A.G. Cullis (eds.), Les Editions de Physique, Les Ulis, France, p. 519, 1986.

[5] J. Stoemenos, B. Aspar, and J. Margail, "Mechanisms of SIMOX synthesis and related microstructural properties," in *Silicon-On-Insulator Technology and Devices,* S. Cristoloveanu (ed.), The Electrochemical Society, Pennington, NJ, p. 16, 1994.

[6] J. Stoemenos, C. Jaussaud, M. Bruel, and J. Margail, "New conditions for synthesizing SOI structures by high dose oxygen implantation," *J. Crystal Growth*, vol. 73, p. 546, 1985.

[7] Y. Li, J.A. Kilner, R.J. Chater, P.L.F. Hemment, A. Nejim, A.K. Robinson, K.J. Reeson, C.D. Marsh, and G.R. Booker, "The effects of dose and target temperature on low energy SIMOX layers," *J. Electrochem. Soc.*, vol. 140, p. 1780, 1993.

[8] L. Meda, S. Bertoni, G.F. Cerofolini, and C. Spaggiari, "Thin buried oxide in implanted silicon," *Nucl. Instr. Methods Phys. Res.*, vol. B80/81, p. 813, 1993.

[9] G.K. Celler, P.L.F. Hemment, K.W. West, and J.M. Gibson, "High quality Si-on-SiO_2 films by large dose oxygen implantation and lamp annealing," *Appl. Phys. Lett.*, vol. 48, p. 532, 1985.

[10] J. Stoemenos, "Microstructure of SIMOX buried oxide, mechanisms of defect formation and related reliability issues," *Microelectron. Eng.*, vol. 22, p. 307, 1993.

[11] P.L.F. Hemment, K.J. Reeson, J.A. Kilner, R.J. Chater, C. Marsh, G.R. Booker, J.R. Davis, and G.K. Celler, "Novel dielectric/silicon planar structures formed by ion beam synthesis," *Nucl. Instr. Methods Phys. Res.*, vol. B21, p. 129, 1987.

[12] J.R. Davis, A. Robinson, K. Reeson, and P.L.F. Hemment, "Fully isolated SOI islands by masked oxygen implantation," *IEEE SOS/SOI Technol. Conf. Proc.*, p. 71, 1987.

[13] H.J. Hovel, "A review of characterization methods for SOI substrates," in *Silicon-On-Insulator Technology and Devices,* S. Cristoloveanu (ed.), The Electrochemical Society, Pennington, NJ, p. 133, 1994.

[14] J. Macia, T. Jawhari, A. Pérez-Rodriguez, and J.R. Morante, "On the optical characterization of SIMOX materials," in *Silicon-On-Insulator Technology and Devices,* S. Cristoloveanu (ed.), The Electrochemical Society, Pennington, NJ, p. 148, 1994.

[15] N. Guillemot, D. Tsoukalas, C. Tsamis, J. Margail, A.M. Papon, and J. Stoemenos, "Suppression mechanisms for oxidation stacking faults in silicon on insulator," *J. Appl. Phys.*, vol. 71, p. 1713, 1992.

[16] L.F. Giles, A. Nejim, F. Cristiano, and P.L.F. Hemment, "Nucleation of oxidation induced stacking faults (OISF) in ion implanted SIMOX structures," in *Silicon-On-Insulator Technology and Devices,* S. Cristoloveanu (ed.), The Electrochemical Society, Pennington, NJ, p. 98, 1994.

[17] P.K. McLarty, D.E. Ioannou, and H.L. Hughes, "Deep states in silicon-on-insulator substrates prepared by oxygen implantation using current deep level transient spectroscopy," *Appl. Phys. Lett.*, vol. 53, p. 871, 1988.

[18] S. Cristoloveanu, J. Pumfrey, E. Scheid, P.L.F. Hemment, and R.P. Arrowsmith, "Thermal donor and new donor generation in SOI material formed by oxygen implantation," *Electron. Lett.*, vol. 21, p. 802, 1985.

[19] S. Cristoloveanu, "Oxygen-related activity and other specific electrical properties of SIMOX," *Vacuum*, vol. 42, p. 371, 1991.

[20] S. Cristoloveanu, S. Gardner, C. Jaussaud, J. Margail, A.J. Auberton-Hervé, and M. Bruel, "Silicon on insulator material formed by oxygen implantation and high-temperature annealing: carrier transport, oxygen activity, and interface properties," *J. Appl. Phys.*, vol. 62, p. 2793, 1987.

[21] J. Wyncoll, K.N. Kang, S. Cristoloveanu, P.L.F. Hemment, and R.P. Arrowsmith "Temperature dependence of Hall mobility and electrical conductivity in SIMOX films," *Electron. Lett.*, vol. 20, p. 485, 1984.

[22] S. Cristoloveanu, "A review of the electrical properties of SIMOX substrates and their impact on device performance," *J. Electrochem. Soc.*, vol. 138, p. 3131, 1991.

[23] P. Normand, D. Tsoukalas, N. Guillemot, and P. Chenevier, "A pile-up phenomenon during arsenic diffusion in silicon-on-insulator structures formed by oxygen implantation," *J. Appl. Phys.*, vol. 66, p. 3585, 1989.

[24] P.M. Lenahan, "Electron spin resonance and instabilities in metal-insulator-semiconductor systems," *Microelectron. Eng.*, vol. 22, p. 129, 1993.

[25] A.G. Revesz, G.A. Brown, and H.L. Hughes, "Bulk electrical conduction in the buried oxide of SIMOX structures," *J. Electrochem. Soc.*, vol. 140, p. 3222, 1993.

[26] M. Delfino, M. Jaczynski, A.E. Morgan, C. Vorst, M.E. Lunnon, and P. Maillot, "Gettering of copper in silicon-on-insulator structures formed by oxygen ion implantation," *J. Electrochem. Soc.*, vol. 134, p. 2027, 1987.

[27] J. Jablonski, Y. Miyamura, M. Imai, and H. Tsuya, "Heavy metal gettering in SIMOX wafers," in *Silicon-On-Insulator Technology and Devices,* S. Cristoloveanu (ed.), The Electrochemical Society, Pennington, NJ, p. 28, 1994.

[28] P. Dimitrakis, G.J. Papaioannou, and S. Cristoloveanu, "Electrical characterization of interface between buried oxide and silicon substrate," in *Silicon-On-Insulator Technology and Devices,* S. Cristoloveanu (ed.), The Electrochemical Society, Pennington, NJ, p. 230, 1994.

[29] M. Levy, E. Scheid, and S. Cristoloveanu, "Depth profiles of the optical properties of buried oxides (SIMOX) by ellipsometry," *Thin Solid Films*, vol. 148, p. 127, 1987.

[30] K. Mitani and U.M. Gösele, "Wafer bonding technology for silicon-on-insulator applications: a review," *J. Electron. Materials*, vol. 21, p. 669, 1992.

[31] T. Abe, T. Takei, A. Uchiyama, K. Yoshizawa, and Y. Nakazato, "Surface impurities encapsulated by silicon wafer bonding," *Jpn. J. Appl. Phys.*, vol. 29, p. L2311, 1990.

[32] P.B. Mumola, G.J. Gardopee, P.J. Clapis, C.B. Zarowin, L.D. Bollinger, and A.M. Ledger, "Plasma thinned SOI bonded wafers," *IEEE Int. SOI Conf. Proc.*, p. 152, 1992.

[33] Q.Y. Tong and U. Gösele, "VLSI SOI fabrication by SIMOX wafer bonding (SWB)," *IEEE Int. SOI Conf. Proc.*, p. 72, 1992.

[34] W.P. Maszara, G. Goetz, T. Caviglia, A. Czerhati, J. Johnson, and J.B. McKitterick, "Wafer bonding for SOI," *Mat. Res. Soc. MRS'87 Conf. Proc.*, p. 259, 1990.

[35] K. Mitani, V. Lehmann, R. Stengl, D. Feijoo, U.M. Gösele, and H.Z. Massoud, "Causes and prevention of temperature dependent bubbles in silicon wafer bonding," *Jpn. J. Appl. Phys.*, vol. 30, p. 615, 1991.

[36] S. Bengtsson and O. Engström, "Interface charge control of directly bonded silicon structures," *J. Appl. Phys.*, vol. 66, p. 1231, 1989.

[37] J. Haisma, E.M.L. Alexander, T.M. Michielsen, J.A. Pals, G.A. Spierings, and J.W.A. van der Velden, "Novel trends in SOI technology: wafer bonding and thinning," *Eur. SOI Conf. Proc.*, p. E2, 1988.

[38] K.Y. Ahn, R. Stengl, T.Y. Tan, U. Gösele, and P. Smith, "Stability of interfacial oxide layers during silicon wafer bonding," *J. Appl. Phys.*, vol. 65, p. 561, 1989.

[39] F.P. Widdershoven, J. Haisma, and J. Naus, "Boron contamination and antimony segregation at the interface of directly bonded silicon wafers," *J. Appl. Phys.*, vol. 68, p. 6253, 1990.

[40] W.P. Maszara, G. Goetz, A. Caviglia, and J. McKitterick, "Bonding of silicon wafers for silicon-on-insulator," *J. Appl. Phys.*, vol. 64, p. 4943, 1988.

[41] S. Bengtsson, A. Jauhiainen, and O. Engström, "Oxide degradation of wafer bonded metal oxide semiconductor capacitors following Fowler–Nordheim electron injection," *J. Electrochem. Soc.*, vol. 139, p. 2302, 1992.

[42] J.B. Lasky, "Wafer bonding for silicon-on-insulator technology," *Appl. Phys. Lett.*, vol. 48, p. 78, 1986.

[43] L.J. Spangler and K.D. Wise, "A technology for high-performance single-crystal silicon-on-insulator transistors," *IEEE Electron Device Lett.*, vol. 8, p. 137, 1987.

[44] J.D. Beasom, "120V AC, 250V DC monolithic power supply made with bonded wafer technology," *IEEE Int. SOI Conf. Proc.*, p. 66, 1992.

[45] K. Petersen, J. Brown, T. Vermeulen, P. Barth, J. Mallon, and J. Bryzek, "Ultra-stable, high temperature pressure sensors using silicon fusion bonding," *Sensors Actuators*, vol. A21, p. 96, 1990.

[46] P.W. Barth, "Silicon fusion bonding for fabrication of sensors, actuators and microstructures," *Sensors Actuators*, vol. A23, p. 919, 1990.

[47] M. Shimbo, K. Furukawa, F. Fukuda, and K. Tanzawa, "Silicon-to-silicon direct bonding method," *J. Appl. Phys.*, vol. 60, p. 2987, 1986.

[48] M.D. Church, "A comparative study of functional 16 K bipolar PROM circuits fabricated on bonded, oxide isolated and junction isolated substrates," *IEEE Int. SOI Conf. Proc.*, p. 175, 1989.

[49] K. Aubuchon, J. Pinter, M. Matloubian, M. Barger, R. McClain, and O. Marsh, "Initial evaluation of CMOS/SOI fabricated on plasma-thinned bonded silicon wafers," *IEEE Int. SOI Conf. Proc.*, p. 154, 1992.

[50] M. Haond, D.P. Vu, A.M. Aguirre, and S. Perret, "Electrical performances of devices made on SOI films obtained by lamp ZMR," *IEEE Circuits Devices*, vol. 3, p. 27, 1987.

[51] B.Y. Tsaur, "Zone-melting-recrystallization silicon-on-insulator technology," *IEEE Circuits Devices*, vol. 3, p. 13, 1987.

[52] J.A. Knapp, "Silicon-on-insulator structures formed by aline-source electron beam: experiment and theory," *J. Appl. Phys.*, vol. 58, p. 2584, 1985.

[53] P. Seegebrecht, "3–D project and current status of 3–D technology in West Germany," *5th Int. Workshop on Future Electron Devices Proc.*, Japan, p. 19, 1988.

[54] L. Pfeiffer, K.W. West, D.C. Joy, J.M. Gibson, and A.E. Gelman, "Reduced subboundary misalignment in SOI films scanned at low velocities," in *Semiconductor-on-Insulator and Thin Film Transistor Technology,* A. Chiang, M.W. Geis, and L. Pfeiffer (eds.), North-Holland, MRS Symp. Proc., vol. 53, p. 29, 1986.

[55] D. Dutartre, M. Haond, and D. Bensahel, "Microscopy of thin Si films during lamp zone melting," in *Semiconductor-on-Insulator and Thin Film Transistor Technology,* A. Chiang, M.W. Geis, and L. Pfeiffer (eds.), North-Holland, MRS Symp. Proc., vol. 53, p. 89, 1986.

[56] J.S. Im, C.K. Chen, C.V. Thompson, M.W. Geis, and H. Tomita, "Liquid–solid interface morphologies and defect structures in zone-melting-recrystallized silicon-on-insulator films," in *Silicon-on-Insulator and Buried Metals in Semiconductors,* J.C. Sturm, C.K. Chen, L. Pfeiffer, and P.L.F. Hemment (eds.), MRS Symp. Proc., vol. 107, p. 169, 1988.

[57] P.M. Zavracky, D.P. Vu, L. Allen, W. Henderson, H. Guckel, J.J. Sniegowski, T.P. Ford, and J.C.C. Fan, "Large diameter SOI wafers by zone-melting-recrystallization," in *Silicon-on-Insulator and Buried Metals in Semiconductors,"* J.C. Sturm, C.K. Chen, L. Pfeiffer, and P.L.F. Hemment (eds.), MRS Symp. Proc., vol. 107, p. 213, 1988.

[58] M.W. Geis, C.K. Chen, R.W. Mountain, N.P. Economou, W.T. Lindley, and P.L. Hower, "Use of zone melting recrystallization to fabricate a three-dimensional structure incorporating power bipolar and field-effect transistors," *IEEE Electron Device Lett.*, vol. 7, p. 41, 1986.

[59] C.I. Drowley and M. Hammond, "Conditions for uniform selective epitaxial growth," *Solid State Technol.*, p. 135, 1990.

[60] A. Ogura and Y. Fujimoto, "Novel technique for Si lateral epitaxial overgrowth: tunnel epitaxy," *Appl. Phys. Lett.*, vol. 55, p. 2205, 1989.

[61] G.W. Neudeck, "Three-dimensional CMOS integration," *IEEE Circuits Devices*, vol. 6, p. 32, 1990.

[62] P.J. Schubert and G.W. Neudeck, "Confined lateral selective epitaxial growth of silicon for device fabrication," *IEEE Electron Device Lett.*, vol. 11, p. 181, 1990.

[63] R.P. Zingg, B. Höfflinger, and G.W. Neudeck, "Stacked CMOS inverter with symmetric device performance," *IEDM Techn. Digest*, p. 909, 1989.

[64] R. Bergmann, "Model for defect-free epitaxial lateral overgrowth of Si over SiO_2 by liquid phase epitaxy," *J. Crystal Growth*, vol. 110, p. 823, 1991.

[65] S.S. Tsao, "Porous silicon techniques for SOI structures," *IEEE Circuits Devices*, p. 3, 1987.

[66] G. Bomchil, A. Halimaoui, and R. Herino, "Porous silicon: the material and its applications to SOI technologies," *Microelectron. Eng.*, vol. 8, p. 293, 1988.

[67] G. Bomchil, A. Halimaoui, and R. Herino, "Porous silicon: the material and its applications in silicon-on-insulator technologies," *Appl. Surface Sci.*, vol. 41/42, p. 604, 1989.

[68] K. Ehara, H. Unno, and S. Muramoto, "1.5 μm FIPOS/CMOS VLSI process with low wafer warpage and Si deposit-defect-free," *Electrochem. Soc. Ext. Abs.*, vol. 85, p. 457, 1985.

[69] S. Cristoloveanu, "Advanced silicon on insulator materials: processing, characterization and devices," in *Semiconductor Silicon*, G. Harbeke and M.J. Schulz (eds.), Springer, Berlin, vol. 13, p. 223, 1989.

[70] H.M. Manasevit and W.I. Simpson, "Single-crystal silicon on a sapphire substrate," *J. Appl. Phys.*, vol. 35, p. 1349, 1964.

[71] G.W. Cullen and C.C. Wang, *Heteroepitaxial Semiconductors for Electronic Devices*, Springer, Berlin, 1978.

[72] S. Cristoloveanu, "Silicon films on sapphire," *Rep. Prog. in Phys.*, vol. 50, p. 327, 1987.

[73] A.C. Ipri, "The properties of silicon-on-sapphire substrates, devices and integrated circuits," in *Applied Solid State Science: Silicon Integrated Circuits*, D. Kahng (ed.), Academic, New York, p. 253, 1981.

[74] T. Englert, G. Arbstreiter, and J. Pontcharra, "Determination of existing stress in silicon films on sapphire substrate using Raman spectroscopy," *Solid-State Electron.*, vol. 23, p. 31, 1980.

[75] I. Golecki, "The current status of silicon-on-sapphire and other heteroepitaxial silicon-on-insulator technologies," in *Comparison of Thin Film Transistor and SOI Technologies*, H.W. Lam and H.J. Thompson (eds.), North-Holland, Amsterdam, p. 3, 1984.

[76] T. Yoshii, S. Taguchi, T. Inoue, and H. Tango, "Improvement of SOS device performance by solid-phase epitaxy," *Jpn. J. Appl. Phys.*, vol. 21 (Suppl. 21–1), p. 175, 1982.

[77] S.S. Lau, S. Matteson, J.W. Mayer, P. Revesz, J. Gyulai, J. Roth, T.W. Sigmon, and T. Cass, "Improvement of crystalline quality of epitaxial Si layers by ion-implantation techniques," *Appl. Phys. Lett.*, vol. 34, p. 76, 1979.

[78] Y. Yamamoto, T. Sugiyama, A. Hara, and T. Inada, "Crystalline quality improvement of silicon on sapphire films by oxygen implantation and subsequent thermal annealing," *J. Appl. Phys.*, vol. 53, p. 793, 1982.

[79] D.J. Smith, L.A. Freeman, R.A. McMahon, H. Ahmed, M.G. Pitt, and T.B. Peters, "Characterization of Si-implanted and electron-beam-annealed silicon-on-sapphire using high-resolution electron microscopy," *J. Appl. Phys.*, vol. 56, p. 2207, 1984.

[80] Y. Kobayashi, T. Ikeda, M. Nakamura, and T. Suzuki, "Improvement of characteristics in SOS/MOSFETs by laser annealing," *Jpn. J. Appl. Phys.*, vol. 21 (Suppl. 21–1), p. 181, 1982.

[81] R.T. Smith and C.E. Weitzel, "Influence of sapphire substrate orientation on SOS crystalline quality and SOS/MOS transistor mobility," *J. Crystal Growth*, vol. 58, p. 61, 1982.

[82] L. Jastrzebski, J. Lagowski, G.W. Cullen, and J.I. Pankove, "Hydrogenation induced improvement in electronic properties of heteroepitaxial silicon-on-sapphire," *Appl. Phys. Lett.*, vol. 40, p. 713, 1982.

[83] J.H. Lee, S. Cristoloveanu, and A. Chovet, "Non-homogeneous electrical transport through silicon-on-sapphire thin films: evidence of the internal stress influence," *Solid-State Electron.*, vol. 25, p. 947, 1982.

[84] P. Gentil and S. Chaussé, "Low frequency noise measurements on silicon-on-sapphire (SOS) MOS transistors," *Solid-State Electron.*, vol. 20, p. 935, 1977.

[85] F. Balestra, J. Brini, and P. Gentil, "Deep depleted SOI MOSFETs with back potential control: a numerical simulation," *Solid-State Electron.*, vol. 28, p. 1031, 1985.

[86] A.M. Goodman, "An investigation of the Si–sapphire interface using the MIS capacitance method," *IEEE Trans. Electron Dev.*, vol. 22, p. 63, 1975.

[87] P. Gentil, "Simple method for evaluating electronic properties at silicon–sapphire interface," *Electron. Lett.*, vol. 16, p. 617, 1980.

[88] M. Dufour and S. Cristoloveanu, "Hot carrier effects in double injection phenomena," *Appl. Phys. A*, vol. 29, p. 87, 1982.

[89] R. Rios, R.K. Smeltzer, and G.A. Garcia, "Radiation effects in fully depleted CMOS/SOS," *IEEE Int. SOI Conf. Proc.*, p. 44, 1991.

[90] D. Pribat, L.M. Mercandalli, J. Siejka, and J. Perriere, "Interface oxidation of epitaxial silicon deposits on (100) yttria stabilized cubic zirconia," *J. Appl. Phys.*, vol. 58, p. 313, 1985.

Chapter 3

SOI Devices

Silicon-on-insulator devices and circuits have progressed at a remarkable rate during the past several years. The theoretical advantages of SOI over bulk silicon, summarized in Chapter 1, have been fully demonstrated in terms of performance and reliability. Recent interest has turned to possible use of fully depleted SOI films for deep submicron circuits. Other areas in which the capability of SOI has been identified include low-power and mixed-technology circuits. In this chapter, we shall review a number of typical SOI devices, circuits, and related applications. Among them are the radiation-hardened CMOS circuits, the advanced CMOS ULSI, high/low voltage, high temperature devices, three-dimensional circuits, and sensors. Some innovative and speculative components will be presented in the last section.

3.1 Advanced CMOS and Bipolar Devices

Ultra-large-scale integration (ULSI) is the mainstream of the microelectronics industry. It is essentially in this high-priority arena that the very attractive arguments of SOI technology have to be materialized. Thin-film CMOS, BiCMOS, and complementary bipolar structures can overcome the severe technological limitations that bulk silicon is expected to encounter in scaling below $0.25\,\mu$m: the failure of LOCOS isolation, threshold-voltage roll-off, increased subthreshold slope, hot-electron degradation, soft errors, and latch-up. In addition to being a high-speed, low-power process, new innovative structures and SOI–dedicated design can further increase the performance advantages over

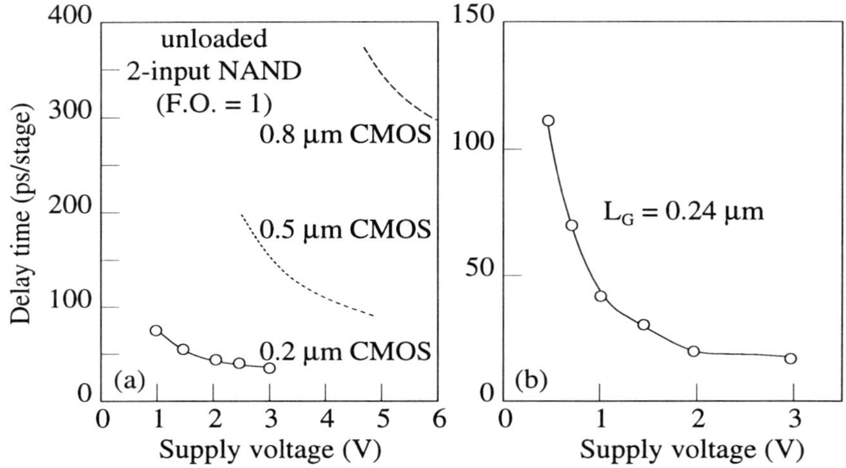

Figure 3.1 Gate delay versus supply voltage for (a) a two-input NAND ring-oscillator (fan-out =1, unloaded) formed on a gate array and (b) a 0.24 μm inverter (after Tsuchiya et al.[1]).

equivalent bulk-type devices. Moreover, SIMOX-based BiCMOS or complementary bipolar ICs may have circuit density comparable to that of MOS circuits. They will enjoy higher speed due to a denser layout and lower parasitic capacitances, and the simpler fabrication process will render them more cost effective.

3.1.1 Advanced CMOS Devices and Circuits

The superior performance and capability of fully depleted CMOS components was first demonstrated on long-channel transistors, and since then has been confirmed on several generations of submicron devices. It is admitted that CMOS on SOI is one generation ahead. For example, 0.35 μm SOI MOSFETs perform as well as 0.25 μm bulk-silicon transistors, which implies a huge economy in processing equipment. Although repeatedly demonstrated, this theorem has not been sufficiently convincing for SOI to be widely used because, in the meantime, bulk-Si technology was able to maintain an extremely rapid rate of development by overcoming its intrinsic drawbacks. An especially critical area, for bulk Si (but not for SOI) is the domain of sub–0.15 μm channel lengths,

where short channel effects, power consumption, and complex isolation schemes become prohibitive.

Recently, Tsuchiya et al.[1] reported a subquarter-micron, ultrathin-film, CMOS technology on low-dose, high-quality SIMOX wafers (3–5 $\times 10^{17}\,\mathrm{cm}^{-2}$ oxygen dose, 90-nm-thick BOX, 50-nm-thick film, and less than $300\,\mathrm{cm}^{-2}$ dislocations). The device characteristics are compatible with low-power and high-performance ULSI applications: no kink effect, sharp subthreshold slope, enhanced current drivability, and reduced short-channel effects. By using an N^+-polysilicon-gate for n-MOSFETs and a P^+-polysilicon-gate for p-MOSFETs, good symmetrical subthreshold characteristics, with 73 mV/decade swing, were achieved for both types of MOSFETs.

Figure 3.1a shows the variation of delay time with supply voltage for a two-input NAND ring-oscillator.[1] The delay time is 45 ps/stage at 2 V and 53 ps/stage at 1.5 V, which is about three times faster than the 0.5 μm bulk CMOS devices operating at 3.3 V. An inverter ring-oscillator fabricated on a gate array with the same 0.24 μm CMOS technology showed a delay time of 30 ps/stage at 2 V and 45 ps/stage at 1 V (Fig. 3.1b). Furthermore, multigigahertz phase-locked loop (PLL) LSIs operate at 4 GHz and dissipate 19 mW power for 2.5 V, which is the highest speed ever reported for silicon-based PLLs.

Subquarter-micron channel length SIMOX technology, with enhancement-mode n- and p-channel MOSFETs and depletion-mode n-channel MOSFETs, is equally feasible on thicker buried oxides (400 nm BOX with a 50–160-nm-thick silicon layer).[2] High-performance circuits operating at very low power were realized by using a fully depleted process with both LDD and non-LDD devices. A 101–stage depletion-mode NMOS inverter ring oscillator, fabricated on a 50 nm SIMOX film with gate oxide thickness of 7 nm and effective gate length of 0.2 μm, had a propagation delay of 14–18 ps/stage at low supply voltage (1.5 V) and room temperature. High-performance, fully depleted p- and n-channel CMOS circuits have also been demonstrated on 0.1 μm plasma-assisted chemical-thinned (PACE) bonded silicon wafers.[3]

Very impressive mesoscopic MOSFETs with 100 nm gate length have been fabricated on SIMOX.[4] The thicknesses of the buried oxide, silicon film, and gate oxide were, respectively, 80 nm, 30 nm, and 7 nm. The film doping was $3 \times 10^{17}\,\mathrm{cm}^{-3}$ for n-channels and $5 \times 10^{16}\,\mathrm{cm}^{-3}$ for p-channels, and linear-graded LDD structures were used. Relatively high values of field-effect mobilities were measured: $440\,\mathrm{cm}^2/\mathrm{Vs}$ for electrons and 140 cm^2/Vs for holes. 0.1-μm-long n–MOSFETs show a remarkable immunity to short-channel effects, which is attributed to the thinness of the BOX. This is confirmed in Figure 8.19b (Section 8.5) by direct comparison of transistors fabricated on SIMOX wafers with either thin (80 nm) or thick (480 nm) buried oxides.

Almost at the boundary between fiction and SOI technology, 85-nm-long transistors were processed on 8-nm-thick silicon films.[4] If we keep in mind that an MOS inversion layer is typically 1–10 nm thick, the 8 nm transistor appears to be formed by peeling off the inversion layer from the wafer surface! The characteristics of 85-nm-long n- and p-channel transistors are well behaved. The subthreshold swing is remarkably small (95 mV/decade) in spite of the huge value of series resistances ($10^5 \Omega/\square$).

Back to reality: subquarter-micron SOI MOSFETs still need improvement. The weakness of low drain breakdown voltage, caused by parasitic bipolar action, will automatically be erased by the ineluctable reduction in supply voltage below 2.5 V. A matter of primary concern is the high value of series resistances, which tend to obscure the excellent intrinsic performance of ultrathin-film MOSFETs. One potential solution is the silicidation of source and drain regions.[5] A self-aligned $TiSi_2$ silicide process, which reduces the sheet resistance, is feasible but rather complicated by the limited available thickness.[6] Series resistances of 250 Ω-μm have recently been achieved in 0.12-μm-long, 60-nm-thin MOSFETs by partial cobalt silicidation.[7] The $CoSi_2$ silicide is made by two-step rapid thermal annealing of Ti/Co laminates such as not to fully consume the silicon film.

Another alternative is the *local thinning* of the transistor body. In a standard SIMOX wafer, the starting thickness of the silicon film is around 200 nm. Selective LOCOS–like oxidation, which uses a mask that exposes only the gate regions, and subsequent etching leave a nonuniform (recessed) surface structure.[8, 9] Once the poly-Si gate has been deposited and etched, source and drain implants are performed in the thick regions. The final transistor has a 50-nm-thick body and 130-nm-thick source and drain. Such a difference in thickness represents a gain by a factor of six in terms of series resistance (less than 50 $\Omega/\square$), which yields three times better current drivability as compared to conventional full wafer thinning.[8]

Other points under consideration are connected with (1) optimization of film and BOX thicknesses,[10] (2) engineering (or suppression) of LDD regions, (3) tailoring of the channel doping profile, (4) choice of adequate

Table 3.1 Transistor parameters for various SOI bipolar devices

SOI film	Bonded wafer	ELO	SIMOX
N_D (cm^{-3})	5×10^{16}	10^{17}	10^{17}
Epi-thickness (μm)	None	0.4–1.2	0.2–1.2
Base-thickness (μm)	0.12	0.12	0.15
Current gain (β)	16	30	120
BV_{CEO} (V)	4.7	2.8–7.5	3–9.5
BV_{CBO} (V)	16	8–18	
R_E	23	8	
C_{EB} (fF)	49.1	32	
C_{CB} (fF)	86.7	21	
Peak f_T (GHz)	4.5	10–20	4.5
Area (μm^2)	0.15x40	0.15x30	0.17x 40
Reference	[15]	[16]	[17]

metal and doping for the gate, (5) association of inversion-mode and accumulation-mode MOSFETs, (6) customization of the design rules for SOI, (7) dynamic (transient) floating body effects, etc.

3.1.2 Bipolar Devices

Early bipolar junction transistors (BJTs) on SOI have suffered from their low performance, caused by the width and high resistance of the base.[11, 12] Nevertheless, both NPN and PNP lateral BJTs having a base width shorter than 0.5 μm were fabricated on thin (100 nm) SOI films, and showed current gains of 75 and 40, respectively.[13] The minority carrier lifetime in the base was 1 μs, and the leakage current was in the sub-pA range. The device fabrication is compatible with the SOI CMOS process.

The fabrication of *vertical* BJTs, using bulk-Si concepts, requires thick silicon films that are available by wafer bonding or by epitaxy over SIMOX. In thin SOI films, there is not enough space to accommodate the buried collector region. A pure SOI solution consists of simply omitting the collector so that the base region extends to the BOX. The inversion channel, activated by the back gate, plays the role of collector.[14]

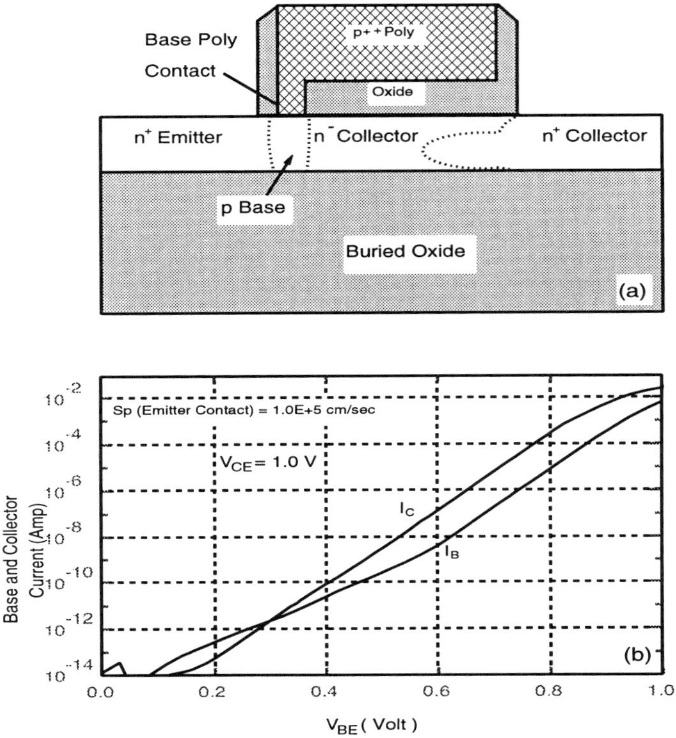

Figure 3.2 (a) Cross-sectional view of a lateral SOI BJT and its base doping contours and (b) Gummel plot (after Shahidi et al.[16]).

Progress in SOI processes has led to the development of a number of novel lateral double-diffused BJTs with thin base width using bonded wafer,[15] epitaxial lateral overgrowth,[16] and SIMOX[17] technologies. The transistor parameters for these SOI BJTs are summarized in Table 3.1. Higaki[15] proposed the bipolar device, which was fabricated on a 150-nm-thick, 5×10^{16} cm^{-3} phosphorus-doped Si film on wafer-bonded substrate. It has the common emitter breakdown voltage of 4.7 V, a peak cutoff frequency of 4.5 GHz, and current gain of 16 over a wide current range.

The SIMOX BJT exhibits a high current gain (120) and a breakdown voltage of 3–9.5 V, and can be configured as a hybrid MOS/BJT in which both MOS channel and BJTs can drive high currents (see Fig. 3.19). The device implemented in the ELO SOI has a current gain between 10 and 30 and a breakdown voltage between 2.8 and 7.5 V according to the N^- collector width. It also has a higher peak cutoff frequency (20 GHz), due to the low base resistance through the direct sidewall P^+ contact.[16] By contrast, the base contacts of others are located at the edges of the device, which results in high base resistances inducing current crowding near the base contacts. This renders much of the device width useless and reduces the cutoff frequency.

Figure 3.2a shows the cross-sectional view of a lateral BJT whose structure is analogous to a vertical $N^+PN^-N^+$ bipolar device.[16] The p-base contact is to the P^+ poly contact on the top of the oxide through the P^+ poly side wall. The device operation is dependent on the front- and back-gate bias, the Si film thickness, and the usual BJT design parameters like doping and width of base and collector. The BJT has a heavily doped base layer to compensate for the high N^--collector concentration. The Gummel plot (Fig. 3.2b) shows the base and collector current for a collector width of 0.4 μm and V_{CE} =1.0 V. For VLSI circuit applications, a fully functional 16 K SOI bipolar PROM circuit has been demonstrated using bonded wafers and lateral trench isolation.[18]

BiCMOS technology has attracted considerable attention because of its high speed, high packing density, and low power dissipation,[19] but its implementation on bulk silicon is very complex. The use of thin-film SOI for lateral bipolar devices permits much simpler circuit integration. Lateral bipolar transistors on SOI have very small junction capacitance, easy isolation, and a CMOS-like device area.[20] These features are particularly suitable in a dense high-performance BiCMOS technology. A double polysilicon bipolar technology, developed by Hiramoto et al.,[21] uses bonded SOI substrates to improve the α-particle immunity, as well as scaled CMOS memory cells to reduce the power consumption and to increase the packing density. The cutoff frequency of the bipolar transistors is as high as 27 GHz, and the area of the CMOS memory cell is 58 μm^2 only, which is attractive for ultra-high-speed, high-density ECL-CMOS SRAM applications.

3.2 Radiation-Hardened Circuits

Among the SOI technologies, SIMOX is considered to be the most promising technology to inherit from SOS the field of radiation-hard applications. The primary reason for this is that SOS proved impractical to reach submicron lithography levels, whereas SIMOX is fully compatible with current VLSI fabrication. In addition, the properties of the buried oxide formed by oxygen implantation can be engineered for improved protection against permanent radiation damage. For example, electron traps are deliberately introduced in order to compensate for the build-up of positive charges during radiation.

In comparison to the quasi-infinite sapphire substrate in SOS devices, the SIMOX BOX is relatively thin. The biasing of the back gate (i.e., the internal field in the BOX) may prove useful for minimizing the generation of the defects as well as their electrical influence. Back-gate bias effects are discussed in Section 10.2. Sometimes, it is suitable to isolate the front-channel operation from BOX and back-interface defects. This is achieved by overimplanting the bottom of the film or by synthesizing self-shielding layers at the buried interface.

SIMOX technology meets the main requirements for spatial and defense integrated circuits (ICs): (i) radiation-hardness and reliability under harsh environments, (ii) high-speed and low-power operation, and (iii) capability of operating over a wide temperature range (-270 °C to 350 °C). SIMOX-based ICs are intrinsically hardened to transient radiation upsets such as gamma radiation and single-event upsets (SEU). Rad-hard CMOS test circuits fabricated on 0.1 μm SIMOX films exhibit performance superior to bulk-silicon counterparts. The subthreshold swing is less than 70 mV/decade, and the ring oscillators have gate delays of 95 and 68 ps and power-delay products of 8.3 and 54 fJ for supply voltages of 3 or 5 volts, respectively.[22] Tests under total-dose gamma radiation showed that even after 1–10 Mrad, these devices are still operational.

The key area of applications for the rad-hard SOI parts are in static memories. The 64 K and 256 K SRAMs formed on partially depleted or fully depleted SIMOX wafers are in production at several semiconductor companies, while 1 Mbit SRAM with 0.8 μm feature sizes has been successfully demonstrated on SIMOX.[23]

Hite et al.[24] reported a novel SRAM cell that makes use of a gated resistor/transistor in the cross-coupling to obtain a high level of resistance to upset over temperature while maintaining a minimum worst-case WRITE time of less than 20 ns at +125 °C. The cell was implemented in a 256 K SRAM fabricated with 1 μm technology on SIMOX wafers with 0.3 μm Si film and 0.4 μm buried oxide. The source/drain implants were driven to the BOX interface so that the depletion region was only

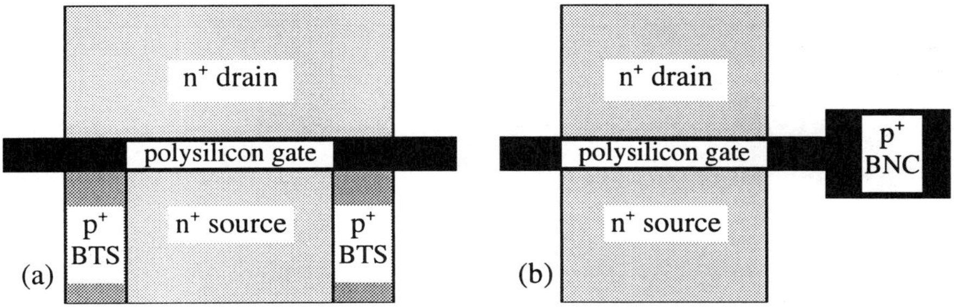

Figure 3.3 (a) SOI double Body-Tied-to-Source (BTS) transistor, and (b) Body Node Contact (BNC) transistor (after Hite et al.[24]).

at the drain/body interface beneath the gate. The body nodes of all transistors, except the pass transistors, were connected to the associated source node by body-tied-to-source straps, as shown in Figure 3.3a. For the n-channel transistor, the body tie is a P^+ source/drain diffusion that contacts the P^+ body region. The silicide process then connects the P^+ body tie to the N^+ source.

Pass gates, such as the access transistors in the memory cell, have the body node connected to a supply voltage by a mesa connection at the edge (Fig. 3.3b). These body contacts help to reduce parasitic bipolar effects, but do not completely eliminate them because of distributed resistance from the contact to portions of the body node.

The 8x32 K SRAMs were characterized for SEU using nickel, iodine, bromine, and gold ions at fluence levels of 2×10^6 particles/cm^2. The tests included measurements in the continuous READ mode, at room temperature and 125 °C, with variable power supply and angle of incidence. A SEU of less than 1×10^{-10} errors/bit-day over temperature was achieved. The cross section versus effective LET (Fig. 3.4a) shows a small increase in sensitivity from 25 °C to 125 °C, caused by the decrease of node restoration drive current of the p-channel load transistors. The cross section does not reach a saturation value at the highest LET of 163 MeV/(mg/cm^2). SEU tests found the 8x32 K SRAMs to be more sensitive to upset with increasing supply voltage, the opposite of usual expectations. The influence of temperature and supply voltage on the access time is shown in Figure 3.4b.

Studies of Mrad(Si) irradiation effects have been conducted on novel

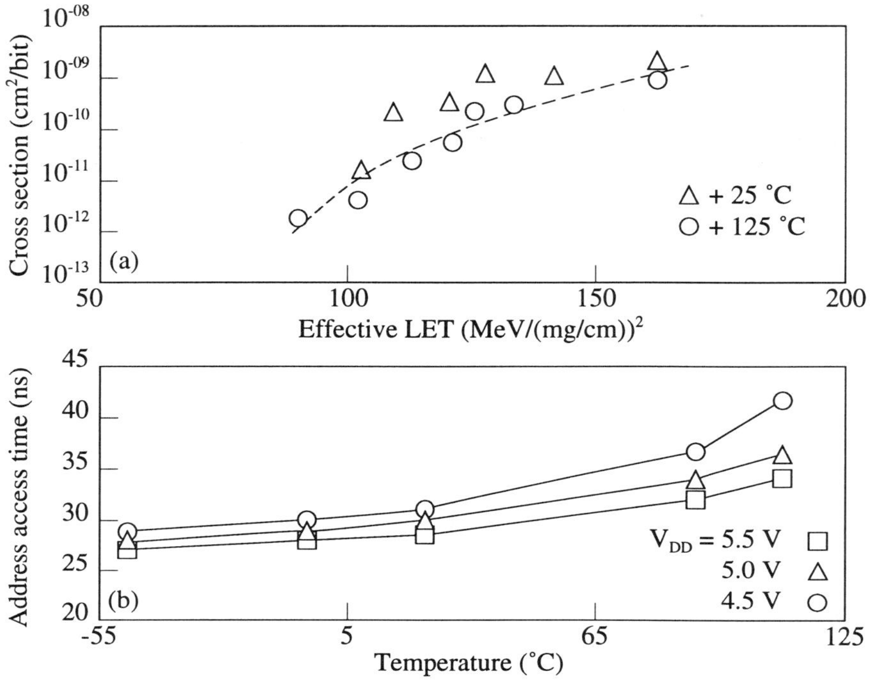

Figure 3.4 (a) Cross section versus LET for SOI 8x32 K SRAM at 5.5 V, and (b) address access time versus temperature (after Hite et al.[24]).

double-gate MOS transistors.[25, 26] The results showed that Gate-All-Around (GAA) MOSFETs offer a higher radiation tolerance than the single-gate SOI transistors. This may be attributed to the fact that, in GAA devices, the active polysilicon gate is present at both the front and the back interfaces of the MOSFET. The test reveals that the maximum transconductance is decreased by about 40% for the 50 nm gate-oxide devices, and by 10% only for the 30 nm oxides, at 10 Mrad(Si) level. The subthreshold slope degradation after exposure indicates the creation of interface traps. At a total dose of 30 Mrad(Si), the gate-referred noise, which was extremely small before irradiation, is increased by two orders of magnitude, both in linear operation and in saturation. The GAA and, in general, thin double-gate SOI MOSFETs look very promising for rad-hard applications up to 10 Mrad(Si) level.

The successful fabrication of a fully functional 1 Mbit SRAM by Lu et al.[23] is a significant milestone in the demonstration of the viability of ULSI on SOI. This SRAM was implemented on commercially available single-implanted SIMOX wafers. The BOX thickness was 500 nm, whereas the epitaxial Si film was 300 nm thick. The SRAM is based on mesa-isolated CMOS SOI technology with triple-level metal and rather conventional bulk CMOS design rules. Features and performance for this 1 Mbit SRAM include 0.8 μm channel length, six-transistor cell, bond pad selection of both the organization (x1, 4x, or x8) and the I/O compatibility (CMOS or TTL), and an access time of 20 ns at 5 V and 25 °C operation. The cell and chip areas are 7.2x14.4 μm^2 and 12.2x17 μm^2, respectively, with total moat area of 110 mm^2.

The design took advantage of the SOI technology feature of allowing abutted N^+ and P^+ drain regions in the layout of a full six-transistor memory cell. The cell includes gated resistors in the cross-coupling for increased immunity to upset by cosmic rays. A trade-off of increased area and reduced drive current exists if the body node voltage is controlled, versus increased parasitic bipolar transistor effects if the body node voltage is allowed to float. All transistors for which the source and drain are defined by the logic have the body node voltage tied to the source. For selected pass-gate transistors, such as the bit-line equalization transistors, the body node is allowed to float. One selection criterion for use of a floating body was that there be zero voltage across the transistor in standby condition.

3.3 High-Voltage Devices

Monolithic high-voltage and power-integrated circuits have wide applications in automotive, aircraft, television, robotics, motors, etc. They associate, on a single die, power transistors or switches with CMOS control and logic functions. The critical step in the fabrication of such "smart power" components is the isolation of subcircuit parts that handle very different voltage levels. The conventional p–n junction isolation used in bulk-silicon technology can be advantageously replaced by SOI, which also brings the additional capability of high speed and density.[27] Some typical SOI devices, mostly fabricated on SIMOX and bonded wafers, will be described in this section.

Double-diffused MOS transistors (DMOS) are key devices for power-integrated circuit technology. In comparison to bipolar transistors, their

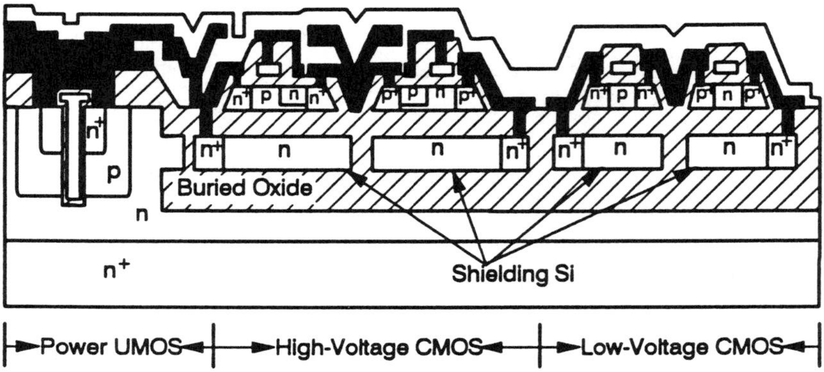

Figure 3.5 Cross-sectional view of an intelligent power device with double buried oxide (after Ohno et al.[29]).

advent offered high input impedance, low drive currents, high switching speed, gate biases compatible with standard CMOS, and simplicity of gate drive circuits. In *vertical* DMOS transistors, the drain contact is located at the bottom of the bulk-silicon wafer (Figures 3.5 and 3.6) and makes impossible the independent biasing of two DMOSFETs. The implementation of DMOS–based circuits on SOI demands a buried-oxide step structure (interrupted BOX). The *top-drain* vertical DMOS is a variant made on thick silicon films. A highly conductive layer (silicides or high-dose implants) is formed at the film–BOX interface and serves as buried drain contact.[28]

The structure best suited for SOI is the *lateral* DMOS transistor (Figs. 3.7 and 3.8). The distinguishing feature is the drift region between the active, gate-controlled channel and the drain. This lateral buffer zone is depleted in the off-state (reverse biased p–n junction), whereas in the on-state it contributes an additional resistance in series with the channel. At large gate bias, it is the drift region that dominates the current value, on-resistance, and power dissipation.[28] Several alternative approaches have been proposed for the optimization of lateral DMOS transistors on plain SOI or on interrupted BOX.[29, 30]

A unique feature of SIMOX is the electric-field shielding layer, which can be formed by either back-interface engineering (low-temperature

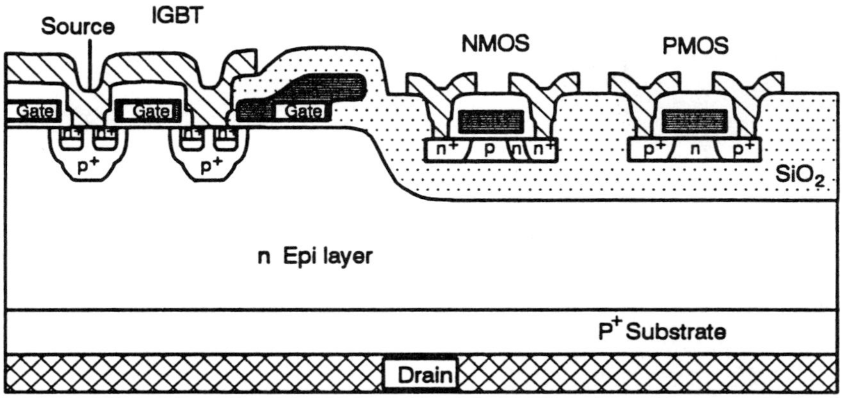

Figure 3.6 A 600 V and 25 A intelligent vertical IGBT with CMOS circuitry formed by SIMOX technology (after Vogt[32]).

postimplantation annealing) or synthesis of double buried oxide. The configuration of an intelligent power circuit, with localized double BOX structure, is shown in Figure 3.5. It consists of a vertical power MOSFET, located in the bulk-silicon area of the chip, and high- and low-voltage CMOS circuits that are electrically protected by the BOX.[31] The high-voltage CMOS transistors are provided with offset gates and field-plates that achieve high source-drain breakdown voltage and make up a high-voltage gate drive circuit for the vertical power transistor. The crystalline silicon layer between the double buried oxide is connected to the source by metallization and serves as a buried back gate that can effectively shield the CMOS from varying substrate potential. All components are individually and completely isolated by oxide layers to ensure latch-up-free operation.

Tests on this device yield promising results: (i) 63 V breakdown voltage and 0.3 ohm on-resistance at $V_G = 12$ V for the power transistor, (ii) stability of the threshold voltages in the CMOS transistors and of the inverter characteristics, even if the substrate potential varies from -100 to +100 V, (iii) low source-drain leakage currents ($\sim 10^{-14}$ A/μm at V_D=5 V), and (iv) an on/off ratio of around 7 at T = 125 °C. This circuit offers two basic functions for automotive applications: control of battery feeding to an external load and overcurrent protection for the

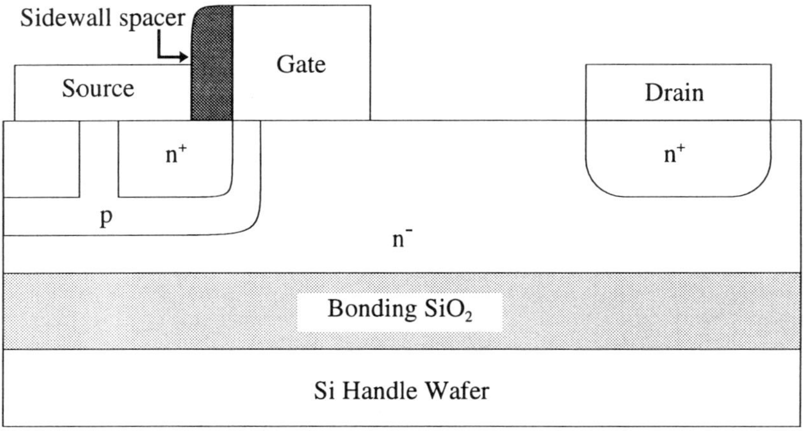

Figure 3.7 The cross-sectional view of a lateral DMOS power transistor on 0.4 μm SOI film (after O'Connor et al.[34]).

power transistor.

Another high-voltage, high-power SIMOX circuit has been reported recently by Vogt.[32] Local SIMOX is used to isolate low-power CMOS for analog/digital circuits from a vertical 600–1200 V insulated-gate bipolar transistor (IGBT) (Fig. 3.6). The dielectric isolation helps to master the high-current (> 25 A) switching, despite the strong minority carrier injection within the IGBT. In addition to many other applications, these devices can provide miniaturized single-chip power supplies for smart sensor systems.

The merit of wafer bonding is the almost infinite flexibility for adjusting the thicknesses of the silicon overlayer and buried oxide. A 120 V AC and 250 V DC monolithic power supply has been processed with bonded wafer technology.[33] The 90 V–1.3 A power DMOS of Figure 3.7 was fabricated on a 400-nm-thick SOI film.[34] The power devices are isolated from each other and from the control circuitry by LOCOS oxidizing through the thin SOI film, and both high- and low-side drivers can be combined on a single chip.

The thin SOI layer virtually eliminates step coverage problems with interconnects and avoids the complicated planarization schemes often needed for dielectrically isolated power devices. The DMOS has the source contact self-aligned to the gate, while the drain is offset from

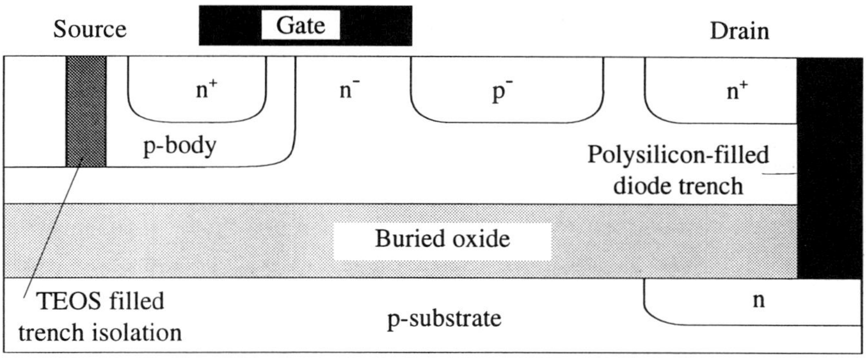

Figure 3.8 Cross-sectional view of an SOI lateral DMOS with buried diode and surface p-layer (after Pein et al.[35]).

the gate to facilitate a high breakdown voltage. The source, drain, and gate contacts are formed by a self-aligned cobalt silicide process. The maximum values of the saturation current and breakdown voltage are 1.3 A and 90 V, respectively. The device on-resistance (4.1 ohms) can be traded off against saturation voltage by varying the length of the drain-drift region.

Pein et al.[35] have reported a high-voltage lateral DMOS transistor that has the drain connected to the substrate via a reverse-biased buried diode (Fig. 3.8). This structure requires relatively thick buried oxide (1 μm) and Si layers ($\sim 5\mu$m) that are simpler to fabricate on bonded and etched-back SOI wafers. It is similar to junction-isolated devices in that most of the potential drop in the vertical direction occurs in the lightly doped p-substrate. A P^- layer is included along the surface of the drift region and enables the on-resistance of the device to be lowered.[35] The buried diodes are fabricated by sequential trench etching through the Si layer and underlying buried oxide followed by ion implantation of the buried diode. Contacts to the diode are made by filling the trench with heavily doped n-type polysilicon followed by an etchback planarization. Lateral isolation of the devices is achieved through trenches, filled with TEOS oxide and subsequently planarized.

Operation of the IGBT and lateral DMOS as high-side switches has been confirmed experimentally. The combination of low on-resistance and high switching speed obtained with SOI devices is favorably compared to equivalent bulk Si components. This combination promises enhanced functionality and performance of single-die power integrated circuits on SOI.

3.4 High-Temperature Devices

Many applications are envisaged for high-temperature integrated circuits rated beyond the standard limit of 125 °C. Operation of bulk MOS devices is usually limited to about 150–200 °C, due essentially to the increase of the junction leakage current. Additional problems are the reduction of carrier mobility and the shift of threshold voltage with increasing temperature.

The temperature range of silicon components may be substantially extended by using SOI devices, since they possess three main advantages over bulk-silicon counterparts. First, the small junction area in SOI MOSFETs reduces the high-temperature leakage currents by as much as 3 to 4 orders of magnitude. Second, the threshold voltage variation with temperature is 2 to 3 times smaller than in bulk, and the output conductance of SOI MOSFETs is actually improved. Finally, the thermally activated latch-up is absent in SOI. These features enable the fabrication of analog and digital SOI CMOS circuits operating beyond 300 °C with little performance degradation. Several types of SOI MOS devices have been investigated in a wide temperature range.[36, 37]

Submicron fully and partially depleted MOSFETs were fabricated, respectively, in 80-nm-thick and 200-nm-thick silicon films on SIMOX substrates. The leakage currents were found to increase by about six orders of magnitude between 25 °C and 350 °C. They show an activation energy corresponding to the silicon bandgap and, therefore, are dominated by the mechanism of carrier diffusion.[36] The subthreshold swing increases to about 200 mV/decade at 200 °C and 260 mV/decade at 300 °C. The threshold voltage decreases almost linearly with temperature, the slope being around -2 mV/°C for partially depleted and -1.5 mV/°C for fully depleted MOSFETs. The carrier mobility follows the typical $T^{-1.5}$ law of inversion layers. When the opposite interface is maintained in accumulation, a reduced temperature dependence of the mobility, $\mu \sim T^{-1}$, is observed in fully depleted transistors.

Reliability tests were conducted at various temperatures on p-channel devices, which primarily suffer from electron injection into the SIMOX

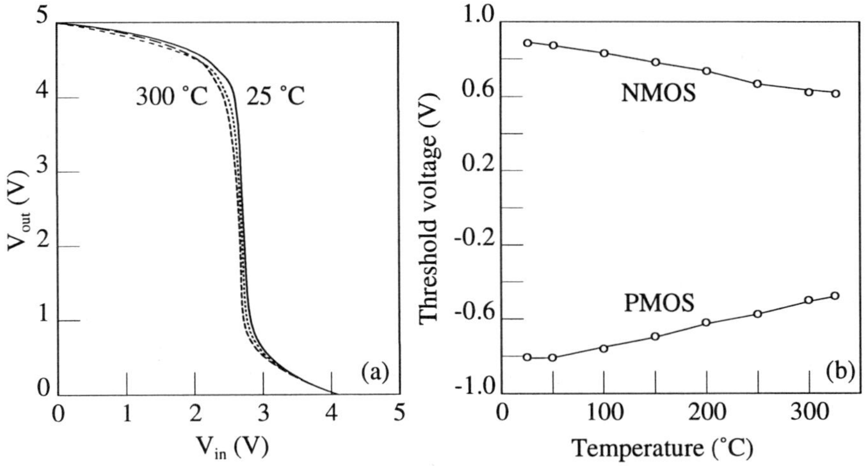

Figure 3.9 (a) The transfer characteristics of a GAA inverter at 25, 100, 200, and 300 °C, and (b) variation of the n- and p-channel threshold voltages with temperature for the GAA MOSFETs (after Francis et al.[37]).

buried-oxide (see also Section 10.1). After front-channel stress, the shift in back-channel threshold voltage increases with temperature in the 20–130 °C range, and decreases above.

The aging attenuation observed at high temperature is presumably due to the shortening of the electron mean free path with temperature that reduces the fraction of hot carriers injected into the BOX.[36] This argument also explains the lower degradation of the back-channel transconductance at high temperatures. It is concluded that the BOX is still reliable at 350 °C.

Francis et al.[37] have studied the high-temperature behavior of SOI and gate-all-around (GAA) CMOS circuits ranging from simple inverters and logic gates to frequency dividers. Although none of the SOI and GAA circuit designs was optimized, the major causes of failure in bulk-CMOS logic at high temperature, i.e., excess power consumption and degradation of logic levels and noise margin, were found to be greatly alleviated. CMOS inverters exhibit full functionality and very little change in static characteristics for temperatures up to 320 °C.

Figure 3.9a shows the transfer characteristics of a GAA inverter operating at 25, 100, 200, and 300 °C. The results reveal that switching voltage remains stable, which is attributed to the extremely weak and symmetrical variation of the n- and p-threshold voltages with temperature, as shown in Figure 3.9b. The output voltage range is reduced by only a few millivolts due to the slightly increased leakage current of the OFF device and the reduced carrier mobility of the ON device.

In logic gates with series transistors, such as AND and NAND gates, the increase of standby supply current with temperature (about one order of magnitude between 20 °C and 320 °C) remains even more limited than expected. This is due to the fact that in SOI circuits, the drain leakage current of each individual transistor flows towards its source, and thus into the following transistor, unlike bulk circuits, where all drain leakage currents are collected by the substrate. The circuit speed was tested on toggle-chain frequency dividers. The results showed that these SOI circuits were still functional for temperature up to 320 °C at a maximum input frequency of 100 MHz, which is about half the speed achieved at room temperature.[38]

3.5 Low-Power Applications

The rapidly growing interest in low-power integrated circuits is justified by the huge demand for battery-operated equipment. Low voltages are also mandatory for subquarter-micron CMOS ULSI circuits, in terms of short-channel effects, hot-carrier reliability, and power dissipation. However, the lowering of the operating voltage results in a loss of performance in bulk-silicon devices. This trend is a major opportunity—and a challenge—for SOI. Fully depleted devices are preferred over partially depleted ones due to current drive enhancement, steeper subthreshold slope, and easier scaling with low threshold voltage.

Low-voltage ($\leq 2\,\mathrm{V}$) operation implies a small gap voltage between off- and on-states—in other words, low threshold voltages and steep subthreshold slopes. This requirement corresponds to the domain of excellence of ultrathin fully depleted SOI MOSFETs, which offer a subthreshold slope close to the ideal value of 60 mV/decade. Threshold voltages around 0.3–0.5 V are targeted because they lead to reasonable values of standby currents and are easily achievable: the thinner the film, the lower the threshold voltage. Floating-body effects and hot-carrier-induced degradation tend to disappear at very low voltage. The lifetime

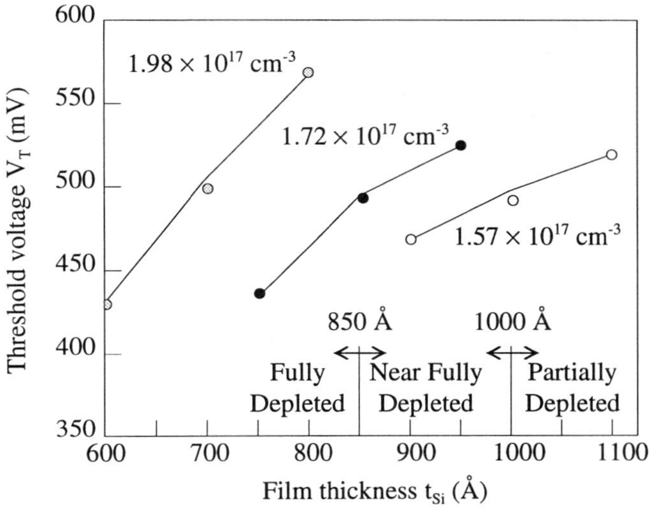

Figure 3.10 Threshold voltage sensitivity versus film thickness for three different doping concentrations in the silicon film (after Wilson et al.[40]).

of 0.1 μm SOI MOSFETs, extrapolated from stressing tests, exceeds 10 years for 2.1–2.5 V operating voltage.[1] In spite of some uniformization of the speed, which is the consequence of velocity saturation, a gain of 25 % over bulk devices is still expected for 0.2 μm SOI transistors. Short-channel effects need to be reduced further by tailoring of the film doping and use of very thin buried oxides. In the meantime, self-heating will be attenuated as a benefit of both low power and thin BOX.

The improvement in low-voltage performance in SOI over its bulk counterparts is due primarily to the much-reduced junction capacitance, leakage, and body effects. The relatively small junction area for SOI circuits provides the possibility of very low standby currents. Houston et al.[39] have demonstrated a 1 μm CMOS/SOI 64 K SRAM with a 10 nA standby current and 150 mW active power at 25 MHz, while for 256 K SRAMs the standby current is less than 50 nA.

Keys to the success of SOI technology also depend greatly on the availability and cost of high-quality wafers that meet the severe requirements of low power/voltage circuits. The practicality of realizing a fully depleted device is subjected to the constraints of Si film thickness, threshold voltage, leakage current targets, and source/drain silicidation. Since the circuit performance and the parametric control at low supply voltage are very sensitive to the variation of threshold voltage, uniformity of silicon film thickness is a priority.

The threshold-voltage dependence on film thickness is compared in Figure 3.10 for three different doping levels. Fully depleted transistors are far more sensitive to thickness variations than partially depleted MOSFETs. Therefore, to ensure acceptable threshold-voltage control, an extremely tight film thickness distribution is required. Further progress is expected in the uniformity of SOI wafers. A temporary compromise would be to design nearly fully depleted devices, at the limit where realistic Si film thickness uniformity can presently be achieved.[40] Advantages such as improved capacitance, body effect, short-channel effect, device density, and subthreshold slope can still be retained. The primary trade-off is some loss of current drive/mobility and potential hot-carrier injection.

3.6 Three-Dimensional Devices

Three-dimensional (3–D) integrated circuits are attractive for at least two reasons: (1) high packing density and (2) outstanding performance, multifunctionality, and versatility due to direct interconnects between the successively stacked active layers. SOI partisans enjoy saying that 3–D integration cannot even be imagined in bulk silicon. For simplicity, the same SOI technology should be used over the whole process. This leaves the recrystallization technique as a unique contender. It is important, during the recrystallization of a new layer, not to damage the already completed underlying devices. Laser or electron-beam recrystallized ZMR offers, in this respect, the best security margins for shallow in-depth penetration of thermal gradients.

The *mezzanine* structure is an intermediate solution for the fabrication of 3–D smart power circuits. It combines the merits of bulk silicon for power devices and of SOI for advanced CMOS circuits. ZMR SOI is formed in selected locations and therefore does not cover the whole wafer. A power DMOS transistor is integrated in the SOI-free bulk-silicon area, whereas the CMOS processor uses the SOI section of the chip.[41]

Full *double-layer* transistors, fabricated with ELO technology,[42] are shown in Figure 3.11. The left-side transistor has dual-gate configu-

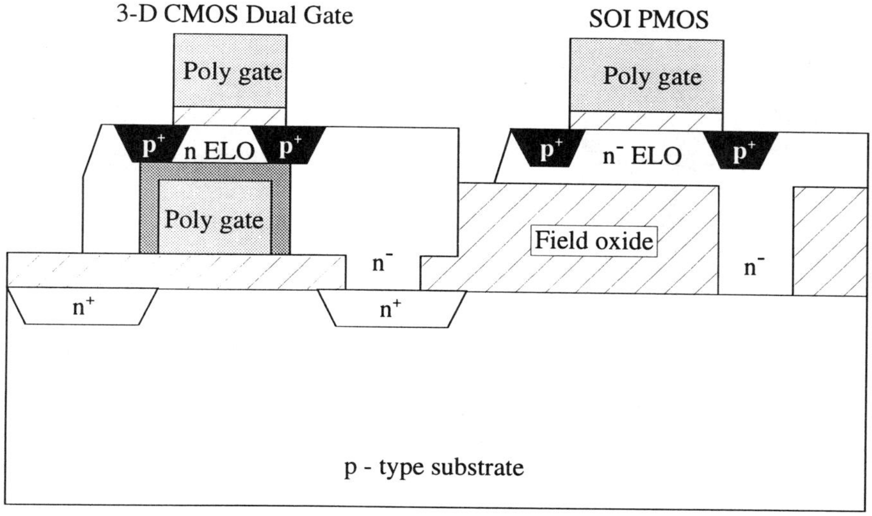

Figure 3.11 Three-dimensional dual-gate stacked p–MOSFET (left) and fully compatible p–MOSFET (right) fabricated on ELO SOI (after Zingg et al.[42]).

ration. The buried polysilicon gate acts as input to both the stacked inverted p–MOSFET and the bottom bulk-Si n–MOSFET. The shared gate has reduced parasitic capacitance, and the latch-up paths are eliminated.[43]

Double-layer image processors were processed by laser recrystallization of unseeded ZMR.[44] Antireflecting silicon nitride patterns served to control the location of grain boundaries. Photodiodes were integrated in the active upper layer just above the signal-processing circuit located in the original bulk-Si wafer. Electrical connections between the two stacked layers were made through via-holes. Static RAMs were processed following the same guidelines. The memory cells were situated in the upper layer and the input/output buffer circuits in the base wafer.[45]

The above technology has been extended to *triple-level* structures.[46] After processing the bulk-Si devices, a 0.7-μm-thick LPCVD oxide was planarized by reactive ion etching. A polysilicon layer 0.5 μm thick was deposited, followed by the patterning of parallel antireflecting Si_3N_4 stripes. After laser recrystallization, the single crystal regions, which were photolithographically defined, were etched into device islands. The

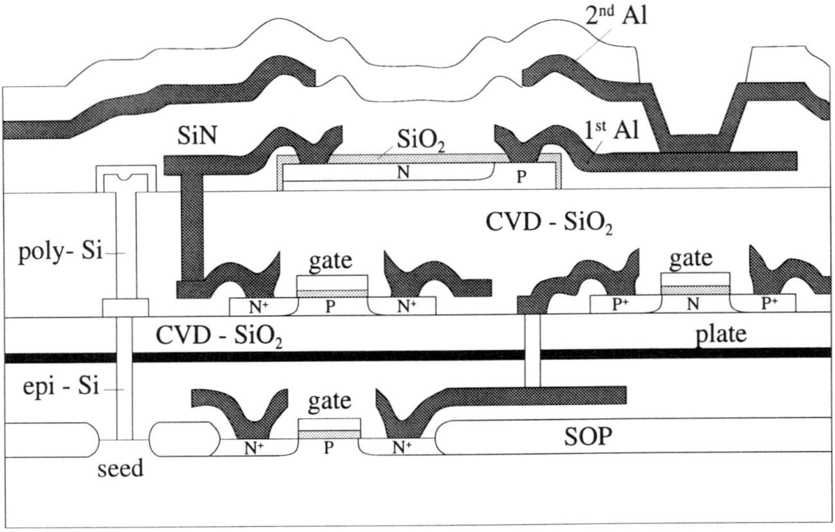

Figure 3.12 Schematic cross-section of a three-layer 3–D image signal processor (after Nishimura et al.[47]).

NMOS circuits were processed with heat treatments limited at 950 °C.

Phosphorus-doped polysilicon interconnects were used to guarantee endurance to successive thermal treatments. The whole sequence was duplicated, with slight changes, for the third level. The three stacked circuits were fully functional. Only a tolerable degradation of electron mobility occurred in the SOI levels: 735 cm^2/Vs in bulk, 528 cm^2/Vs in the first SOI layer, and 549 cm^2/Vs in the top SOI film. Tensile stress was suspected to enhance the mobility of the third-level transistor as compared to the second-level MOSFET.

A high-performance image signal processor was fabricated with the same generic process, except that the laser recrystallization technique was improved by seeding.[47] Small seed openings ($2 \times 2\,\mu m^2$ size and $100\,\mu m$ interdistance) were filled with single-crystal silicon for the first SOI layer, and with polysilicon for the second SOI level. As illustrated in Figure 3.12, the image processor is composed of (i) photodiode arrays in the upper SOI level, (ii) A/D converters in the intermediate SOI layer, and (iii) arithmetic units and shift registers in the bottom bulk-Si level. The size of the pixel was about 1 mm^2, and the whole chip area was $8 \times 8\,\mathrm{mm}^2$. The light sensing, quantization, and data processing were conducted simultaneously, allowing extremely fast parallel processing ($3\,\mu$s for one image frame).

These successful attempts fully demonstrate the feasibility of 3–D SOI circuits. However, it is unlikely that 3–D mass production will start soon. There is still enough space indeed for ULSI circuits to be implemented in a single layer.

3.7 Transducers

The role of a transducer (or sensor) is to transform a nonelectrical excitation into an electrical signal. The development of appropriate sensors made possible the penetration of microelectronics into traditional nonelectrical fields of application. For each category of input signals (mechanical, chemical, thermal, optical, or magnetic), there were successful attempts to use SOI instead of bulk Si.

3.7.1 Mechanical Sensors

A *pressure* sensor, schematically reproduced in Figure 3.13a, consists of a thin diaphragm fabricated by etching a cavity into the wafer substrate. Piezodevices integrated on the top of the membrane sense any deformation that results from pressure variations. The absolute value of the pressure can also be detected with sensors whose cavity was sealed under vacuum.[1] The piezodevices are simple resistors or MOS transistors that come by pairs in order to minimize thermal effects. For calibration and signal processing, the sensing element is duplicated outside the diaphragm.

The obvious advantage of SOI is that the buried oxide, formed by wafer bonding or low- (or high-) dose oxygen implantation, provides an ideally selective etch-stop layer. Ultrathin membranes are manufacturable and allow very low pressures to be detected. The thickness of the membrane can be adjusted by epitaxy for applications in almost any ranges of pressure. Diaphragms with well-controlled shape and size are achieved by anisotropic etching of Si substrate.[32] To this purpose, a silicon nitride mask is deposited on the back of the wafer, and rectangular or circular openings are processed.

[1] Such a scelling procedure can be found at the origin of wafer bonding technology.

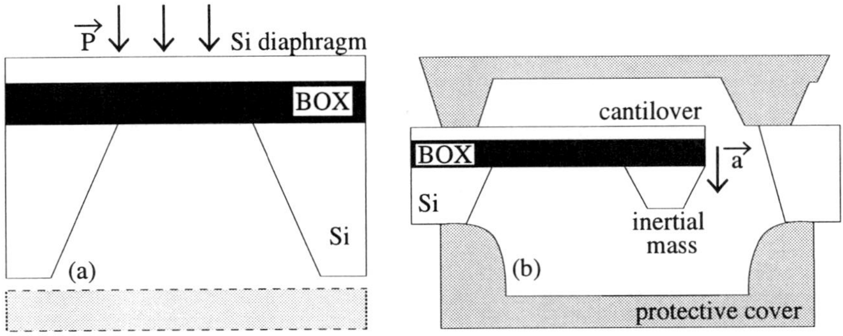

Figure 3.13 Cross-sections of two mechanical microtransducers fabricated on SIMOX: (a) Pressure sensor (the dotted base of the device is optional and serves for the measurement of absolute pressure); (b) accelerometer with cantilever structure.

The *accelerometer* of Figure 3.13b has a similar cantilever structure, except that one side of the cavity is mobile and serves as inertial mass. A vertical acceleration (or fluctuation) modifies the force to which the inertial mass is submitted. The resulting bending of the cantilever is merely converted in piezoresistance.

Piezoelectric layers (ZnO) represent an alternative to piezoresistors. Incorporated in the gate insulator of MOS transistors, their polarization will modify the oxide charge and threshold voltage according to the magnitude of the incoming pressure or acceleration.

Single SIMOX[48] or double SIMOX[49] structures can be used as optical waveguides. Once a cavity was etched off in the substrate, the displacement of the diaphragm or cantilever is sensed through the change in light propagation properties. The diaphragm can be implemented into a system of differential capacitors such that its bending modifies the electrode interdistance—hence the capacitance.

3.7.2 Magnetic Sensors

Van der Pauw (Fig. 5.3) and MOS–Hall devices (Fig. 9.1) are conventional sensors for the detection of the magnetic fields B perpendicular to the wafer surface. The low carrier mobility in silicon eliminates any possibility of magnetoresistance applications: $\Delta R/R \sim 2\mu^2 B \Delta B$.

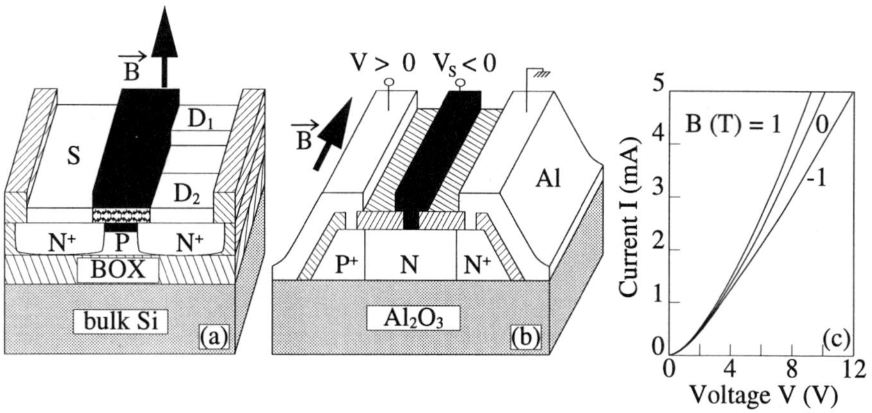

Figure 3.14 Magnetic sensors integrated on SOI: (a) dual-drain magnetotransistor; (b) SOS magnetodiode with optional Schottky contact; (c) I–V characteristics of the SOS magnetodiode under magnetic field (after Mohaghegh et al.[53]).

MOS transistors with multiple drains (Fig. 3.14a) operate as a three-terminal Hall device. A normal magnetic field induces a current imbalance by allowing one of the drains to receive more carriers.[50] The optimum geometry of the dual-drain magnetotransistor is square-like ($L \simeq W$) with roughly a 35% gap between the drains. Based on similar principles, lateral bipolar magnetotransistors with dual collectors have been fabricated in SIMOX.[32]

A *magnetodiode* is a very unique SOI transducer.[51, 52] This device is simply a long p–i–n diode biased in the forward mode to achieve double-injection conditions (Fig. 3.14b). The current magnitude is governed by the high-density electron-hole plasma that is supplied by the N^+ and P^+ reservoirs. A quadratic current–voltage law, $I \sim V^2$, is typically obtained, the current being strongly dependent on the carrier recombination rates in the volume and at the interfaces of the film.

When a magnetic field is applied in the plane of the wafer, perpendicular to the current, the electron-hole pairs experience a vertical deflection by the Lorentz force. According to the polarity of the magnetic field, the pairs will be shifted towards either the top or the bottom film interfaces. The recombination rate is enhanced at the interface receiving the carriers and lowered at the opposite interface.

Most interesting is the case of interfaces with very contrasting properties, as in SOS or in low-temperature annealed SIMOX. Carrier pairs deflected towards the defective Si–Al_2O_3 interface (in SOS) or back Si–SiO_2 interface (LTA–SIMOX) are very rapidly recombined; hence, the average carrier density and the current are reduced (Fig. 3.14c). For opposite polarity of the magnetic field, carriers are shifted toward the high-quality Si–SiO_2 interface and therefore escape the recombination at the bottom interface. The current increases accordingly.

For the magnetodiode effect to work properly, the device must be 10–20 times longer than the ambipolar diffusion length. Even more crucial is the film thickness, which should be comparable to or slightly smaller than the diffusion length.[52] The latter requirement is extravagant for bulk silicon, but can be fulfilled in SOI. The revenge of lower-quality SOI technologies is to offer ideal conditions for magnetodiode operation: poor carrier lifetime and asymmetric interfaces. Excellent sensitivities have been achieved in SOS magnetodiodes: $\Delta V/\Delta B \simeq 10\,\mathrm{V/T}$ and $\Delta I/\Delta B \simeq 150\,\mathrm{mA/T}$.[53]

A more elaborated device has a Schottky contact added onto the surface of the magnetodiode. The reverse current of the Schottky diode is proportional to the surface carrier concentration (underneath the contact), which again is modulated by the magnetodiode effect. The surface concentration of electron-hole pairs is even more sensitive to magnetic-field variations than their average concentration. The more complex processing of Schottky magnetodiodes is rewarded by a gain in sensitivity: $\Delta V/\Delta B \simeq 30\,\mathrm{V/T}$ in SOS.[53]

Three-dimensional detection of magnetic fields can be achieved by combining two magnetodiodes perpendicular to each other with a Hall sensor. The magnetic sensors can also be used as *indirect* detectors of pressure and position. They simply measure the variations of the field delivered by a small permanent magnet placed on a membrane or any other mobile element.

3.7.3 Chemical Sensors

The ion-sensitive field-effect transistor (ISFET) is designed to determine the presence and concentration of a particular chemical element in a solution. A special membrane, selectively sensitive to the ion to be detected, is deposited on the top oxide and plays the role of a transistor gate. When the ISFET is immersed in a solution, the absorption of ions modifies the balance of electrical charges. This is equivalent to a gate bias effect, which governs the transistor characteristics. The ion concentration is monitored through changes in drain current, or better, in the body voltage that is necessary to maintain the drain current constant.

The advantage of SOI comes from its specific dual-gate configuration. Even if one gate is missing, being replaced by the membrane, the opposite gate can still be operated for additional control and flexibility. SIMOX ISFETs have been processed by etching a cavity, as is the case for pressure sensors, and depositing a pH-sensing layer on the back of the BOX. Sensitivities very close to the theoretical peak value (60 mV/pH) have been reported.[32]. The ISFET opens wide opportunities for applications in the food, medical, and chemical industries.

3.7.4 Thermal Sensors

All silicon devices suffer from temperature variations. Viewed in another way, they are remarkably good temperature sensors. For instance, the voltage drop across a p–n diode, measured at constant forward current, is a quasi-linear function of temperature: $V = (kT/q)\ln(1 + I/I_0)$. Second-order nonlinearities are induced by the temperature dependence of the reverse current and can easily be compensated for. In a bipolar transistor, the base-emitter voltage, measured at constant collector current, behaves similarly. The practical problem is that the silicon material surrounding the device heats up as well. Its thermal inertia hampers the rapidity and frequency of the measurement.

At this point, SOI becomes very attractive because the silicon film is thin and, most importantly, the silicon substrate can be etched off. Owing to the poor thermal conductivity of oxide and air (100 times lower than for Si), the thermoresistors, diodes, or transistors, integrated on a thin SOI diaphragm, behave as if they were virtually isolated from the surrounding silicon mass.

The applications cover a broader domain than the standard thermometry. A *thermopile* for infrared detection (Fig. 3.15a), with sensitivity above 200 V/W and time constant of 20 ms, has been built on SIMOX.[32] The radiation signal is dissipated in an absorbing layer. The resulting increase in temperature is sensed by an integrated thermocouple, located between the Si film and Al metallization. The same concept serves for thermal *microwave power* detectors.

The *flow* sensor of Figure 3.15b associates thermometers with a ther-

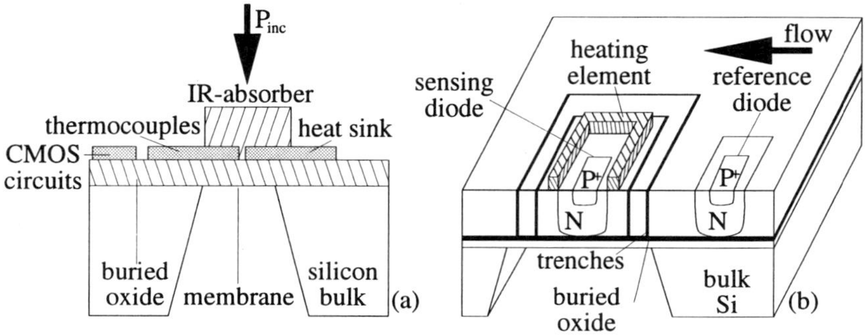

Figure 3.15 (a) Infrared thermopile and (b) flow sensor fabricated on SIMOX (after Vogt[32]).

mally isolated heating element. The local temperature is more or less reduced according to the lateral flow of gas or liquid.

3.7.5 Optical Sensors

Although photoconductors and phototransistors exist, the photodiode is the typical optical sensor in silicon. Photogenerated electron-hole pairs are separated by the junction field and give rise to a photocurrent. The light intensity is determined from the photocurrent magnitude at reverse bias, or from the change in diode voltage measured in open-circuit conditions or for a fixed forward current. Zener photodiodes make use of the internal multiplication effect in the avalanche region. Silicon photodiodes are p–n or p–i–n type and can have vertical or lateral configurations.

The main interest and originality of SOI photodiodes are in the context of three-dimensional integration (Section 3.6) of sophisticated image processors.[47] The third active layer of the stacked device presented in Figure 3.12 receives a number of photocell arrays. Each photocell consists of a shallow N^+P junction; the junction depth is 0.2 μm, and the film thickness is 0.7 μm. Ultrathin SOI layers are not suitable because light absorption in the Si film is too weak. The responsivity of SOI photodiodes at long wavelengths is poorer than for bulk Si, but it is sufficient at 0.55 μm wavelength, which is the preferential domain of sensitivity for the human eye. Other problems are related to optical interference in multilayered SOI and to the defects subsisting at the back interface, which may also hamper the sensitivity.

The input optical signal and the resulting output electrical signal are increased by connecting in parallel hundreds of photodiodes. A photocurrent of 600 nA/mm^2 was measured at bright condition (900 lux).[47] It is expected that stacked diodes, integrated in each active layer of three-dimensional circuits, would present different responsivity ranges, thereby allowing color discrimination.[14]

The output voltage can be remarkably increased by series circuits of photodiodes.[54] An array of 1 cm^2, containing thousands of mesa-isolated SOS photocells, can deliver as much as 10 kV at 10 μA when illuminated by 1 W laser diodes. This makes possible the optical-to-electrical conversion of DC power and opens many applications. Such converters are specific to SOI or SOS because they offer an unrivaled degree of electrical isolation.

3.8 Innovative Devices

3.8.1 Double-Gate and Volume-Inversion Transistors

The dual-gate configuration is the intrinsic feature of SOI devices. Some components, especially those dealing with high voltage and power, have a hard time accommodating this flexibility; hence, their design is intended to avoid the influence of the bulk-silicon substrate. But the general trend is to select a back-gate bias that provides ideal conditions for optimum performance of the front channel.

The *double-gate MOSFET* is one step ahead of the context of interface coupling. The two gate biases are interrelated and simultaneously scanned: $V_{G_2} = KV_{G_1}$, where the coefficient K accounts for the difference in thickness and threshold voltage between the gate and buried oxides. What happens to inversion-mode MOSFETs when increasing $V_{G_{1,2}}$? Nothing, really, in thick, partially depleted films. The two channels grow independently, whereas the undepleted central region of the film is controlled by majority carriers. The total current comes from the parallel combination of two transistors.

The case of fully depleted films is far more exciting. The potential in the middle of the film tends to follow the two surface potentials as they increase from accumulation to strong inversion (see also Fig. 8.1). In ultrathin (< 100 nm) and undoped ($< 10^{15}$ cm^{-3}) layers, biased in the near-threshold range, the parabolic potential distribution across the film

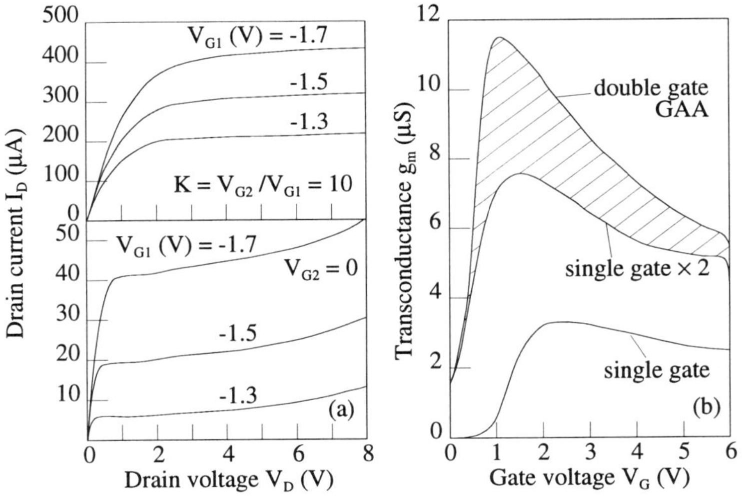

Figure 3.16 (a) Drain characteristics of a fully depleted SIMOX MOSFET operated in volume-inversion mode ($V_{G_2} = 10V_{G_1}$) and in standard front-channel mode ($V_{G_2} = 0$). (b) Transconductance curves in gate-all-around GAA–MOSFET and reference single-gate SOI MOSFET (after Cristoloveanu[49] and Colinge[14]).

becomes almost flat, with less than 5 mV concavity in the central region. Even if the concavity is enlarged with increasing gate bias, the key point is that the whole film can reach strong inversion. The device has received the name VI–MOSFET or *volume-inversion* MOS transistor.[2]

The minority carriers are no longer confined at interfaces, and many of them enjoy flowing in the silicon-film volume. They do not have to fear so much the presence of interface traps, oxide charges, surface roughness, and other surface-related unpleasant scattering events. Increased values of carrier mobility and transconductance are expected. Furthermore, the film volume is much thicker than a surface inversion layer and can contain a large number of minority carriers, allowing the drain current to increase.

The concept of volume inversion has duly been verified by operating SIMOX transistors in both the VI-mode (with $K = 10$) and the conventional front-channel mode ($K = 0$).[2] The VI–MOSFET offers enhanced transconductance and steeper subthreshold slope. Figure 3.16a

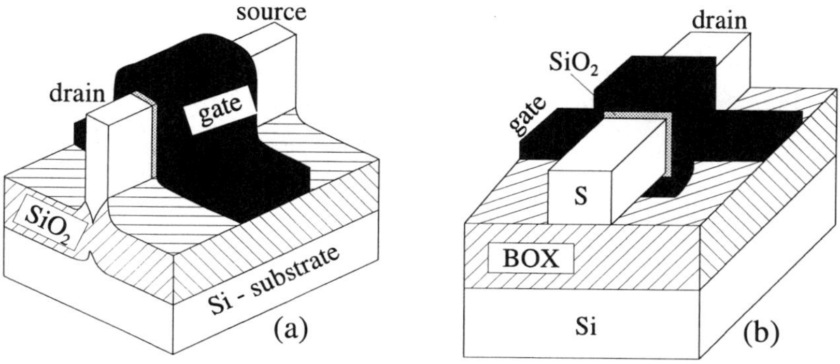

Figure 3.17 Configuration of (a) DELTA[56] and (b) Gate-All-Around (GAA) transistors.[14]

shows that, in the saturation region, the drain current is, as expected, much higher in the VI–MOSFET. The kink effect is suppressed, and the breakdown voltage is clearly improved (from 10 V for $K = 0$ to 15 V for $K = 10$). The linear region is extended because a higher voltage is needed for volume pinch-off. Similar benefits have been obtained in the case of depletion-mode transistors with *volume accumulation*.

The question about how to bias the back gate and avoid the impractical use of high back-gate voltages is partially answered by the synthesis of ultrathin buried oxides. The structure of Figure 3.21a has been envisaged to provide independent secondary gates to each transistor.[49] The miracle of fiction-to-reality conversion happened once more in the magic SOI kingdom, and this structure has been effectively fabricated for the benefit of the smart power circuit shown in Figure 3.5. The silicon islands buried in the BOX were formed by double oxygen implantation.[31]

The DELTA transistor of Figure 3.17a is a vertical single-gate version of the horizontal double-gate MOSFET. Very narrow ($< 0.2\,\mu$m) silicon islands are anisotropically etched on the bulk-Si wafer. Subsequent LOCOS-like oxidation proceeds underneath the islands and forms an SOI structure. The lateral sides of the island act as front and back channels and share the same gate voltage (i.e., $K = 1$) and gate oxide. The DELTA transistor exhibits nearly ideal subthreshold slope (62 mV/decade), high current drivability, and immunity to short channel effects down to $0.15\,\mu$m lengths.[56]

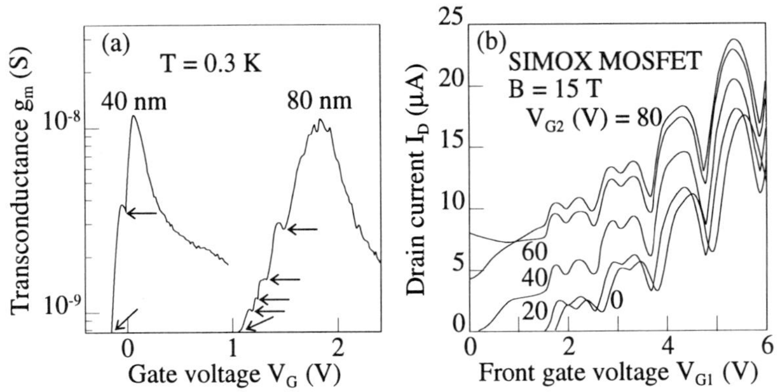

Figure 3.18 (a) Quantum structures on the low-temperature transconductance curves of GAA transistor (arrows indicate the filling of various subbands, after Colinge et al.[57]); (b) Shubnikov–de Haas oscillations at a high magnetic field, showing the influence of the back-gate bias on the front-channel characteristics of a fully depleted SIMOX MOSFET (after Ouisse et al.[59]).

The *gate-all-around* (GAA) transistor (Fig. 3.17b) is fabricated by etching a cavity into the BOX, underneath the silicon-device island.[14] The resulting silicon "bridge" is thermally oxidized and wrapped into a LPCVD polysilicon gate. A continuous MOS structure, with constant oxide thickness, covers the whole periphery of the transistor body. If the film doping and thickness are small enough, the GAA–MOSFET becomes an ideal specimen of volume-inversion transistors.

In early days, it was suspected that the outstanding performance of the VI-MOSFET was an artefact induced by the large bias applied on the back gate. Such an argument cannot be applied to double-gate symmetrical structures. Nevertheless, the GAA transistor persists in exhibiting transconductance and current values that exceed *twice* the values measured in a reference single-gate MOSFET (Fig. 3.16b). The interface coupling is, of course, more than the sum of two channels. It can give rise to original quantum effects: oscillations in transconductance $g_m(V_G)$ curves (Fig. 3.18a)[57] and an increase in the threshold voltage as soon as the film thickness is reduced below a certain critical value ($\sim 10\,\text{nm}$).[58]

As discussed in Section 8.1, the VI–MOSFET can be a test vehicle

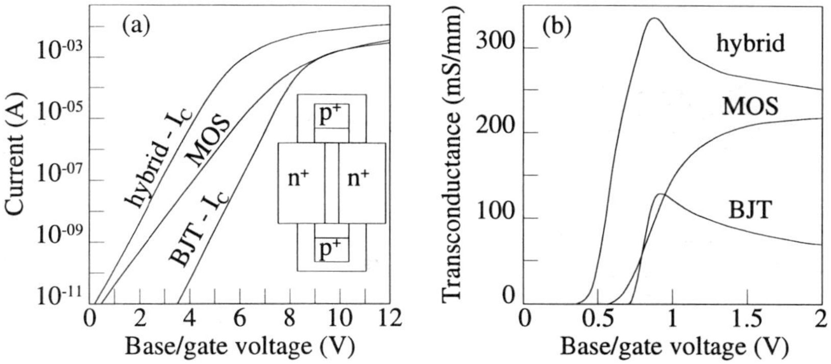

Figure 3.19 Quarter-micron hybrid bipolar–MOSFET SOI transistor (insert). (a) Subthreshold Gummel and (b) transconductance characteristics for the hybrid and individual modes of operation (after Parke et al.[17]).

for achieving 2–D to 3–D transitions. Figure 3.18b shows Shubnikov–de Haas experiments performed at a high magnetic field in a fully depleted SIMOX MOSFET (160-nm-thick film and $2 \times 10^{16}\,\text{cm}^{-3}$ doping level).[59] Current oscillations occur if the Landau-level separation exceeds the thermal energy and if the relaxation time is longer than the cyclotron orbit ($\mu B > 1$). As the population of the inversion layer grows, the Landau levels are successively filled, and a conductivity peak is observed whenever the Fermi level crosses a Landau level. The periodicity of the oscillations with increasing front-gate bias gives evidence for the 2–D character of the SOI transistor and corresponds to the gate oxide thickness. The periodicity does not change when the back gate is driven to strong inversion, which means that the bidimensionality of the system is not perturbed. Much thinner and lower-doped layers would be necessary for 2–D to 3–D transitions.

It is obvious that volume inversion hardly occurs in highly-doped ($> 10^{17}\,\text{cm}^{-3}$) films, which are frequently used for conventional submicron MOSFETs. The film thickness is even more stringent a requirement, but is not an obstacle anymore, since SOI films thinner than 50 nm are currently being manufactured.

Nowadays, the double-gate symmetrical structure is the subject of modeling efforts,[60, 60] while its representatives, GAA and DELTA, are

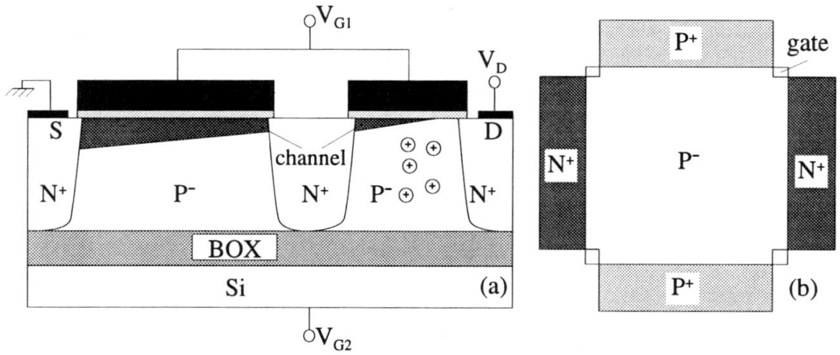

Figure 3.20 Configuration of the (a) twin-gate MOSFET[63] and (b) single-device inverter with cross–MOS structure (after Gao et al.[64]).

undergoing experimental exercises. They show attractive quantitative properties (reduced noise,[62] improved cryogenic operation, etc.) as well as exciting qualitative features (quantum wire operation, tolerance to radiation and hot-carrier effects, etc.).

3.8.2 Other Novel Devices

Since the lateral bipolar structure is inherent to all MOSFETs, their functions can readily be combined. This is of interest in deep submicron devices, where the base (or body) is short enough for the bipolar transistor to achieve reasonable gains.

The *hybrid* mode of operation implies that surface channel and buried bipolar conduction occur simultaneously. The bipolar–FET transistor, originally called a *voltage-controlled bipolar–MOS device*,[14] is formed by connecting the floating body to the gate. A 0.2-μm-long device has been processed in partially depleted SIMOX material.[17] The bipolar current dominates in the subthreshold region, whereas at high gate voltage the MOSFET gains control. The hybrid BJT–FET device outperforms the individual transistors: (i) the subthreshold slope is 60 mV/decade instead of 93 mV/decade for the MOSFET (Fig. 3.19a), (ii) the current gain is over 10^4 as compared to only 15 for the BJT, and (iii) the current is 1.5 larger than for the MOSFET. Figure 3.19b compares the transconductance curves for each mode of operation. The device looks especially

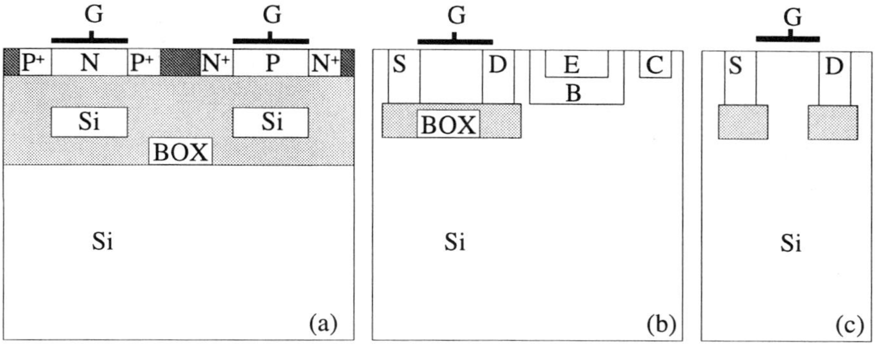

Figure 3.21 Schematics of novel SOI devices: (a) buried silicon islands in double SIMOX for independent biasing of the back gates in volume inversion and power devices, (b) combined MOS and bipolar/power devices in interrupt SIMOX material, and (c) intermittent SOI for body-to-substrate current path.

promising for low-voltage integrated circuits.

The *twin MOS transistor* of Figure 3.20a is intended to suppress the kink and other undesirable floating-body effects that occur in partially depleted SOI.[63] Two n-channel MOSFETs with a common gate are connected in series and measured as a single transistor. The floating N^+ region shared by the two transistors acts as a natural barrier separating the bodies of the two transistors. Channel pinch-off and impact ionization do occur in the drain-side transistor, but the majority of carriers cannot reach the source-side transistor. It follows that the body potential near the actual source is not raised, which inhibits the onset of the kink effect. The device works properly only if the source-side transistor is longer and dominates the output characteristics. The test conducted on $2 + 1$-μm-long channel shows kink-free characteristics and enhanced breakdown voltage. During hot-carrier stress, the defects are generated on the drain side and do not hamper the operation of the dominant transistor. This yields very good immunity to aging for the whole device.[63]

The cross–MOSFETs structure shown in Figure 3.20b operates as a *single device inverter*.[64] The P^- silicon film is surrounded by pairs of P^+ and N^+ contacts and serves as common body for the inversion-mode n-channel and the accumulation-mode p-channel Siamese transistors. The unique front gate controls both devices. When the front interface is

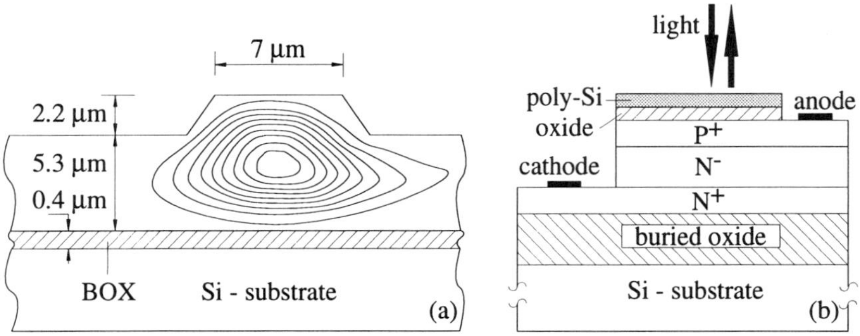

Figure 3.22 (a) Optical rib-waveguide in thick SIMOX film and field distribution (after Schmidtchen et al.[68]). (b) Infrared optical modulator on SOI (after Xiao et al.[69]).

inverted for n-channel conduction, the film is fully depleted and the p–MOSFET is off. Symmetrically, when the p-channel is on (accumulation or partial depletion), the n–MOSFET is blocked. This functionality mirrors the operation of CMOS inverters. Another type of single-device inverter, with dual-gate control, was conceived to make use of the back-gate bias.[65]

Many other devices were envisaged in relation to the interrupt SOI structure. Figure 3.21b schematically shows the association of low-power CMOS circuits integrated in the SOI region with devices, such as vertical power and bipolar transistors, which are better acclimatized to bulk silicon. This is just the straightforward version of the mezzanine structure, the pioneer of 3–D devices. In Figure 3.21c, the small openings in the buried oxide, which are natural in the ELO and patterned SIMOX processes, are used to relieve the MOSFETs from floating-body effects. The benefit of SOI isolation is nevertheless maintained in the junction regions.[66]

Outstanding *microwave* capabilities have been demonstrated for SOI MOSFETs.[67] The SIMOX process has been implemented on float-zone silicon wafers with extremely high resistivity. A key to success is the resistivity of the silicon film, which exceeds $10^4\,\Omega\,\mathrm{cm}$ even after high temperature annealing. The resistance of source/drain and polysilicon gate regions has been reduced using titanium silicide and gold processes. These features, combined with the general-purpose capabilities of SOI (excellent isolation and low junction-to-body capacitances), yield a maximum frequency of operation as large as 32 GHz for quarter-micron n-channel MOSFETs.[67]

The potential of SOI is acclaimed in domains other than microelectronics. The dream, of course, is to associate—on the same chip—high-frequency electronic processing with optical functions (connections, switching, modulation, etc.). SOI *optical waveguides* have been conceived that take advantage of the low optical absorption of silicon film and of the large difference in the refractive indexes of Si, SiO_2, and air. The rib-waveguide of Figure 3.22a has been processed on SIMOX by epitaxial growth of a 7-μm-thick film and anisotropic etching.[68] The tests indicate that the waveguide losses are below 0.5 dB/cm. *Optical modulators* have been demonstrated on SOI. The device consists of a vertical p–i–n diode and a Fabry–Perrot cavity, which is actually the capped silicon film (Fig. 3.22b).[69] Incoming infrared light bounces into the cavity. The carriers injected in the base of the forward biased diode induce a modulation of the light, a shift of the resonance frequency of the cavity, and finally a modification of the reflectance.[14]

Bibliography

[1] T. Tsuchiya, T. Ohno, and Y. Kado, "Present status and potential of subquarter-micron ultra-thin-film CMOS/SIMOX technology," in *Silicon-On-Insulator Technology and Devices*, S. Cristoloveanu (ed.), The Electrochemical Society, Pennington, NJ, p. 401, 1994.

[2] J. Chen, S. Parke, J. King, F. Assaderaghi, P. Ko, and C. Hu, "A high-speed SOI technology with 12 ps/18 ps gate delay operating at 5 V/1.5 V," *IEDM'92 Tech. Digest*, p. 35, 1992.

[3] K. Aubuchon, J. Pinter, M. Matloubian, M. Barger, R. McClain, and O. Marsh, "Initial evaluation of CMOS/SOI fabricated on plasma-thinned bonded silicon wafers," *Int. SOI Conf. Proc.*, p. 154, 1992.

[4] Y. Omura, S. Nakashima, K. Izumi, and T. Ishii, "0.1 μm gate, ultrathin-film CMOS devices using SIMOX substrate with 80 nm thick buried oxide layer," *IEEE Trans. Electron Devices*, vol. 40, p. 1019, 1993.

[5] K. Suzuki, T. Tanaka, Y. Tosaka, T. Sugii, and S. Andoh, "Source/drain contact resistance of silicided thin-film SOI MOSFETs," *IEEE Trans. Electron Devices*, vol. 41, p. 1007, 1994.

[6] Y. Yamaguchi, T. Nishimura, Y. Akasaka, and K. Fujibayashi, "Self-aligned silicide technology for ultra-thin SIMOX MOSFETs," *IEEE Trans. Electron Devices*, vol. 39, p. 1179, 1992.

[7] L.T. Su, M.J. Sherony, J.E. Chung, and D.A. Antoniadis, "Optimization of series resistance in sub–0.2 μm SOI MOSFETs," *IEEE Electron Device Lett.*, vol. 15, p. 145, 1994.

[8] O. Faynot and B. Giffard, "High performance ultrathin SOI MOSFETs obtained by localized oxidation," *IEEE Electron Device Lett.*, vol. 15, p. 175, 1994.

[9] M. Chan, F. Assaderaghi, S.A. Parke, C. Hu, and P.K. Ko, "Recessed-channel structure for fabricating ultrathin SOI MOSFET with low series resistance," *IEEE Electron Device Lett.*, vol. 15, p. 22, 1994.

[10] L.T. Su, J.B. Jacobs, J.E. Chung, and D.A. Antoniadis, "Deep-submicrometer channel design in silicon-on-insulator (SOI) MOSFETs," *IEEE Electron Device Lett.*, vol. 15, p. 183, 1994.

[11] M. Rodder, "Silicon-on-insulator bipolar transistor," *IEEE Electron Device Lett.*, vol. 4, p. 193, 1983.

[12] B.Y. Tsaur, "Fully-isolated lateral bipolar-MOS transistors fabricated in zone melting recrystallized Si films on SiO_2," *IEEE Electron Device Lett.*, vol. 4, p. 269, 1983.

[13] J.P. Colinge, "Half-micrometer base lateral bipolar transistors made in thin silicon-on-insulator films," *Electron. Lett.*, vol. 22, p. 886, 1986.

[14] J–P. Colinge, *Silicon-on-Insulator Technology: Materials to VLSI*, Kluwer, Boston, 1991.

[15] N. Higaki, T. Fukano, and A. Fukuroda, "A thin-base lateral bipolar transistor fabricated on bonded SOI," *Symp. VLSI Tech. Proc.*, p. 53, 1991.

[16] G.G. Shahidi, D.D. Tang, and B. Davari, "A novel high-performance lateral bipolar transistor formed on SOI substrate," *IEDM'91 Tech. Digest*, p. 633, 1991.

[17] S.A. Parke, C. Hu, and P.K. Ko, "Bipolar–FET hybrid-mode operation of quarter-micrometer SOI MOSFETs," *IEEE Electron Device Lett.*, vol. 14, p. 234, 1993.

[18] M.D. Church, "A comparative study of functional 16 K bipolar PROM circuits fabricated on bonded, oxide isolated and junction isolated substrates," *IEEE Int. SOI Conf. Proc.*, p. 175, 1989.

[19] H.J. Shin, C.L. Chen, E.D. Johnson, Y. Taur, S. Ramaswamy, and G. Bourdon, "Full-swing complementary BiCMOS logic circuits," *IEEE BCTM Conf. Proc.*, p. 229, 1989.

[20] G.G. Shahidi, "SOI: opportunities and challenges for sub-0.25 μm VLSI," *IEEE BCTM Conf. Proc.*, p. 255, 1992.

[21] T. Hiramoto, N. Tamba, M. Yoshida, T. Hashimoto, T. Fujiwara, K. Watanabe, M. Odaka, M. Usami, and T. Ikeda, "A 27 GHz double polysilicon bipolar technology on bonded SOI with embedded 58 μm^2 CMOS memory cells for ECL-CMOS SRAM applications," *IEDM'92 Tech. Digest*, p. 39, 1992.

[22] J.L. Leray, E. Dupont-Nivet, J.F. Peret, Y.M. Coïc, M. Raffaelli, A.J. Auberton-Hervé, M. Bruel, B. Giffard, and J. Margail, "CMOS/SOI hardening at 100 Mrad (SiO_2)," *IEEE Trans. Nucl. Sci.*, vol. 37, p. 2013, 1990.

[23] H. Lu, E. Yee, L. Hite, T. Houston, Y. Sheu, R. Rajgopal, C.C. Shen, J.M. Hwang, and G. Pollack, *IEEE Int. Solid State Circuits Conf. Digest*, p. 182, 1993.

[24] L.R. Hite, H. Lu, T.W. Houston, D.S. Hurta, and W.E. Bailey, "An SEU resistant 256 K SOI SRAM," *IEEE Trans. Nucl. Sci.*, vol. 39, p. 2121, 1992.

[25] J.P. Colinge, M.H. Gao, A. Romano-Rodriguez, H. Maes, and C. Claeys, "Silicon-on-insulator gate-all-around device," *IEDM'90 Tech. Digest*, p. 595, 1990.

[26] E. Simoen, U. Magnusson, I. Born, J. Vlummens and C. Claeys, "Mrad(Si) irradiation effects in GAA SOI nMOSFETs," in *Silicon-On-Insulator Technology and Devices*, S. Cristoloveanu (ed.), The Electrochemical Society, Pennington, NJ, p. 375, 1994.

[27] E. Arnold, S. Merchant, M. Amato, S. Mukherjee, H. Pein, and A. Ludikhuize, "Comparison of junction isolated and SOI high voltage devices in source follower mode," *4th Int. Symp. Power Semiconductor Dev. Proc.*, p. 242, 1992.

[28] E. Arnold, "Silicon-on-insulator devices for high voltage and power IC applications," *J. Electrochem. Soc.*, vol. 141, p. 1983, 1994.

[29] B. Edholm, J. Olsson, and A. Söderbärg, "A self-aligned lateral bipolar transistor realized on SIMOX material," *IEEE Trans. Electron Devices*, vol. 40, p. 2359, 1993.

[30] I.J. Kim, S. Matsumoto, T. Sakai, and T. Yachi, "Breakdown voltage improvement for thin-film SOI power MOSFETs by a buried oxide step structure," *IEEE Electron Device Lett.*, vol. 15, p. 148, 1994.

[31] T. Ohno, S. Matsumoto, and K. Izumi, "An intelligent power IC with double buried-oxide layers formed by SIMOX technology," *IEEE Trans. Electron Devices*, vol. 40, p. 2074, 1993.

[32] H. Vogt, "Advantages and potential of SOI structures for smart sensors," in *Silicon-On-Insulator Technology and Devices*, S. Cristoloveanu (ed.), The Electrochemical Society, Pennington, NJ, p. 430, 1994.

[33] J.D. Beasom, "120 V AC, 250 V DC monolithic power supply made with bonded wafer technology," *Int. SOI Conf. Proc.*, p. 66, 1992.

[34] J.M. O'Connor, V.K. Luciani, and A.L. Caviglia, "High voltage DMOS power FETs on thin SOI substrates," *Int. SOI Conf. Proc.*, p. 167, 1990.

[35] H. Pein, E. Arnold, H. Baumgart, R. Egloff, T. Letavic, S. Merchant, and S. Mukherjee, "SOI high voltage LDMOS and LIGBT transistors with a buried diode and surface p-layer," *Int. SOI Conf. Proc.*, p. 146, 1992.

[36] T. Ouisse, G. Reichert, S. Cristoloveanu, O. Faynot, and B. Giffard, "Analysis of SIMOX MOS transistors operated in the high temperature range," *Mat. Sci. Eng. B*, 1995.

[37] P. Francis, A. Terao, B. Gentinne, D. Flandre, and J.P. Colinge, "SOI technology for high-temperature applications," *IEDM'92 Tech. Digest*, p. 353, 1992.

[38] D. Flandre, C. Jacquemin, and J.P. Colinge, "Design techniques for high-speed and low-power and high-temperature digital CMOS circuits on SOI," *Int. SOI Conf. Proc.*, p. 164, 1992.

[39] T.W. Houston, H. Lu, P. Mei, T.G.W. Blake, L.R. Hite, T. Sundaresan, M. Matloubian, W.E. Bailey, J. Liu, A. Peterson, and G. Pollack, "A 1 μm CMOS/SOI 64 K SRAM with 10 nA standby current," *IEEE SOS/SOI Technol. Conf. Proc.*, p. 137, 1989.

[40] S.R. Wilson, B.Y. Hwang, J. Foerstner, T. Wetteroth, M. Racanelli, J. Tsao, and M. Huang, "Requirements for TFSOI materials and their effects on devices," in *Silicon-On-Insulator Technology and Devices*, S. Cristoloveanu (ed.), The Electrochemical Society, Pennington, NJ, p. 413, 1994.

[41] H. Achard, J.L. Mermet, H. Bono, J.P. Joly, and A. Monroy, "Contribution to the use of laser recrystallization on top of existing devices," *Eur. SOI Conf. Proc.*, p. B–03, 1988.

[42] R.P. Zingg, J. Friedrich, G.W. Neudeck, and B. Hoefflinger, "Three-dimensional stacked MOS transistors by localized silicon epitaxial overgrowth," *IEEE Trans. Electron Devices*, vol. 37, p. 1452, 1990.

[43] G.W. Neudeck, "Three-dimensional CMOS integration," *IEEE Circuits Devices*, vol. 6, p. 32, 1990.

[44] S. Hirose, T. Nishimura, K. Sugahara, S. Kusunoki, Y. Akasaka, and N. Tsubouchi, "A 10–bit linear image sensor fabricated in double active layers," *1985 Symp. VLSI Tech. Conf. Proc.*, p. 34, 1985.

[45] T. Nishimura, Y. Inoue, K. Sugahara, M. Nakaya, Y. Horiba, and Y. Akasaka, "A three dimensional static RAM," *1985 Symp. VLSI Tech. Conf. Proc.*, p. 30, 1985.

[46] K. Sugahara, T. Nishimura, S. Kusonoki, Y. Akasaka, and H. Nakata, "SOI/SOI/bulk-Si triple-level structure for three-dimensional devices," *IEEE Electron Device Lett.*, vol. 7, p. 193, 1986.

[47] T. Nishimura, Y. Inoue, K. Sugahara, S. Kusonoki, T. Kumamoto, S. Nakagawa, M. Nakaya, Y. Horiba, and Y. Akasaka, "Three dimensional IC for high performance image signal processor," *IEDM'87 Conf. Proc.*, p. 111, 1987.

[48] R. Watts, A.L. Robinson, and R.A. Soref, "Electromechanical optical switching and modulation in micromachined silicon-on-insulator waveguides," *Int. SOI Conf. Proc.*, p. 62, 1991.

[49] S. Cristoloveanu, "Advanced silicon on insulator materials: processing, characterization and devices," in *Semiconductor Silicon*, G. Harbeke and M.J. Schulz (eds.), Springer, Berlin, p. 223, 1989.

[50] S. Cristoloveanu, "Integrated magnetic sensors: an overview," *J. Korean Inst. Electron. Eng.*, p. 86, 1986.

[51] P. Lilienkamp and H. Pfleiderer, "An EFSI–SOS magnetodiode," *Phys. Status Solidi (a)*, vol. 43, p. 479, 1977.

[52] S. Cristoloveanu, "Magnetic field and surface influences on double injection phenomena in semiconductors: the magnetodiode effect theory," *Phys. Status Solidi (a)*, vol. 64, p. 683 and vol. 65, p. 281, 1981.

[53] A. Mohaghegh, S. Cristoloveanu, and J. de Pontcharra, "Double injection phenomena under magnetic field in SOS films. A new generation of magnetosensitive micro-devices," *IEEE Trans. Electron Devices*, vol. 28, p. 237, 1981.

[54] W.B. Dubbelday, L.D. Flesner, G.A. Garcia, G.P. Imthurm, and R.J. Hirschi, "Very high voltage photocell arrays," *Int. SOI Conf. Proc.*, p. 84, 1991.

[55] F. Balestra, S. Cristoloveanu, M. Benachir, J. Brini, and T. Elewa, "Double-gate silicon on insulator transistor with volume inversion: a new device with greatly enhanced performance," *IEEE Electron Device Lett.*, vol. 8, p. 410, 1987.

[56] D. Hisamoto, T. Kaga, and E. Takeda, "Impact of the vertical SOI 'DELTA' structure on planar device technology," *IEEE Trans. Electron Devices*, vol. 38, p. 1419, 1991.

[57] J.P. Colinge, X. Baie, and V. Bayot, "Evidence of two-dimensional carrier confinement in thin n-channel SOI gate-all-around (GAA) devices," *IEEE Electron Device Lett.*, vol. 15, p. 193, 1994.

[58] Y. Omura, S. Horiguchi, M. Tabe, and K. Kishi, "Quantum-mechanical effects on the threshold voltage of ultrathin-SOI nMOSFETs," *IEEE Electron Device Lett.*, vol. 14, p. 569, 1993.

[59] T. Ouisse, S. Cristoloveanu, and D.K. Maude, "Experimental investigation of silicon-on-insulator metal-oxide-semiconductor field-effect transistors at high magnetic field and low temperature," *J. Appl. Phys.*, vol. 74, p. 408, 1993.

[60] P. Francis, A. Terao, D. Flandre, and F. Van de Wiele, "Modeling of ultra-thin double-gate nMOS/SOI transistors," *IEEE Trans. Electron Devices*, vol. 41, p. 715, 1994.

[61] K. Suzuki, T. Tanaka, Y. Tosaka, H. Horie, and Y. Arimoto, "Scaling theory for double-gate SOI MOSFETs," *IEEE Trans. Electron Devices*, vol. 40, p. 2326, 1993.

[62] E. Simoen, U. Magnusson, and C. Claeys, "A low-frequency noise study of gate-all-around SOI transistors," *IEEE Trans. Electron Devices*, vol. 40, p. 2054, 1993.

[63] M.H. Gao, J–P. Colinge, L. Lauwers, S–H. Wu, and C. Claeys, "Twin-MOSFET structure for suppression of kink and parasitic bipolar effects in SOI MOSFETs at room and liquid helium temperatures," *Solid-State Electron.*, vol. 35, p. 505, 1992.

[64] M.H. Gao, S.H. Wu, J–P. Colinge, C. Claeys, and G. Declerck, "Single device inverter using SOI cross–MOSFETs," *Int. SOI Conf. Proc.*, p. 138, 1991.

[65] J.F. Gibbons and G. Declerck, "Single device inverter using SOI cross–MOSFETs," *IEEE Electron Device Lett.*, vol. 3, p. 191, 1982.

[66] T.I. Kamins, P.J. Marcoux, J.L. Moll, and L.M. Roylance, "Patterned implanted buried-oxide transistor structures," *J. Appl. Phys.*, vol. 60, p. 423, 1986.

[67] A.K. Agarwal, M.H. Hanes, T.W. O'Keeffe, J.R. Szedon, H.M. Hobgood, T.J. Smith, R.R. Siergiej, C.D. Brandt, M.C. Driver, R.N. Thomas, and H.C. Nathanson, "MICROX^{TM}: an all silicon microwave technology," *Int. SOI Conf. Proc.*, p. 144, 1992.

[68] J. Schmidtchen, A. Splett, B. Schüppert, and K. Petermann, "Low loss integrated-optical rib-waveguides in SOI," *Int. SOI Conf. Proc.*, p. 142, 1991.

[69] X. Xiao, J.C. Sturm, P.V. Schwartz, and K.K. Goel, "Vertical 1.3 μm optical modulator in silicon-on-insulator," *Int. SOI Conf. Proc.*, p. 171, 1990.

Chapter 4

Wafer-Screening Techniques

4.1 The Basis for Wafer Screening

Wafer screening refers to the evaluation of crystal quality, doping uniformity, defect and impurity concentrations, thicknesses of the silicon film and buried oxide, and other electrical and physical properties (carrier lifetime, interface trapping, etc.) of the starting wafers prior to the fabrication of SOI devices and circuits. Fluctuations in quality and uniformity of SOI materials have great impact on the performance, yield, and reliability of the VLSI circuits. Therefore, it is essential that diagnostic tools, including both destructive and nondestructive techniques, be available for the initial assessment of the key physical and electrical parameters of the starting SOI materials.

Among the wide variety of characterization methods, transmission electron microscopy (TEM) has been used extensively for analyzing the microstructure and crystal quality of SOI materials. Since TEM is a destructive and time-consuming technique, it cannot be employed on a routine basis. Nondestructive tools such as spectroscopic ellipsometry, optical reflectance, and dual-beam S-polarized reflectance can offer a quick evaluation of the top Si film thickness and quality. Contactless dual-beam optical modulation and surface photovoltage techniques have been developed for mapping the lifetime and diffusion-length profiles in the SOI wafers. In addition, the nondestructive optical modulation technique has also been utilized to determine the interface recombination velocities in SOI wafers. Most electrical characterization proceeds from capacitance and conductance measurements and requires the fabrication

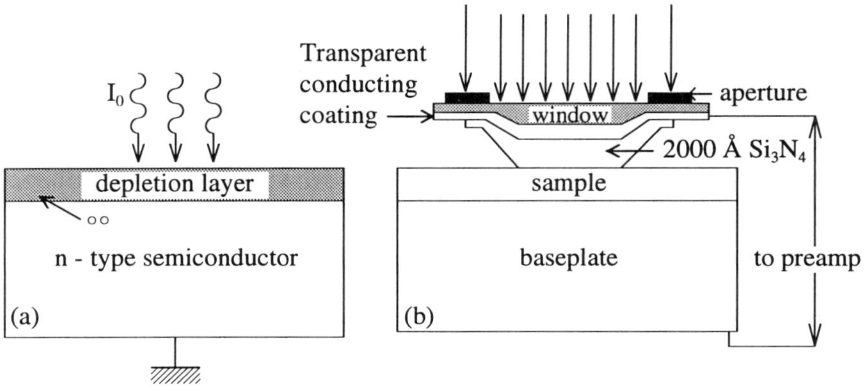

Figure 4.1 (a) Cross-sectional view of a semiconductor specimen for SPV measurement, and (b) an SPV capacitively coupled probe (after Goodman[2]).

of p–n junctions and capacitor structures; hence, it is more suitable for the assessment of processed devices. However, the special SOI structure makes possible the implementation of new methods, totally unknown in bulk silicon.

In the first part of this chapter, several nondestructive optical techniques will be proposed for screening the quality of starting SOI wafers. The second part will be dedicated to electrical characterization by quick turnaround capacitance measurements and the point-contact pseudo-transistor, which is probably the most specific and surprising method in SOI.

4.2 Surface Photovoltage

The surface photovoltage (SPV) technique, first reported by Johnson[1] back in 1957, was applied to investigate the minority carrier diffusion lengths and lifetimes in a variety of semiconductor materials such as GaAs,[2] Si,[3] InP,[4] and SIMOX[5]. A steady-state photoexcitation is used to generate a constant surface photovoltage at different wavelengths. The SPV method is suitable for processing control and material acceptance tests.

4.2.1 Principle and Experimental Setup

SPV measurements can be performed on large SOI wafers. One side of the wafer is chemically treated to induce a surface depletion region (Fig. 4.1a). The surface depletion region is created by boiling n-Si in water for one hour. For p-Si, a one-minute etch in 20 ml concentrated HF + 80 ml SiO_2 is recommended. Another surface treatment for Si samples is a standard Si clean/etch, removing any residual SiO_2 in buffered HF and treating n-Si in an aqueous solution of $KMnO_4$. The $KMnO_4$ step is omitted in p-Si. The surface with the induced depletion layer is illuminated by a chopped monochromatic light with energy slightly greater than the energy bandgap of the semiconductor ($h\nu \geq E_g$), while the opposite surface of the wafer is kept in the dark. Electron-hole pairs are generated and diffused to the surface depletion region, where they are separated by the electric field to produce an SPV.

The SPV signal is capacitively coupled (MIS-like capacitor structure) to a lock-in amplifier for amplification and measurement (see Fig. 4.1b). The light intensity is adjusted to produce a constant surface photovoltage at different wavelengths of light illumination. The light intensity required for such a constant SPV signal is drawn as a function of the reciprocal absorption coefficient and results in a linear plot. Extrapolation to the negative horizontal axis, at zero light intensity, yields the minority carrier diffusion length. SPV equipment is commercially available from *Semiconductor Diagnostics Inc.* (USA).

4.2.2 Model and Parameter Extraction

The general theory of the SPV method was described by Johnson[1] and Goodman.[2] A simplified version, applicable to SOI materials, is presented next. The SPV signal developed at the illuminated surface of the semiconductor specimen is a function of the excess minority carrier density injected into the surface space-charge region. The excess carrier density is in turn dependent on the incident light intensity, optical absorption coefficient, and minority carrier diffusion length. Since the surface photovoltage is proportional to the excess carrier concentration, the SPV signal for an n-type semiconductor can be expressed as[4]

$$V_{spv} = C_1 \Delta p \tag{4.1}$$

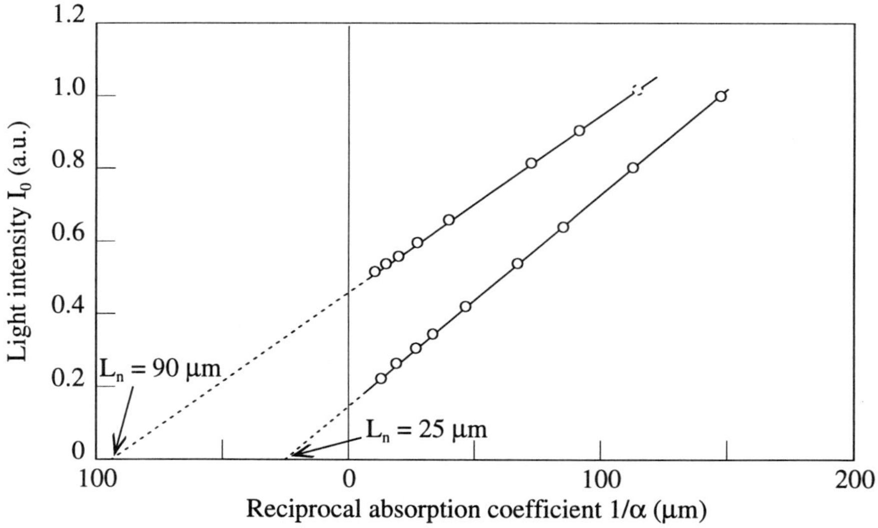

Figure 4.2 Relative light intensity I_0 versus inverse absorption coefficient for p-silicon samples. The arrows indicate the extrapolation of the diffusion length (after Schroder[6]).

where C_1 is a proportionality constant and

$$\Delta p = \frac{\eta I_o(1-R)}{(D_p/L_p + S_1)} \frac{\alpha L_p}{(1+\alpha L_p)} \tag{4.2}$$

is the excess hole density. η is the quantum efficiency, I_o is the light intensity, R is the reflection coefficient, D_p is the hole diffusion coefficient, S_1 is the front-surface recombination velocity, α is the optical absorption coefficient, and $L_p = \sqrt{D_p \tau}$ is the hole diffusion length. The above equations hold if $\alpha^{-1} \gg L_p$, $\Delta p \gg n$, and $\alpha d > 1$ (where d is the thickness of the specimen).

If η and R are assumed constant over the measured wavelength range, the incident light intensity I_o required to produce a constant SPV signal is directly proportional to the reciprocal absorption coefficient α^{-1}:

$$I_o = C_2(\alpha^{-1} + L_p) \tag{4.3}$$

where C_2 is a constant, independent of the photon wavelength.

The linear plot of I_o versus α^{-1} is extrapolated to the zero light intensity and intercepted at the negative horizontal axis, which yields the effective hole diffusion length (i.e., $L_p = -1/\alpha$). Figure 4.2 shows the relative photon intensity I_o versus the inverse absorption coefficient α^{-1} for two p-type samples. The negative intercepts yield $L_n = 25\ \mu$m and 90 μm, respectively.

The SPV method has been validated in silicon wafers where the minority carrier diffusion lengths ranged from a few tens to a few hundred micrometers.

In order to determine the minority carrier diffusion lengths from the SPV method, an accurate knowledge of the optical absorption coefficient versus wavelength is required. Any error in the absorption coefficient data adversely affect the diffusion length and carrier lifetime evaluations. For the stress-relieved silicon wafers, the absorption coefficient α (in cm^{-1}) versus wavelength (in μm) can be described by[1, 2]

$$\alpha = \left(\frac{84.732}{\lambda} - 76.417\right)^2 \tag{4.4}$$

Equation (4.4) is also valid in SOI materials for wavelengths from 0.7 to 1.1 μm.

SPV experiments have been performed on the *back side* of SOI wafers for investigating the minority carrier diffusion lengths and lifetimes in the silicon substrates. In particular, the diffusion length is well correlated with the concentration of heavy metals and, therefore, stands as a reliable monitor of wafer contamination. A diffusion length of 200 μm typically corresponds to 10^{17} Fe/cm^3, which is beyond the detection limit of SIMS and spark spectroscopy.[5] A longer diffusion length is synonymous with a lighter contamination.

Unfortunately, the SPV method fails to determine the minority carrier diffusion lengths in the silicon film because the optical absorption occurs mostly in the Si substrate of the SOI wafers.

Front-side probing is only possible after removal of the silicon film and buried oxide. Measurements in as-implanted SIMOX show that the diffusion length (from 40 to above 220 μm) is similar on both sides of the wafer, indicating that the recombination centers are quasi-uniformly distributed in the silicon substrate.[5] Alternative methods such as the DBOM technique should be used for measuring the excess carrier lifetimes in the top silicon film of SOI materials.

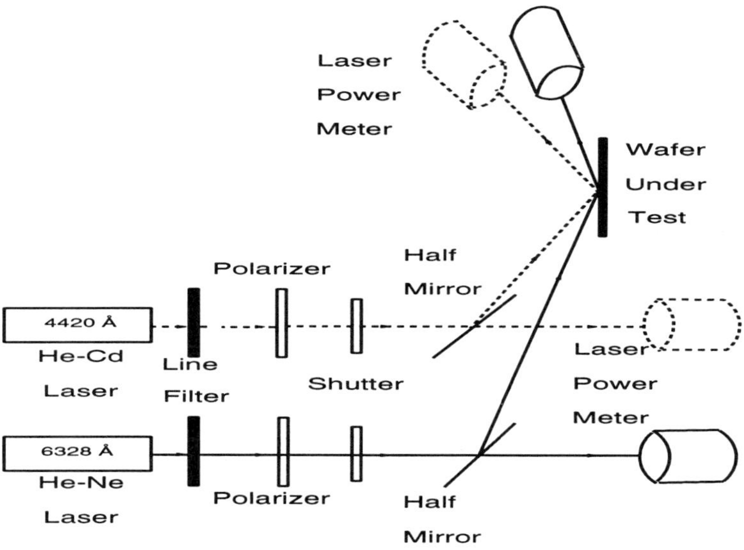

Figure 4.3 The schematic diagram of the dual-beam S-polarized reflectance (DBSPR) experimental setup.

4.3 Dual-Beam S-Polarized Reflectance

The dual-beam S-polarized reflectance (DBSPR) technique is a contactless nondestructive optical method for determining the thicknesses of both the top silicon film and buried oxide layer in SOI wafers.

4.3.1 Principle and Experimental Setup

The DBSPR technique, developed recently by Chang and Li,[7] is based on the S-polarized reflectance of a visible pump-beam on the SOI wafer. A theoretical model allows the mapping of the silicon film and buried oxide thicknesses across the entire wafer. The results obtained for several SIMOX samples were found to be in good agreement with the values determined by spectroscopic ellipsometry and reflectance.

Figure 4.3 shows the schematic diagram of the experimental setup. A 14 mW He–Cd CW blue laser ($\lambda = 0.442\,\mu$m) and a 4 mW He–Ne CW red laser ($\lambda = 0.6328\,\mu$m) are used in the experiment. Both lasers are focused to a beam size of about 2 mm diameter and polarized by

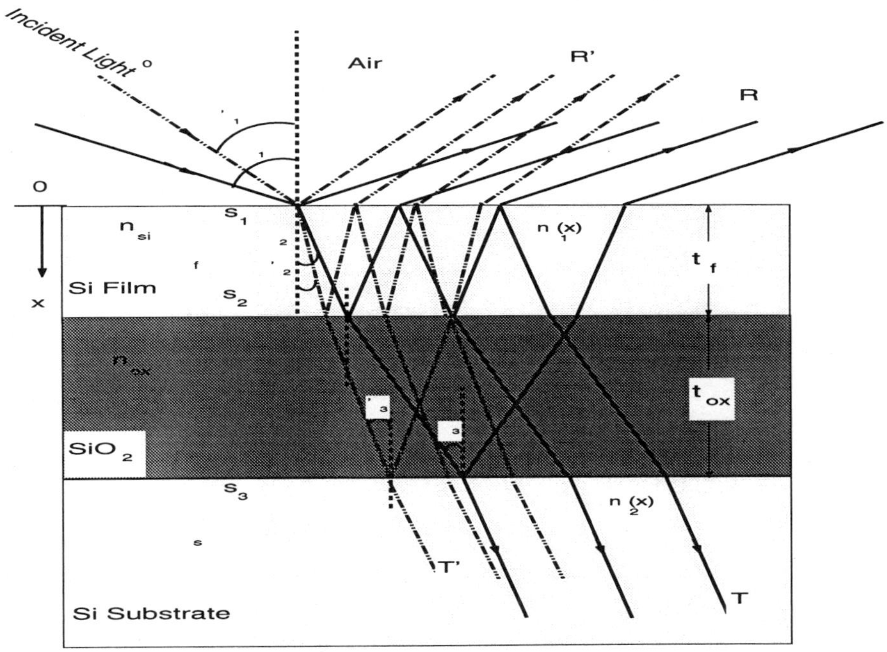

Figure 4.4 The schematic diagram of the incident probe beam and pump beam impinging on an SOI wafer.

the dichroic linear polarizers (with extinction ratios of $< 1.6\text{x}10^{-4}$ for $\lambda = 0.442\,\mu\text{m}$ and $< 7.6x10^{-5}$ for $\lambda = 0.6328\,\mu\text{m}$). The two laser beams are reflected by half-silver coated mirrors and then focused at the same spot on the surface of the measured sample. The S-polarized reflectances are monitored with a laser power meter. Error in reflectance measurement due to the fluctuation of laser power can be reduced by monitoring the incident laser power intensity and reflected power intensity simultaneously.

In the experiment, the blue laser beam was used to determine the Si film thickness t_{si}, whereas the red laser beam served to measure the buried oxide thickness t_{ox}. The multiple-angle reflectance measurement was first performed using the blue laser beam, and the same procedure was then repeated using the red laser beam on the same spot of the wafer. Next, the single-angle reflectances of the blue and red laser beams were measured by scanning the laser beam across the entire SIMOX wafer.

4.3.2 Model and Parameter Extraction

We briefly describe the DBSPR theory for determining the thicknesses of the Si film and buried oxide layer in a SOI wafer using the S-polarized reflectance measurements. As shown in Figure 4.4, an S-polarized laser beam is impinging at an angle of incidence θ_1 on the surface of a SIMOX wafer. The total reflection coefficient can be expressed as

$$r = \frac{r_0 + r_1 e^{-i\Delta_1} + r_2 e^{-i(\Delta_1+\Delta_2)} + r_0 r_1 r_2 e^{-i\Delta_2}}{1 + r_0 r_1 e^{-i\Delta_1} + r_0 r_2 e^{-i(\Delta_1+\Delta_2)} + r_1 r_2 e^{-i\Delta_2}} \tag{4.5}$$

where

$$\Delta_1 = \frac{4\pi}{\lambda} n_{si} t_{si} \cos\theta_2 \tag{4.6}$$

$$\Delta_2 = \frac{4\pi}{\lambda} n_{ox} t_{ox} \cos\theta_3 \tag{4.7}$$

r_0, r_1, and r_2 are the Fresnel reflection coefficients at the air/Si-film, Si-film/buried-oxide, and buried-oxide/Si-substrate interfaces, respectively; n_{si} and n_{ox} are the refractive indexes of Si and SiO_2. Δ_1 and Δ_2 are the phase changes that result from the double travel of light in the top Si film and buried-oxide layer.

Silicon Film Thickness. The thickness of the top Si film in an SOI wafer is determined by shining the blue laser to obtain the S-polarized reflectance and measuring the total reflectance with a laser power meter. Equation (4.5) may be expressed as

$$r \simeq \frac{r_0 + r_1 e^{-i\Delta_1}}{1 + r_0 r_1 e^{-i\Delta_1}} \tag{4.8}$$

by neglecting the back reflectances below the buried-oxide/Si-substrate interface. Since the blue light is mainly absorbed in the top Si film ($\alpha \simeq 3 \times 10^4\,\text{cm}^{-1}$),[5] the approximation holds for both thick and thin films for certain angles of incidence. Therefore, this technique is adequate for measuring the film thickness in SOI wafers with $t_{si} > 0.03\,\mu\text{m}$.[7]

The film thickness is determined from the DBSPR measurements using the expression

$$R_m - R = 0 \tag{4.9}$$

where R_m is the measured reflectance and R is the approximated reflectance given by Equation (4.8). Most of the blue light being absorbed

Table 4.1 Measured Si film thickness for different SIMOX samples.

Thickness (μm)	Sample 1		Sample 2		Sample 3	
Maximum	0.2048		0.5443		1.6484	
Minimum	0.1970		0.5246		1.5536	
Average	0.1988*	0.2004†	0.5389*	0.5300†	1.6162*	1.6500†
Standard Deviation	0.0015		0.0055		0.0436	

⋆ Measured by S-polarized reflection method.
† Measured by reflection interference spectroscopy.

in the Si film, Equation (4.9) can be written as

$$X + \xi + 2l\pi \pm \cos^{-1}\left(\frac{e^b(R_m - R_0)}{2\sqrt{R_0 R_1}\sqrt{A^2 + B^2}}\right) = 0 \tag{4.10}$$

where $l = 0, 1, 2, 3, \ldots$ denotes the different modes; A, B, and ξ are given by

$$\begin{aligned} A &= \cos(\delta_0 - \delta_1) - R_m \cos(\delta_0 + \delta_1) \\ B &= \sin(\delta_0 - \delta_1) + R_m \sin(\delta_0 + \delta_1) \\ \xi &= \tan^{-1}(B/A) \end{aligned} \tag{4.11}$$

$X = 4\pi t_{si} u/\lambda$ and $b = 4\pi t_{si} v/\lambda$, with u and v expressed as

$$\begin{aligned} 2u^2 &= +N_{si}^2 - k^2 - \sin^2\theta_1 + \sqrt{(N_{si}^2 - k^2 - \sin^2\theta_1)^2 + 4N_{si}^2 k^2} \\ 2v^2 &= -N_{si}^2 + k^2 + \sin^2\theta_1 + \sqrt{(N_{si}^2 - k^2 - \sin^2\theta_1)^2 + 4N_{si}^2 k^2} \end{aligned} \tag{4.12}$$

Note that u and v are the real and imaginary parts of $n_{si}\cos\theta_2$, respectively. In the DBSPR measurement, only two parameters, namely, the film thickness t_{si} and the real part refractive index N_{si} in the film, are unknown. Thus, t_{si} and N_{si} are found by applying Equation (4.10) to experiments made at multiple angles of incidence.[7]

Since the above method is tedious and time consuming, it is highly desirable to use a single-angle of incidence along with numerical calculation of Equation (4.9). The multiple-angle step is only performed in

Table 4.2 Measured buried oxide thickness for different SIMOX samples.

Thickness (μm)	Sample 1		Sample 2		Sample 3	
Maximum	0.3944		0.4032		0.4808	
Minimum	0.3629		0.3410		0.4065	
Average	0.3891*	0.3479†	0.3663*	0.3900†	0.4484*	0.3900†
Standard Deviation	0.0062		0.0127		0.0112	

⋆ Measured by S-polarized reflection method.
† Measured by spectroscopic ellipsometry.

the initial measurement to obtain the value of t_{si}, which is followed by single-angle measurements over the entire surface. This allows a quick determination of film thickness fluctuations in SOI wafers.

Buried Oxide Thickness. A red He–Ne CW laser is employed to obtain the S-polarized reflectance data. Using the value of t_{si} obtained above, the buried-oxide-layer thickness t_{ox} is calculated from the following two equations:

$$R_m - R_{total}(\theta_1, \lambda, t_{si}, t_{ox}, n_{ox}, n_{si}) = 0 \tag{4.13}$$

and

$$R_{total}(\theta_1, \lambda, t_{si}, t_{ox}, n_{ox}, n_{si}) = \frac{C_1}{C_2} \tag{4.14}$$

where R_{total} is the total reflectance from the incidence of the red laser beam and $C_{1,2}$ are constants.[7] The value of t_{ox} is determined by using the multiple angle of incidence and Equation (4.13).

Similar to the Si film thickness measurements, the uniformity of buried-oxide thickness in SIMOX wafers can be evaluated from the measured reflectance at the angle of incidence θ_1 and from the numerical solution of Equation (4.13) using values of t_{ox} obtained from the multiple-angle reflectance measurements.

Based on the DBSPR method depicted above, the thicknesses of the Si film and buried oxide have been mapped across three SIMOX wafers, and the results are summarized in Tables 4.1 and 4.2, respectively. For comparison, the average thicknesses of film and buried oxide in these wafers, determined by spectroscopic reflection and ellipsometry, are also included in Tables 4.1 and 4.2.

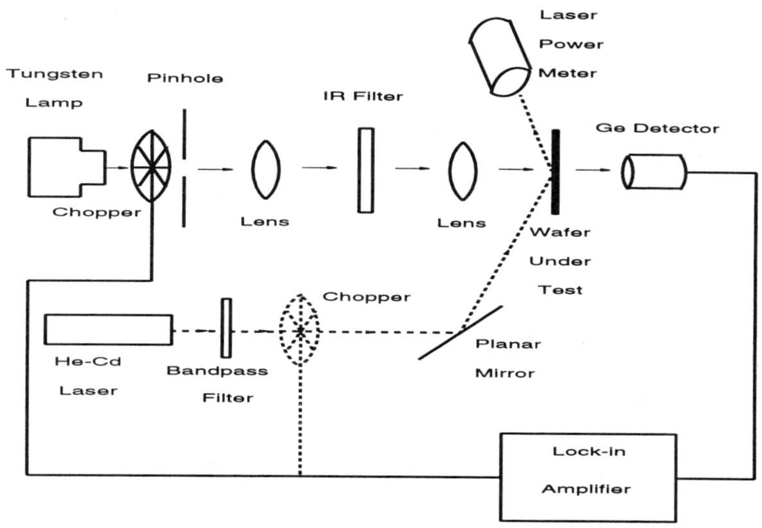

Figure 4.5 The schematic diagram of the dual-beam optical modulation (DBOM) experimental setup.

4.4 Dual-Beam Optical Modulation

The dual-beam optical modulation (DBOM) technique is a contactless, nondestructive optical method for mapping the excess carrier lifetimes in the Si film and the substrate of SOI wafers.

4.4.1 Principle and Experimental Setup

The DBOM technique is based on the modulation of the transmission intensity of an infrared probe beam by a visible pump beam ($h\nu \geq E_g$) via free-carrier absorption in the SOI wafers. The theoretical model, derived by Yang and Li, makes possible the mapping across the SOI wafer of the excess carrier lifetime in the silicon film τ_f and in the substrate τ_s.[9, 10]

Figure 4.5 shows a DBOM experimental setup: a tungsten lamp is used as the infrared (IR) probe beam with a beam diameter of 4 mm. The IR filter is a low-doped silicon wafer with a cutoff wavelength of about 1.1 μm. A He–Ne CW laser ($\lambda = 0.6328\,\mu$m, $P_0 = 5\,$mW) or a Cd-He CW laser ($\lambda = 0.442\,\mu$m, $P_0 = 13\,$mW) is used as a pump beam. The He–Ne and Cd–He lasers are respectively suitable for lifetime measurements in the substrate and in the film. To eliminate three-dimensional effects due to nonequilibrium electrons and holes, the laser beam is reduced to 2 mm in diameter and is aligned with the probe beam. Both the probe beam and the pump beam are chopped at 400 Hz for synchronized detection.

In the experiment, the IR transmission intensity I was first measured with the chopper and the tungsten light turned on, and the resultant signals were detected with a lock-in amplifier. The pump-beam signal was measured when the chopper of the IR probe beam was turned off and the chopper of the pump beam turned on. The resultant pump-beam signal is given by ΔI. The reflected power of the pump beam was evaluated with an optical power meter to determine the reflectivity R of the SOI wafer. An optical baffle around the SOI wafer served to minimize the effects of stray light from the laser pump beam.

4.4.2 Theory and Parameter Extraction

Figure 4.6 shows a schematic diagram of the incident probe beam and pump beam impinging on a SOI wafer. The excess electron densities $\Delta n_1(x)$ and $\Delta n_2(x)$, generated by the incident pump beam in the top Si film and in the substrate of a p-type SOI wafer, respectively, as a function of depth are obtained by solving the continuity equations with proper boundary conditions. The photon flux density ϕ_2 at $x = t_{si} + t_{ox}$ (where t_{si} and t_{ox} denote the Si film and buried-oxide thicknesses, respectively) is derived from Fresnel's equation

$$\begin{aligned} \phi_2 &= \phi_0 \exp(-\alpha t_{si}/\cos\theta_2) \left(\frac{2 n_{si} \cos\theta_2}{n_{si}\cos\theta_2 + n_{ox}\cos\theta_3} \right)^2 \\ &\times \left(\frac{2 n_{ox} \cos\theta_3}{n_{ox}\cos\theta_3 + n_{si}\cos\theta_2} \right)^2 \end{aligned} \tag{4.15}$$

If the excess electron and hole densities are assumed to be equal ($\Delta n = \Delta p$), then the change of optical absorption coefficient due to the free-carrier density generated by the short-wavelength pump beam ($h\nu \geq E_g$) can be written as $\Delta\alpha_{fc} = \sigma_{fc}\Delta n$, where $\sigma_{fc} = \sigma_e + \sigma_h$ is the total optical absorption cross section of electrons and holes. The fractional change of the transmission beam intensity $\Delta I/I$ is related to the total excess carrier densities, $\Delta n = \Delta n_1 + \Delta n_2$, generated in the film and substrate:[9, 10]

$$\Delta I/I = \exp\left[-\sigma_{\text{fc}} \left(\int_0^{t_{si}} \Delta n_1(x)dx + \int_{t_{si}+t_{ox}}^{d} \Delta n_2(x)dx \right) \right] - 1 \tag{4.16}$$

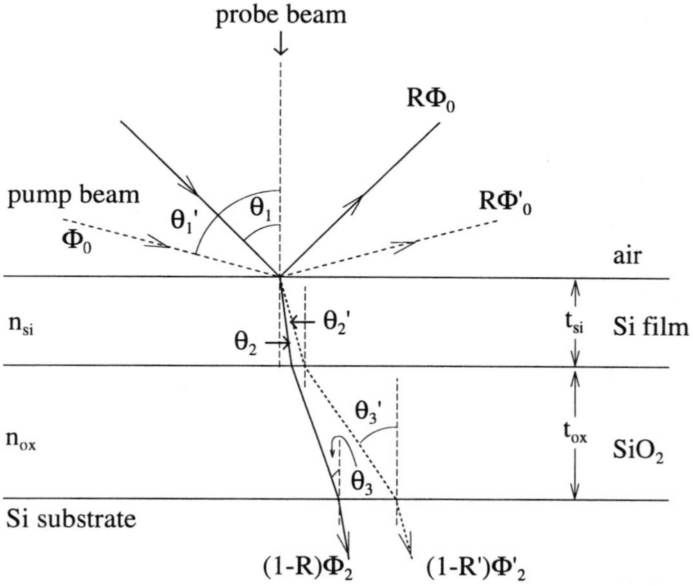

Figure 4.6 The schematic diagram of the incident probe beam and pump beam impinging on an SOI wafer.

where d is the total thickness of the SOI wafer. For thin silicon film with t_{si}/L_f, $\alpha t_{si} \ll 1$, Equation (4.16) can be expressed as[9]

$$\frac{\Delta I}{I} \simeq -\sigma_{\text{fc}}\alpha\eta\phi_o(1-R)\cos\theta_2\left[\left(\frac{\alpha^2 D_n^2-(D_n/L_f)^2}{D_n(S_1+S_2)+t_{si}[S_1S_2+(D_n/L_f)^2]}\right)\right.$$
$$\left.\times\left(\frac{t_{si}^2}{\alpha^2L_f^2-1}\right)\tau_f+\left(L_s\frac{\alpha L_s^2+S_3\tau_s}{L_s+S_3\tau_s}-\frac{1}{\alpha}\right)\left(\frac{\phi_2/\phi_0}{\alpha^2L_s^2-1}\right)\tau_s\right] \quad (4.17)$$

by approximating $\ln(1+\Delta I/I) \simeq \Delta I/I$.

Note that Equation (4.17) can be further simplified by assuming that (1) $\alpha^2 L_s^2$, $\alpha^2 L_f^2 \gg 1$ (valid for $\alpha > 1\times 10^3\,\text{cm}^{-1}$), (2) $L_s \gg S_3\tau_s$ for small interface recombination velocity at the SiO_2/substrate interface, and (3) $D_n \gg S_1S_2\tau_f$. Since typical surface recombination velocities are $S_1 \simeq 10^2$ cm/s at the air/Si-film interface, $S_2 \simeq 1$–10 cm/s at the Si-film/SiO_2 interface, and $D_n \simeq 33.7\,\text{cm}^2/\text{s}$ at 300 K,[5] assumption (3) is valid for $\tau_f < 10^{-2}$ s. Upon irradiation, S_1 was found to increase up to about 10^6 cm/s; hence, assumption (3) is still valid for $\tau_f < 10^{-6}$ s.

Table 4.3 The film and substrate carrier lifetimes for five SIMOX wafers determined by the DBOM and SPV methods. Sample A1 is a multiple-implant wafer (doses of $(0.5 + 0.5 + 0.8) \times 10^{18}\,\mathrm{cm}^{-2}$, annealed at 1310 °C for five hours). Samples A2–A5 received single oxygen implants of $1.8 \times 10^{18}\,\mathrm{cm}^{-2}$ dose. Annealing: A2 same as A1, A3–A5 at 1285 °C for six hours.

Sample	Si-Overlayer Thickness t_{si} (μm)	Buried-Oxide Thickness t_{ox} (μm)	Lifetimes DBOM τ_f (μs), τ_s (μs)	Lifetime SPV τ_s(μs)
A1	0.187	0.379	0.23, 7.5	—
A2	0.226	0.355	0.18, 5.4	—
A3	0.53	0.39	0.46, 14	—
A4	0.85	0.39	0.58, 15	16
A5	1.65	0.39	0.72, 18	19

Equation (4.17) reduces to

$$\frac{\Delta I}{I} \simeq -\sigma_{\mathrm{fc}} \eta \phi_0 (1 - R) \cos\theta_2 \tau_{eff} \tag{4.18}$$

where

$$\tau_{eff} = \left(\frac{\alpha t_{si}}{1 + \tau_f (S_1 + S_2)/t_{si}} \right) \tau_f + (\phi_2/\phi_0)\tau_s \tag{4.19}$$

Based on the above assumptions, the first term on the right-hand side of Equation (4.19) can be neglected. From Equations (4.15) and (4.19), it is found that

$$\frac{\tau_{eff}}{\tau_s} \simeq \frac{\phi_2}{\phi_0} \propto (1 - \alpha t_{si}/\cos\theta_2) \tag{4.20}$$

Equation (4.20) was obtained by using $e^{-\alpha t_{si}/\cos\theta_2} \simeq (1 - \alpha t_{si}/\cos\theta_2)$ in Equation (4.15). Thus, the excess carrier lifetime, τ_s, is determined with Equation (4.20) from the measured value τ_{eff}. It is noted that the above analysis can be applied equally well for holes in an n-type substrate provided that the proper parameters for electrons are replaced by hole parameters.

To measure excess carrier lifetimes in the silicon film of SOI wafers, a short-wavelength He–Cd CW laser is used as the pump beam so that free carriers are generated and absorbed inside the film. In this case, two incident angles of the pump beam are requested for determining simultaneously the carrier lifetimes in the film (τ_f) and in the substrate (τ_s). It is demonstrated that τ_{eff} can be rewritten as[10]

$$\tau_{eff} \simeq \alpha t_{si}\tau_f + (\phi_2/\phi_0)\tau_s \tag{4.21}$$

where ϕ_2 is given by Equation (4.15). The first term on the right-hand side of Equation (4.21) is independent of the incident angle of the pump beam. Using two incident angles yields

$$\tau_f = \frac{1}{\alpha t_{si}} \frac{\tau_{eff}\phi_2' - \tau_{eff}'\phi_2}{\phi_2' - \phi_2} \tag{4.22}$$

$$\tau_s = \frac{(\tau_{eff}' - \tau_{eff})\phi_0}{\phi_2' - \phi_2} \tag{4.23}$$

where τ_{eff} and τ_{eff}' are the measured effective lifetimes and ϕ_2 and ϕ_2' are the calculated photon-flux densities in the substrate region for the two incident angles. Thus, the film carrier lifetime is simply determined by measuring τ_{eff} as a function of the incident angle of the pump beam.

Based on the DBOM theory depicted above, the excess carrier lifetimes in the top Si film and the substrate of five SIMOX wafers with different film thicknesses and annealing temperatures were measured, and the results are summarized in Table 4.3. The average excess carrier lifetimes in the substrate of two SIMOX wafers were also determined by the SPV method and included in Table 4.3. The two methods show very good agreement. Finally, mapping of the excess carrier lifetimes in the top Si film and the substrate can be made across the entire SOI wafer using the DBOM technique.

4.5 Other Optical Methods

The surface recombination velocities at the upper and lower interfaces of the buried oxide can be measured by *nondestructive optical modulation* (NOM), which is a variant of the DBOM technique.[11] The experimental setup is similar to that of Figure 4.5 and consists of a tungsten lamp (probe beam), an IR filter, and two lasers (pump beams). The two laser beams are both reflected by half-silver coated mirrors and then focused at the same spot on the surface of the measured sample. The substrate carrier lifetime τ_s was evaluated with the red laser pump beam impinging on the rough back surface of SOI wafer. The reflectance of a rough surface is obtained by measuring the rms values of rough surface slope m_0 and surface roughness σ with a Sloan Dektak II surface profiler.[12] The IR transmitted intensity I was first detected by chopping the probe beam, and then the resulting I was read out by a lock-in amplifier. The fractional transmitted intensity ΔI was obtained analogously by chopping the pump beam. The S-polarized reflectances of pump beams were monitored by the laser power meter to determine the total front-surface reflectance R and back-interface transmittance T at the SOI wafer.[11]

The recombination velocity S_3 at the lower oxide interface (buried-oxide/Si-substrate) was determined by repeating the procedures, except that the probe and the red laser pump beam were impinging on the passivated front surface (at the same location on the wafer), and the pump beam was irradiated at different oblique incidences[11]

$$S_3 = \frac{(K-\alpha)D_n}{1-L_sK} \tag{4.24}$$

$$K = \left(\frac{\Delta I/I}{\sigma_{fc}\eta\phi_0\cos\theta_2(1-R-T)} - \frac{\Delta I'/I'}{\sigma_{fc}\eta\phi_0'\cos\theta_2'(1-R'-T')}\right) \times \frac{\alpha}{\tau_s}\left(\frac{T}{1-R-T} - \frac{T'}{1-R'-T'}\right)^{-1} \tag{4.25}$$

where the quotes indicate the parameters measured for the second angle of incidence.

The recombination velocity S_2 at the upper oxide interface (top Si film/buried oxide) was obtained by using the blue laser as a pump beam. In a thin silicon film, where $S_1 + S_2 > t_{si}/\tau_f$ and $S_1 \simeq 10^2\,\mathrm{cm}^2$,[13] we have[11]

$$S_1 + S_2 \simeq \left(\frac{\Delta I/I}{\sigma_{fc}\eta\phi_0\cos\theta_2 T} - \frac{\Delta I'/I'}{\sigma_{fc}\eta\phi_0'\cos\theta_2' T'}\right)^{-1} \times t_{si}(1-e^{-\alpha t_{si}})\left(\frac{1-R-T}{T} - \frac{1-R'-T'}{T'}\right) \tag{4.26}$$

The measurements indicate a lower recombination velocity ($S_3 \simeq 50\,\mathrm{cm/s}$) at the BOX–substrate interface than at the front oxide interface (where $S_1 + S_2 \simeq 150$–$350\,\mathrm{cm/s}$), which is consistent with the difference in carrier lifetime.

In addition to the techniques described above, other nondestructive optical methods, such as infrared, ultraviolet,[14] and visible[15, 16] reflectance measurements as well as Raman spectroscopy, have been used in the characterization of SOI materials.

UV reflectance is a reliable monitor of the crystallinity of the Si overlay in SIMOX wafers and provides three morphological features: (i) the overall reflectivity reduction is related to the presence of contamination in the Si film, (ii) a decrease in reflectivity as a function of inverse wavelength is attributed to the surface roughness, and (iii) reduction of intensity of the reflectance peaks at 0.28 μm and 0.367 μm is due to some amorphization of the Si overlay.[14]

The *spectroscopic ellipsometry* (SE) is a nondestructive and sensitive optical technique for determining accurately the thickness, the interface quality, and the composition of multilayer structures such as SOI materials.[17] By measuring the phase shift and amplitude ratio over a wavelength region ranging from 230 to 930 nm, enough information can be obtained, not only on the thicknesses of multilayers with accuracy better than 1 nm but also on crystallinity, void fraction, interface roughness and composition of the mixed layers. The SE technique is based on the detection of the change of polarization of a light beam reflected by a sample. The main advantage of SE over single-wavelength ellipsometry (with constant[18] or variable[19] angle of incidence) is that the measurement is independent of intensity fluctuations of the light source, since only the phase shifts are measured.

In SOI structures, SE measurements are performed at an incidence angle of 75 °—close to the Brewster angle between SiO_2 and Si, for which the measurement is most sensitive. The optical spectra of as-implanted and annealed SIMOX structures are recorded with a Multilayer Optical Spectrometer Scanner (commercially available from Sopra (France)) in the wavelength region between 300 nm and 850 nm; 100 equidistant points take typically one hour of measurement time. The accuracy of the predefined multilayer model of the SOI structure is evaluated by fitting the simulated and measured SE spectra and by minimizing the error function.

If the layers and interfaces of the SOI structure are well defined, a three-layer model (i.e., native oxide/silicon overlay/buried oxide) is sufficient to fit the measured SE spectra, and convergence of the thickness-finding algorithm is achieved in a matter of minutes. The complete theory of spectroscopic ellipsometry is quite complicated and beyond the scope of this book. However, for readers interested in the details of SE theory, we recommend the book chapter written by Azzam and Bashara.[20]

4.6 Point Contact Pseudo-MOS Transistor

The electrical characterization of SOI wafers is a difficult task due to the thinness of the film and complexity of the stacked structure. This is why the electrical properties are, in general, inferred from the analysis of test MOS devices or integrated circuits. Described below is a simple technique that takes advantage of the specific configuration of SOI and has the potential of being nondestructive. The pseudo–MOS transistor, also called the point-contact transistor or Ψ–MOSFET, is the first transistor that does not require any lithography or technology at all.

4.6.1 Principle

The Ψ–MOSFET is based on the upside-down MOS structure that is inherent in all SOI materials. Figure 4.7a shows that the bulk-Si substrate acts as a gate terminal and can be biased, through the metal support, to induce a conduction channel at the upper interface of the buried oxide. The buried oxide plays the role of a gate oxide and the Si film represents the transistor body. To operate the Ψ–MOSFET in situ (without lithography and metallization), low-pressure probes are placed on the film and form source and drain point contacts.[21, 22]

In spite of the device simplicity and nonparallelism of current lines, very pure MOSFET–like characteristics are produced. As shown in Figure 4.7b, the output $I_D(V_D)$ curves exhibit a nearly ideal behavior up to 20 V. The locus of I_{Dsat} vs. V_{Dsat} (dashed line) indicates that the saturation voltage follows the standard MOS model: $V_{Dsat} \simeq V_G - V_T$.

According to the positive or negative bias applied on the gate, accumulation or inversion channels are activated at the interface. Typical subthreshold $I_D(V_G)$ characteristics are shown for n- and p-channels in Figure 4.7c. Strong inversion and strong accumulation curves are given in Figure 4.7d, whereas Figure 4.7e pictures the corresponding transconductance characteristics. From these diagrams, many useful properties may be determined for electrons and holes.

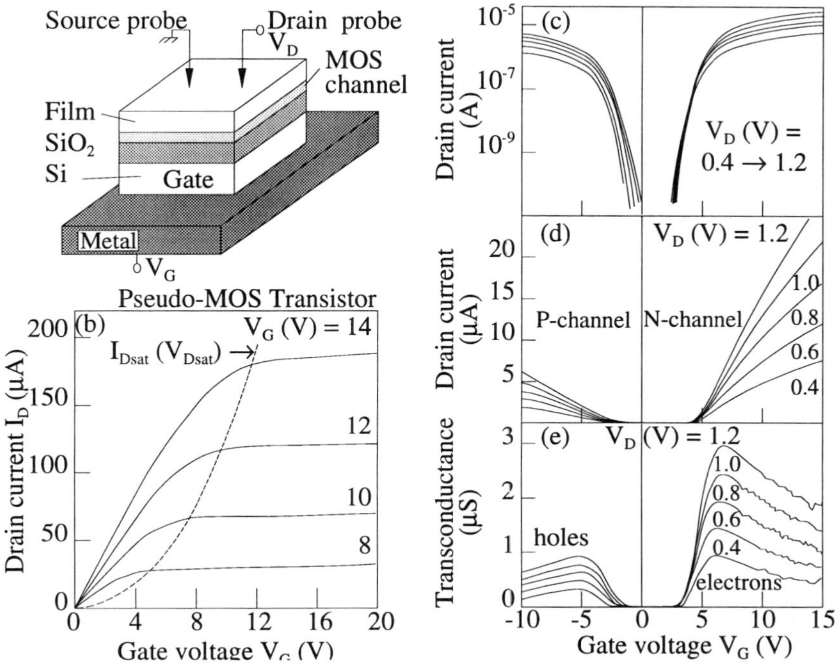

Figure 4.7 (a) Schematic of the pseudo-MOS transistor and typical characteristics: (b) drain current versus drain voltage, (c) subthreshold characteristics, (d) current versus gate voltage in strong inversion and accumulation, (e) transconductance curves.

4.6.2 Experimental Setup

In principle, the Ψ–MOSFET is a wafer-scale experiment. This implies that no leakage current flows through the buried oxide. If the presence of oxide pipes or other shortcuts is suspected, the measurement may be performed on small (1–2 cm^2) pieces. In case of sidewall leakage, the Si film on the sample borders can easily be etched off.

The basic setup is composed of any two-probe system, connected to an HP–4145 or equivalent picoammeter. However, for calibration purposes (see Fig. 4.10a), it is more suitable to use a four-point probe, which currently serves for resistivity measurements. For example, the Jandel system offers tungsten–carbide probes of 0.4 mm diameter and 25 μm tip radius as well as probe pressure control between 10 and 60 g.

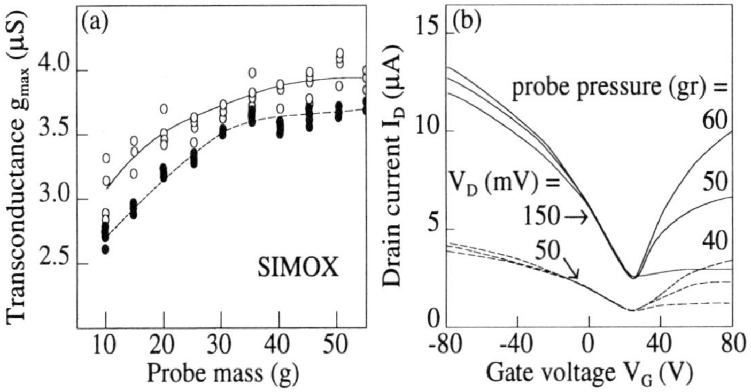

Figure 4.8 Influence of the probe pressure (a) on the transconductance peak, $g_{max} = f_g \mu_0 C_{ox} V_D$ (- - -, from transconductance curves; ——, from $I_D/\sqrt{g_m}$ curves) and (b) on the $I_D(V_G)$ characteristics (after Williams et al.[23]).

Increasing the probe *pressure* gradually reduces the series resistances and provides more accurate (less underestimated) carrier mobility values. In 200-nm-thick fully depleted films, it was found that the transconductance improves by 35 % as the pressure increases from 10 g to 30 g, and then saturates (Fig. 4.8a).[21, 23]

Larger pressures are needed in thicker or more heavily doped films that cannot be totally depleted by applying a substrate bias. Figure 4.8b shows that the influence of the pressure is quite substantial in inversion ($V_G > 0$), where the tip contact with the n-channel is achieved through a p-doped overlay and a depletion zone. In contrast, the contact with the accumulation p-channel is direct and, naturally, the pressure effect is much less relevant. Therefore, the pressure adjustment is a matter of trade-off for each type of SOI wafer. The probes must penetrate the native oxide and preferably reach the depletion region, without producing too much damage. It is remarkable that 10 g pressure induces only a 10 nm probe penetration.[6] Since more than 10 nm depth is consumed during the subsequent circuit processing, the low-pressure method appears to be quasi-nondestructive. The *spacing* between source and drain probes is less critical.[23] Almost identical $I_D(V_G, V_D)$ curves are obtained by interchanging the four probes of the system—for example to vary the source–drain distance from 1 mm to 3 mm.

4.6.3 Model

The drain current originates from the inversion/accumulation channel as well as from the neutral region of the film. The accurate modeling of the Ψ–MOSFET basically requires the solutions of the Poisson and Gauss equations in the film and in the underlying substrate. There are many scenarios according to the various combinations of film/substrate doping levels and types, or interface defects. Fortunately, such a modeling is far less necessary than for capacitance measurements on semiconductor–insulator–semiconductor (SIS) structures (Section 6.1).

We first consider the simpler case of *fully depleted* films. The subthreshold characteristics (Fig. 4.7c) are well behaved and the leakage current is extremely small, which verifies that the current flowing in the "volume" of the film is negligible. For this reason, in Figure 4.7d, there is an off-state bias range. The drain current is therefore governed by either the inversion or the accumulation channel, and the intercepts with V_G–axis roughly correspond to the threshold voltage V_T and flat-band voltage V_{FB}. These characteristics follow the standard MOSFET law

$$I_D = f_g C_{ox} \frac{\mu_0}{1 + \theta(V_G - V_{T,FB})} (V_G - V_{T,FB}) V_D \tag{4.27}$$

where μ_0 is the electron/hole mobility and θ is the mobility attenuation factor

$$\theta = \theta^0 + f_g \mu_0 C_{ox} R_{SD} \gg \theta^0 \tag{4.28}$$

The ideal value, θ^0, is proportional to C_{ox} and can be very small ($\theta^0 \leq 0.01\,\text{V}^{-1}$) for fully processed back-channel SOI MOSFETs (see Section 8.7). In the Ψ–MOSFET, the Schottky nature of the probe contacts results in significant series resistances and larger θ values ($\theta \geq 0.05\,\text{V}^{-1}$).

The geometric coefficient f_g accounts for (1) nonparallel current lines, (2) lack of lateral isolation, and (3) size and shape of probe contacts. It is not possible to define the width, length, and aspect ratio of the Ψ–MOSFET, so the usual relation $f_g = W/L$ does not apply and f_g must be independently determined.

We now discuss the case of a highly doped p-type film that is *partially depleted.* In Figure 4.9a, there are three distinct regions. For positive gate bias ($V_G > V_T$, region I), the interface is inverted and the depletion region depth is a maximum. The total current is the sum of the inversion channel current given by Equation (4.27) and the minimum value of the volume current. For $V_{FB} < V_G < V_T$ (region II), only the volume

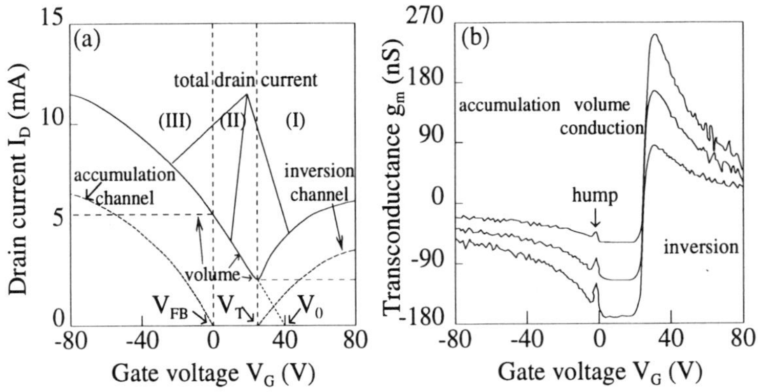

Figure 4.9 (a) Distinct conductance regions in a highly doped film, and (b) transconductance curves showing a dip resulting from the influence of the depleted substrate.

contribution remains, and depends on the extension w_d of the depletion region:

$$w_d = \frac{\epsilon_{si}}{C_{ox}} \left[-1 + \sqrt{1 + \frac{2C_{ox}^2}{q\epsilon_{si}N_A}(V_G - V_{FB})} \right] \tag{4.29}$$

A first-order development is acceptable for most SOI structures. It follows that, as the depletion region shrinks from inversion to accumulation, the volume current increases *linearly* with gate voltage

$$I_D = qf_g\mu_0 N_A V_D(t_{si} - w_d) = f_g\mu_0 C_{ox} V_D(V_0 - V_G) \tag{4.30}$$

where

$$V_0 = V_{FB} + \frac{qN_A}{C_{ox}} t_{si} \tag{4.31}$$

V_0 is the fictitious voltage that would lead to full depletion. It is obtained from the intercept of the linear current in region II with the horizontal axis. In this region, the transconductance is constant.

For lower-gate bias ($V_G < V_{FB}$, region III), the current is given by the accumulation channel (Eq. (4.27)) plus the volume current flowing in the whole film thickness t_{si}.

The above analysis implies that the depth profiles of mobility and doping are constant and the substrate influence is ignored. Most of the time, the interface between the substrate and BOX is accumulated or inverted so that the surface potential is constant and the gate bias variations have full repercussions on the channel. A voltage sharing occurs for a selected range of V_G (from -2 V to 0 V in Figure 4.9b), when the depletion region grows underneath the oxide. The series combination of the oxide and substrate depletion capacitances must here be taken into account, which is equivalent to a thicker buried oxide or a reduction in the effective value of C_{ox}. The hump observed on the transconductance curves of Figure 4.9b is merely a signal from the depletion region that is developing into the substrate. The magnitude and position of the transconductance hump depends on the film and substrate doping levels.

It is important to note that the expressions of V_T and V_{FB} must be recalculated if the channel of the Ψ–MOSFET forms while the substrate is still depleted. For example, for an n-film over a p-substrate, we have[24]

$$V_{FB} = \frac{kT}{q} \ln \frac{N_A N_D}{n_i^2} - \frac{Q_{ox}}{2C_{ox}} - \frac{Q_{ox}^2}{2q\epsilon_{si} N_A} \tag{4.32}$$

where the semiconductor–semiconductor work function was taken into account. This expression is highly different from the one used in bulk Si.

4.6.4 Parameter Extraction

The *geometric coefficient* f_g, which accounts for lateral current spreading, is determined by comparing Ψ–MOSFET data with four-point probe experiments performed with the same system. Applying a current I_{14} between the outer probes, while measuring the voltage drop V_{23} between the inner probes, eliminates the influence of series resistances. This provides well-calibrated values for the sheet resistance $R_\square = V_{23}/I_{14}$, which is V_G–dependent too:

$$I_{14} = 4.53 C_{ox} \frac{\mu_0}{1 + \theta(V_G - V_{T,FB})} (V_G - V_{T,FB}) V_{23} \tag{4.33}$$

A direct comparison between Equations (4.27) and (4.33) can be made by plotting both $I_{14}(V_G)$ and $I_D(V_G)$ curves while keeping the voltage V_{23} constant and equal to V_D. The similarity of the two curves is illustrated in Figure 4.10a. Since I_{14} is almost six times larger than I_D, an empirical value is obtained for the geometrical coefficient $f_g \simeq 0.75$.[21] Experiments on various SOI wafers tend to show that this f_g value is rather universal, although the theoretical reason is not clear yet.

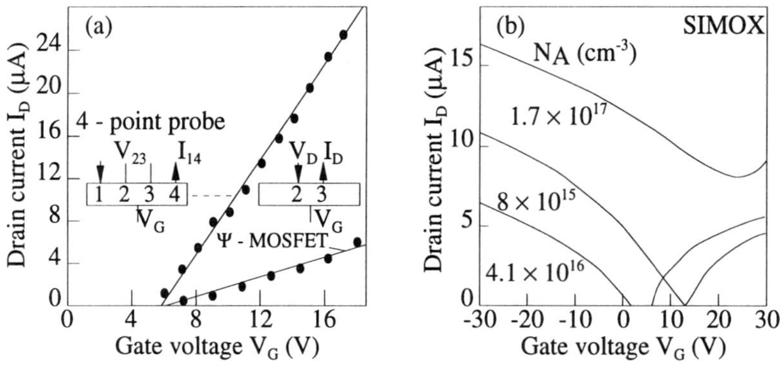

Figure 4.10 (a) Comparison between the variations of the Ψ–MOSFET drain current and the four-point probe current versus gate voltage (the voltage drop between the inner probes was $V_D = V_{23} = 0.2\,\text{V}$).[21] (b) Influence of film doping on $I_D(V_G)$ characteristics.[24]

The *threshold and flat-band* voltages are determined from $I_D(V_G)$ characteristics. The inversion region is discriminated from the accumulation region by a larger sensitivity to probe pressure. In order to cancel first-order series-resistance effects, the values of μ_0 and $V_{T,FB}$ are extracted by plotting the function $I_D/\sqrt{g_m}$ versus V_G (see Figure 4.8a and Section 8.7). The slope, $\sqrt{f_g \mu_0 C_{ox} V_D}$, yields the mobility for electrons or holes, and the intercept gives V_T or V_{FB}.

Coefficient θ is readily obtained from Equation (4.27) by measuring the saturation value, $\theta = I_D/g_m(V_G - V_T)^2 - 1/(V_G - V_T)$, when V_G varies. From coefficient θ, the series resistances are determined using Equation (4.28).

For high doping levels, it is suitable to apply these techniques to the accumulation/inversion channel only. Therefore, we first have to subtract, from the total current, the minimum volume contribution (for V_T extraction) or the maximum volume contribution (for V_{FB} extraction). The voltage V_0, defined in Figure 4.10a, is found by extrapolating to zero the linear portion of the $I_D(V_G)$ characteristic.

A simple look taken at the Ψ–MOSFET $I_D(V_G)$ characteristics of Figure 4.10b is enough to detect if the film behaves as fully or partially depleted. *Doping levels* above $5 \times 10^{15}\,\text{cm}^{-3}$ are quite precisely determined by making use of Equation (4.31) together with the experimental values of V_{FB} and V_0. Qualitative information on the substrate doping (which may change during the synthesis of SOI structures) is delivered by the position of the transconductance dip (Fig. 4.9b): the position of the dip reflects the doping type, while the dip width is related to the doping level.

The characterization of *fixed charges* in the buried oxide proceeds from the analysis of the flat-band voltage. During irradiation tests, Q_{ox} is modified. It is clear that ad hoc SOI expressions for V_T and V_{FB}, as in Equation (4.32), must be used in order to avoid severe and misleading overestimations of fixed-charge build-up.

The density of *interface traps* is determined from the subthreshold swing, $S = (kT/q)(1 + C_d/C_{ox} + qD_{it}/C_{ox})$, shown in Figure 4.7c, or from the difference between V_T and V_{FB}. It should be noted that during radiation exposure, the quality of the *native* oxide is also degraded. It is therefore recommended to etch off the radiated native oxide prior to Ψ–MOSFET tests.

In conclusion, the Ψ–MOSFET stands as a valuable tool to control the quality of as-grown SOI wafers and to anticipate the performance and radiation hardness of integrated circuits. The experiment is straightforward, and a wide set of parameters is easily available. The precise modeling of the semiconductor–insulator–semiconductor structure is not needed unless information on the substrate doping and fixed charge is desired. The main drawback of using pressure contacts comes from series resistances, which primarily affect the transconductance. A specific geometric coefficient has been defined and can be calibrated to correct the mobility values.

4.7 Quick-Turnaround Capacitance

The quick-turnaround (QT) approach is based on the capacitance versus voltage (C–V) and capacitance versus time (C–t, Zerbst) measurements performed on an SIS (silicon–insulator–silicon) capacitor, where the buried oxide serves as the dielectric layer. In the QT approach, a simple process is used to isolate silicon islands in the Si overlay, thus delineating the SIS capacitor.

The QT measurements provide a rapid assessment of the fixed oxide charge and interface-trap densities at both buried-oxide interfaces, and extraction of the average generation lifetimes in the film and substrate. The QT method can also be used to study the effects of postimplant annealing on the oxide charges and interface states in an SOI wafer.[25] A detailed description of the theory, experimental setup, and parameter extraction procedures using the C–V and C–t measurements on an SIS capacitor structure is given in Chapter 6, and only a brief introduction of the QT approach is presented below.

In an SIS capacitor structure, due to charge coupling between the upper (F/O) and the lower (S/O) interfaces of the buried oxide, a significant error can be introduced in parameter extraction when the conventional MOS C–V theory is applied. This can be explained as follows. If the substrate is grounded, the bias voltage applied to the Si film of the SIS capacitor is expressed as

$$V_G = -\psi_{s_2} + \frac{Q_{s_2} + Q_{it_2} + Q_{f_2}}{C_{ox}} + \psi_{s_3} + W_{23} \qquad (4.34)$$

where ψ_s is the surface potential, Q_s is the surface charge, Q_{it} is the interface trapped charge, Q_f is the fixed oxide charge, C_{ox} is the oxide capacitance, and W_{23} is the work function difference between the film and the substrate. The subscripts "2" and "3" refer to the F/O and S/O interfaces, respectively. Charge coupling between the two interfaces can be obtained by differentiating Equation (4.34) with respect to ψ_{s_2} and rearranging the terms

$$C_{s_2} = \left[1 - \frac{d\psi_{s_3}}{d\psi_{s_2}} + (\frac{d\psi_{s_2}}{dV_G})^{-1}\right] C_{ox} - C_{it_2} \qquad (4.35)$$

In Equation (6.20), the $d\psi_{s_3}/d\psi_{s_2}$ term represents the coupling between the charges at the two interfaces. To overcome this difficulty, the parameters are determined by fitting the theoretical high-frequency C–V curve to the measured curve. The surface potential at the S/O interface is swept from inversion to accumulation. Using the requirement of charge neutrality, i.e., equal charges at the F/O and S/O interfaces, the F/O potential is calculated at each point from the given S/O surface potential. The film bias for this charge arrangement is then deduced from Equation (4.34) by assuming that neither the film nor the substrate are heavily doped. A detailed description of the parameter extraction procedure from the C–V measurements is given in Section 6.1.

The QT–SIS capacitor test structure is fabricated by preparing ohmic contacts to the front and back surfaces (in n-type SOI, deposit of phos-

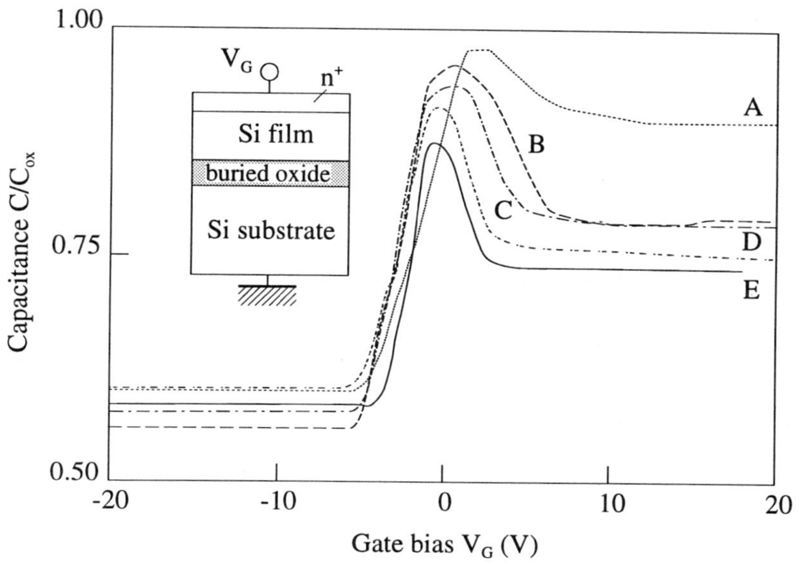

Figure 4.11 Typical C–V curves for buried oxide capacitors in SIMOX wafers with different annealings. Curve A: anneal at 1250 °C for 2 hours. Curves B, C, and D: anneals at 1285 °C for 16, 8, and 2 hours, respectively. Curve E: ideal curve.

phorus-doped glass on both sides, followed by a 950 °C, 15-minute drive-in diffusion). Trenches are then etched in the Si film down to the buried oxide, forming isolated silicon islands on the buried-oxide layer. The typical area of the resultant SIS capacitors is $500 \times 500\,\mu m^2$. Although the plasma etch used to isolate the islands may cause some sidewall damage, the sidewall area is less than 1 % of the total capacitor area, and hence its contribution to the capacitance measurements should be negligible. As compared to the capacitors from the fully processed CMOS test chips, the QT–SIS capacitors are far less influenced by the post-SIMOX processing steps. As a result, the parameters extracted from QT approach are more representative of the quality of the original SOI wafers.

Typical C–V curves are illustrated in Figure 4.11. Although these curves refer to SIMOX wafers implanted with a screen oxide, the trends are quite general.[25] Since both silicon layers are n-type, a positive gate bias depletes the film, while a negative bias depletes the substrate. In general, the presence of a fixed oxide charge rigidly shifts the C–V curve, while interface traps cause a stretch-out. An increase in the donor density increases the value of the minimum capacitance in the C–V plot. Thus, it is seen from Figure 4.11 that the F/O interface is much more sensitive to changes in the annealing parameters than is the O/S interface (i.e., more stretch-out and shift were observed in the C–V curves at the F/O interface than at the S/O interface). For the F/O interface, it was found that lowering the annealing temperature causes an increase in the donor density, while increasing anneal time results in a higher density of the fixed oxide charge.

The *Zerbst capacitance* (C–t) method can be applied to SIS capacitors to assess the average generation lifetime in SOI materials. The C–t measurements consist in monitoring the capacitance transient once the interface under study was pulsed from strong inversion into deep depletion. This condition basically eliminates parasitic transients arising from the other interface (which is kept in strong accumulation) or from interface traps (screened by the inversion layer).[26]

The success of the Zerbst method comes from the strong relationship existing between the quality of SOI wafers/devices and the measured carrier generation lifetimes in the top Si film and the substrate. Further discussion of C–V and Zerbst measurements in SOI samples is given in Chapter 6.

4.8 Pinhole Detection

Buried-oxide imperfections such as pinholes are responsible for leakage paths and premature breakdown of the oxide. For example, in SIMOX, there are nonstoechiometric oxide regions localized where the oxygen implant was masked by surface particles. The characterization of the buried oxide for surface-to-substrate leakage is essential for implantation-process feedback and device-yield optimization.

Traversing pinholes are nondestructively detected using a $CuSO_4$ electrolytic plating apparatus.[27] The standard characterization test consists of an aluminum cathode applied on the back of the wafer and a copper anode on the front of the SIMOX wafer. Inserted between the anode and the wafer surface is a tex-wipe absorbing towel that has earlier been soaked in a copper sulfate solution.

A 25 V potential is applied across the wafer for about 40 minutes.

When the current is allowed to pass through the buried oxide, the electrolytic solution is decomposed. Copper plates form on the absorbing towel at leakage sites. The plates are counted to determine the number of pinholes, which is then divided by the area of the wafer.

A strong correlation was found between the pinhole maps and surface particles counts (with Tensor 4000 surfscan system). Measured pinhole densities in recent SIMOX material are less than $0.1\,\text{cm}^{-2}$ for both standard ($0.4\,\mu$m thick) or thin ($< 0.2\,\mu$m) buried oxides.

4.9 Conclusion

The quick characterization of as-grown SOI materials is a prerequisite for their further optimization and rapid use in microelectronics. It is suitable to determine the quality of the Si overlay, buried oxide, and interfaces prior to circuit fabrication in order to avoid costly processing of poor wafers.

Nondestructive techniques are highly desirable for SOI device and circuit fabrication because of the quality-control problems that exist in the incipient SOI wafers. The SPV technique is widely used by the SOI community for inspection of the minority carrier diffusion lengths and lifetimes in the SOI substrates, while the DBOM and NOM techniques have been developed recently for determining the excess carrier lifetimes and interface recombination velocities in both the film and substrate of the SOI materials. Spectroscopic ellipsometry, dual-beam S-polarized reflectance, and spectroscopic reflectrometry have the capability to provide accurately the thicknesses of the silicon film and buried oxide as well as the interface quality and composition of the multilayered SOI structure.

Of particular interest are the electrical parameters, since they have a direct impact on the performance of integrated circuits. However, the choice of nondestructive techniques for electrical characterization is more limited. In turn, this fact created the opportunity for conceiving unique SOI methods such as the Ψ–MOS transistor, the SIS capacitor, and the electrolytic pinhole detection.

Bibliography

[1] E.O. Johnson, "Measurement of minority carrier lifetime with the surface photovoltage," *J. Appl. Phys.*, vol. 28, p. 1349, 1957.

[2] A.M. Goodman, "A method for the measurement of short minority carrier diffusion lengths in semiconductors," *J. Appl. Phys.*, vol. 32, p. 2550, 1961.

[3] M. Saritas and H.D. McKell, "Diffusion length studies in silicon by the surface photovoltage method," *Solid-State Electron.*, vol. 31, p. 835, 1988.

[4] S.S. Li, "Determination of minority carrier diffusion length in InP by surface photovoltage measurement," *Appl. Phys. Lett.*, vol. 29, p. 126, 1976.

[5] L. Jastrzebski, G. Cullen, and R. Soydan, "Monitoring of heavy metals in as-implanted SIMOX with surface photovoltage," *J. Electrochem. Soc.*, vol. 137, p. 303, 1990.

[6] D.K. Schroder, *Semiconductor Material and Device Characterization*, Wiley, New York, 1990.

[7] Y.S. Chang and S.S. Li, "Measurements of silicon film and buried oxide thickness in SOI wafers by a contactless S-polarized reflectance technique," in *Silicon-On-Insulator Technology and Devices,* S. Cristoloveanu (ed.), The Electrochemical Society, Pennington, NJ, p. 154, 1994.

[8] S.M. Sze, *Physics of Semiconductor Devices,* Wiley, New York, 1981.

[9] P.C. Yang and S.S. Li, "Measurements of substrate carrier lifetimes in silicon-on-insulator wafers by a contactless dual-beam optical modulation technique," *Solid-State Electron.*, vol. 35, p. 927, 1992.

[10] P.C. Yang and S.S. Li, "Measurements of film carrier lifetimes in silicon-on-insulator wafers by a contactless dual-beam optical modulation technique," *Appl. Phys. Lett.*, vol. 61, p. 1408, 1992.

[11] Y.S. Chang and S.S. Li, "Determination of key material parameters in SOI wafers by using a contactless optical technique," *Proc. IEEE SOI Conf. Proc.*, p. 39, 1994.

[12] H. Davies, "The reflection of electromagnetic waves from a rough surface," *Proc. Inst. Elec. Eng.*, vol. 101, p. 209, 1954.

[13] F. Sanii, F.P. Giles, R.J. Schwartz, and J.L. Gray, "Contactless nondestructive measurement of bulk and surface recombination using frequency-modulated free carrier absorption," *Solid-State Electron.*, vol. 35, p. 311, 1992.

[14] G. Harbeke and L. Jastrzebski, "Assessment of the surface quality of SIMOX wafers by UV reflectance," *J. Electrochem. Soc.*, vol. 137, p. 696, 1990.

[15] T.I. Kamins and J.P. Colinge, "Thickness determination for silicon-on-insulator structures," *Electron. Lett.*, vol. 22, p. 1236, 1986.

[16] J.R. Sandercock, "Film thickness monitor based on white light interference," *J. Phys. E: Sci. Instrum.*, vol. 16, p. 866, 1983.

[17] J. Vanhellemont, H.E. Maes, and A. de Veirman, "Spectroscopic ellipsometry studies of SIMOX structures and correlation with cross-section TEM," *J. Vacuum Technol. Appl. Ion Phys.*, vol. 42, p. 359, 1991.

[18] M. Levy, E. Scheid, S. Cristoloveanu, and P. Hemment, "Depth profiles of the optical properties of buried oxides (SIMOX) by ellipsometry," *Thin Solid Films,* vol. 148, p. 127, 1987.

[19] P.K. Dutta and S. Cristoloveanu, "The effects of implantation dose and anneal temperature on the layered structure and electrical properties of oxygen-ion-implanted silicon-on-insulator," *Microelectron. J.*, vol. 22, p. 67, 1991.

[20] R.M. Azzam and N.M. Bashara, *Ellipsometry and Polarized Light,* Elsevier, North-Holland, 1987.

[21] S. Cristoloveanu and S. Williams, "Point-contact pseudo-MOSFET for in-situ characterization of as-grown silicon-on-insulator wafers," *IEEE Electron Device Lett.*, vol. 13, p. 102, 1992.

[22] S.T. Liu, P.S. Fechner, and R.L. Roisen, "Fast turn characterization of SIMOX wafers," in *IEEE SOS/SOI Technol. Conf. Proc.*, p. 61, 1990.

[23] S. Williams, S. Cristoloveanu, and G. Campisi, "Point-contact pseudo-metal/oxide/semiconductor transistor in as-grown silicon on insulator wafers," *Materials Sci. Eng.*, vol. B12, p. 191, 1992.

[24] T. Ouisse, P. Morfouli, O. Faynot, H. Seghir, J. Margail, and S. Cristoloveanu, "A detailed investigation of the pseudo-MOS transistor for in situ characterization of SOI wafers," in *IEEE Int. SOI Conf. Proc.*, p. 30, 1992.

[25] F.T. Brady, S.S. Li, and W.A. Krull, "An electrical study of SIMOX material by a quick-turnaround method," *Solid-State Electron.*, vol. 33, p. 1643, 1990.

[26] J.S. Kang and D.K. Schroder, "The pulsed MIS capacitor: a critical review," *Phys. Stat. Solidi (a)*, vol. 89, p. 13, 1985.

[27] L.P. Allen, A. Genis, R. Dolan, and W. Krull, "Electrolytic analysis of oxide leakage in SIMOX material," *1992 IEEE Int. SOI Conf. Proc.*, p. 22, 1992.

Chapter 5

Transport Measurements

This chapter is dedicated to conventional measurements that provide basic transport parameters such as resistivity, carrier mobility and concentration, scattering process, etc. Although these methods are routinely used in bulk Si, they must carefully be reconsidered for application to thin SOI films. In response to intrinsic problems (large values of the sheet resistance, full depletion, in-depth inhomogeneity, etc.) and new opportunities (substrate bias influence), more refined experimental setup and modeling are necessary.

5.1 Four-Point Probe

The *four–point probe* method is used for rapid inspection of wafer resistivity when the "volume" of the material is not accessible and only the surface may be probed. Two probes are dedicated to current injection, and two other probes measure the voltage drop. This discrimination is necessitated by the existence of parasitic resistances at the metal–semiconductor contact, which are essentially related to the mechanical contact or to the spreading of current lines from the probe tips into the semiconductor.

Among several variants, the most frequent setup consists of four collinear and equidistant probes (Fig. 5.1a). The outer probes are carrying the current, whereas the voltage is sensed between the inner probes. The voltage measurement draws very little current, so the voltage drop across the parasitic resistances may safely be ignored. The sheet resis-

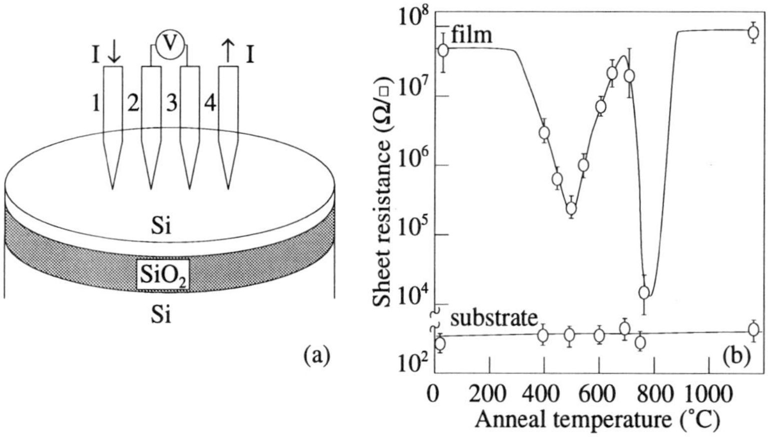

Figure 5.1 (a) Schematics of four-point probing and (b) sheet-resistance variation in SIMOX film and substrate against temperature of successive 30 minutes anneals (after Cristoloveanu et al.[2]).

tance $R_\square$ and the *average* resistivity ρ are given by

$$R_\square = \frac{\rho}{t_{si}} = 4.53 \frac{V_{23}}{I_{14}} \tag{5.1}$$

For the above relation to be valid, two main conditions must be fulfilled:

- The probe spacing (typically $s \simeq 1$–$2\,\text{mm}$) must exceed twice the material thickness ($s > 2t_{si}$). This is not a constraint in SOI structures, where not only the film but also the Si substrate can be probed with the same system.
- The probes must be centered and their distance from the sample borders must be larger than 3 to 4 probe spacings. Otherwise, correction factors, that exist for circular or rectangular samples, must be used.[1]

More or less sophisticated four-point probe systems are commercially available. In thin SOI films, the use of pressure-adjustable probes (Jandel system) is very convenient because the probe penetration into the film can be controlled. In undoped SOI films, the sheet resistance may be extremely large: $R_\square \geq 10^9\,\Omega/\square$, in a $0.1\,\mu$m layer of resistivity 10^2–$10^4\,\Omega$cm. The requirement for a high-impedance detection system may be met by using a differential voltmeter connected to each terminal via Keithley electrometers. The current is selected so as to keep the detected voltage in the mV range. It is suitable to repeat the measurement for several values of the current and to check that the $V_{23}(I_{14})$ characteristic is linear.

There are situations when the electrical influence of interface charges and probe contacts causes the film to be fully depleted, which prevents current injection. Intentional film doping is not a satisfactory solution, since most of the information expected from resistivity data (crystal quality, nature of defects) is erased. The dilemma may be solved by biasing the substrate in accumulation to squeeze the depletion region.

Four-point probe measurements are currently performed on as-grown SOI structures to control the degree of contamination, the homogeneity across the wafer, and the dopant activation. Figure 5.1b illustrates the generation of oxygen-related donors in early SIMOX material.[2] The sheet resistance of the film can drop by orders of magnitude after short anneals at 500 °C and 750 °C. Repeating the experiment on the back of the wafer does not reveal any detectable variation of $R_\square$. This verifies the presence of many oxygen atoms in the Si film subsequent to the implantation process. It is worth noting that such a two-sided experiment offers an elegant way of comparing in situ the resistivities of the film and of the starting bulk-Si wafer.

5.2 Spreading Resistance

In films that show in-depth inhomogeneities of doping and/or mobility, the "average" resistivity measured by four-point probing does not contain enough explicit information. The resistivity profile is determined by spreading resistance. The basic idea is to scan laterally the resistance of a beveled sample and then to convert the data into a vertical profile.

A low-angle ($< 15'$–$3°$) bevel is prepared by mounting the sample on a bevel block and lapping it with diamond paste. Two tungsten probes are moved along the bevel, by steps of a few micrometers, and the resistivity is measured at each location. A vertical resolution of about 35 nm is achieved by $2\,\mu$m steps along a 1° bevel. Higher resolutions are needed in very thin SOI films (50–100 nm). This precludes the use of bevel angles above 30' or scan steps exceeding 1 μm.

The probe interdistance being very small ($\leq$ 0.1 mm), the voltage drop can be maintained below the thermal voltage (25 mV), which guarantees ohmic conduction. The measured resistance is the sum of contact resistance, spreading resistance, and sheet resistance. The contact resistance is constant and does not matter in the differentiation algorithm. The spreading resistance R_{SR}, due to the spreading of current lines from the probe tip into the semiconductor, is given by

$$R_{SR} = \frac{\rho}{\alpha r} \tag{5.2}$$

where r is the probe contact radius and α is a shape-defined coefficient. The parameter (αr) is periodically determined, in particular when the probes are changed or reconditioned. This is done by measuring precalibrated samples or by comparison with four-point probe data.

The resistivity profile is constructed from the resistance values measured at each sample depth. In general, the calculations proceed by determining the resistivity of the first slice of the film, which is located just above the buried oxide. From the second resistance value, the resistivity of the following slice is obtained, and so on. It might be helpful to mark the top surface by depositing an oxide cap. There is no way to measure independently the mobility and doping profiles. Therefore, the doping profile is inferred from the resistivity profile by assuming ideal mobility versus doping laws. This procedure works reasonably well in fully optimized semiconductors, such as bulk Si, where the spreading resistance is routinely used to profile the doping variations through p–n junctions. The method was also extended to resolve two-dimensional profiles by scanning the bevel in various directions.

The critical features of spreading resistance are the probe alignment, calibration, and mechanical scan. Commercially available systems (from Solid State Measurements Inc., USA) are expensive. Also, the preparation of uniform low-angle bevels requires a great deal of skill and experience. Bevel nonuniformities are responsible for artificial structures that appear in the resistivity profile and can be misunderstood as doping inhomogeneities.

Even more serious profiling artifacts arise in undoped thin SOI films. The sheet resistance may just exceed the system sensitivity ($\simeq 10^8\,\Omega$). This usually happens when the probes are stepped in the proximity of the buried oxide. The measurement is confusing, since it is not easy to

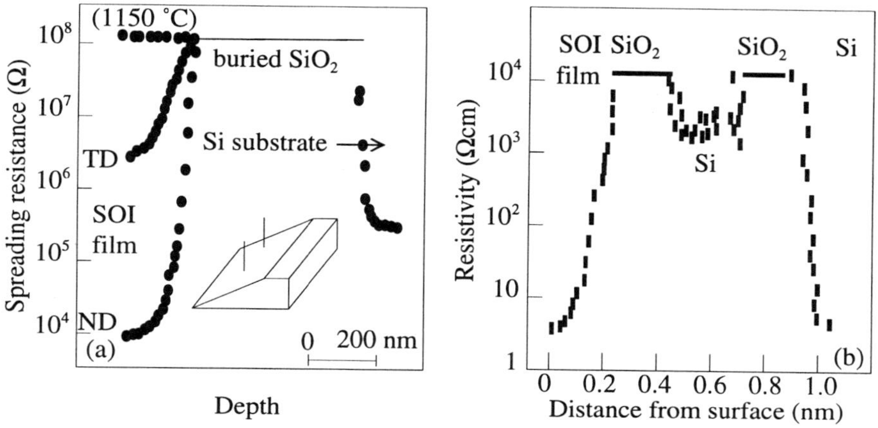

Figure 5.2 Spreading resistance profiles in SIMOX: (a) before and after activation of thermal donors (TD: 450 °C, 30 minutes) and new donors (ND: 750 °C, 10 minutes);[2] (b) double SIMOX structure.

appreciate whether the probes have actually reached the oxide or whether the detection limit has been overcome. The BOX appears to be thicker, and the computation algorithm is corrupted. Only by comparison with calibrated (intentionally doped) samples may this ambiguity be clarified.

The doping profiling, in the bottom region of the film, is not reliable for two other reasons: (1) the resistivity is influenced by oxide charges and (2) the mobility may be highly inhomogeneous, e.g., dominated by crystalline defects and surface scattering rather than by impurity scattering. The use of mobility–doping laws, imported from bulk Si, is indeed a common source of errors.

Despite these inherent weaknesses, the method is very useful and deserves consideration. However, one has to be cautious about the interpretation of resistivity and doping profiles in SOI. Typical spreading resistance profiles of resolution 5 nm are reproduced in Figure 5.2a. No distinction is visible between the Si film and buried oxide in as-grown SIMOX wafer. Thermal activation of oxygen donors at 450 °C and 750 °C highly reduces the resistivity (in agreement with Fig. 5.1b) and reveals the location of the Si layer. No donor formation or any other anneal influence is observed on the other side of the buried oxide.[2]

Figure 5.2b illustrates the resistivity profile in a double SIMOX struc-

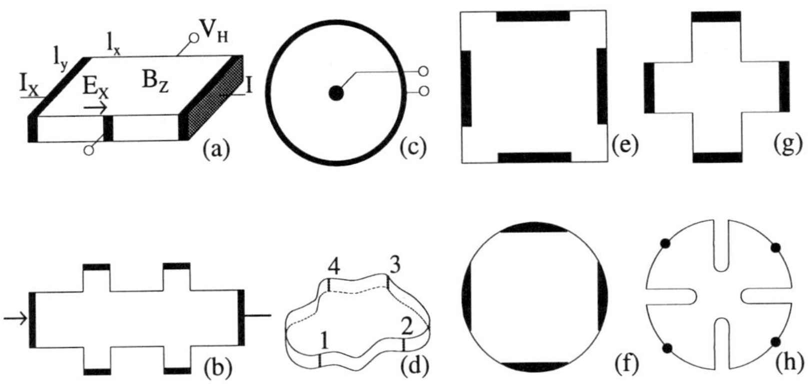

Figure 5.3 (a) Conventional Hall sample, (b) bridge-type device, (c) Corbino disk, (d) van der Pauw sample of arbitrary shape, and (e,f,g,h) alternative van der Pauw shapes.

ture formed by two successive oxygen implants at different energies. Two distinct oxides, separated by a silicon region, are clearly visible. The sandwiched Si layer has poorer quality and higher resistivity than the top Si film. Spreading resistance measurements are currently being dedicated to the analysis of the activation and redistribution of implanted dopants in SOI structures.[3]

5.3 Hall Effect and Magnetoresistance

The Hall effect is a unique tool for the independent determination of carrier concentration and mobility. The conventional theory considers a rectangular slab (Fig. 5.3a). The current is supplied by large end contacts, and a magnetic field B is applied perpendicular to the surface. The carriers tend to be deflected by the Lorentz force. Since no current is allowed in the transversal direction, a Hall electric field develops to oppose the Lorentz force.

The Hall voltage is expressed by

$$V_H = \frac{l_y}{l_x}\mu_H B V_x = R_H \frac{B I_x}{t_{si}} \tag{5.3}$$

where μ_H is the Hall mobility and R_H is the Hall coefficient

$$\mu_H = \frac{R_H}{\rho_B} = \frac{R_H/t_{si}}{\rho_B/t_{si}} = \frac{V_H}{R_\square B I_x} \tag{5.4}$$

This equation shows that the Hall mobility has the privilege of being measurable even in films with unknown thickness, whereas the extraction of R_H and ρ, from V_H and $R_\square$ data, respectively, requires a prior evaluation of t_{si}. The sign of the Hall voltage indicates the semiconductor type and thereby identifies the presence of electrons or holes.

Under the oversimplifying assumption of monokinetic carriers, the Hall mobility reduces to the drift mobility, the Hall coefficient gives the carrier concentration, and there is no magnetoresistance at all:

$$\mu_H = \mu_0 \qquad R_H = \frac{1}{qn} \qquad \frac{\rho_B}{\rho_0} = 1 \tag{5.5}$$

When the carrier energy distribution is taken into consideration, the relaxation time $\tau(\epsilon)$ between collisions must be averaged over the energy range, yielding

$$\mu_H = r_H \mu_0 \qquad R_H = \frac{r_H}{qn} \tag{5.6}$$

with $r_H = \langle \tau^2 \rangle / \langle \tau \rangle^2$. The Hall scattering factor r_H is governed by collisions with acoustic phonons at high temperature ($r_H = 3\pi/8 \simeq 1.18$) and ionized impurities at low temperature ($r_H = 315\pi/512 \simeq 1.93$). A slight magnetoresistance ($\rho_B/\rho_0 = \langle \tau \rangle \langle \tau^{-1} \rangle$, at high magnetic field) arises from fluctuations in carrier velocities around the average velocity.

The above analysis was developed for isotropic semiconductors. In n-type silicon, the isoenergy surfaces are ellipsoidal, and an anisotropy coefficient K accounts for the ratio between the longitudinal and transversal values of effective mass and relaxation time. The Hall and drift mobilities become quite different, which invalidates Equation (5.6). Also, the Hall mobility may depend on sample orientation, especially for high magnetic fields. For example, in n–Si with $I^{(110)} \perp B^{(001)}$ configuration, the scattering coefficient of Equation (5.6) becomes $r_H = 3K(K+2) \langle \tau^2 \rangle / (2K+1)^2 \langle \tau \rangle^2$, with $K \simeq 5$–7.[4]

A proper description of these effects takes into consideration the conductivity tensor of each ellipsoid. Only for selected orientations will the tensor reduce to a scalar. Summing up the contributions of the six ellipsoids accounts for experimentally observed effects. The key aspect is

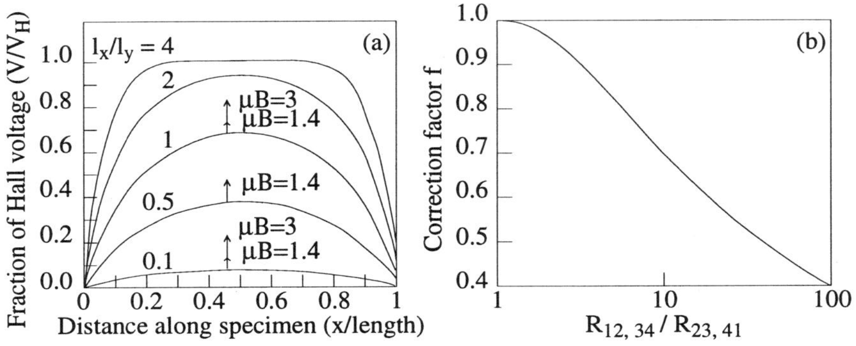

Figure 5.4 (a) Shortening of the Hall voltage along the sample of Figure 5.3a, for various aspect ratios (length/width) and low magnetic fields; the arrows show the attenuation of the shortening effect at higher magnetic fields (units of μB in m^2T/Vs, after Berr[6]). (b) The van der Pauw correction factor f versus asymmetry in pseudoresistances (after Schroder[1]).

that, for a given ellipsoid, the current is not necessarily parallel to the applied electric field (or to the *total* current). A magnetic field parallel to the electric field does deflect the current lines and induces a magnetoresistance. Conversely, such *longitudinal magnetoresistance* measurements are used to detect the anisotropy and stress-induced modifications of energy valleys.[4, 5]

The so-called *geometric magnetoresistance* is a different effect that depends on sample geometry, not on carrier scattering. According to the location of the Hall probes along the sample, the measured Hall voltage may be substantially smaller than the theoretical value given by Equation (5.3). The apparent reduction of Hall mobility is shown in Figure 5.4a as a function of the aspect ratio (l_x/l_y) of the sample.[6] Near the end contacts, the Hall voltage is shorted ($V_H \rightarrow 0$), the Lorentz force is no longer compensated, and a high "geometric" magnetoresistance develops.[1]

The Corbino disk (Fig. 5.3c) is the limiting case for such a geometric

[1]Hall and transversal magnetoresistance are dual effects: when Hall field is suppressed (short-circuited), carrier deflection is enabled, leading to a maximum magnetoresistance. Reciprocally, the magnetoresistance is almost canceled for a maximum Hall field.

magnetoresistance: zero Hall field and $\rho_B/\rho_0 = 1 + \mu_H^2 B^2$. For this reason, the measurement of the pure Hall effect requires *long* rectangular (Fig. 5.3a) or bridge-type (Fig. 5.3b) samples. However, we will see in the next section that van der Pauw samples, with less space-consuming geometry, are more frequently used.

Although Si is referred to as an "extrinsic" semiconductor, there are extreme situations (high temperature or phototransport experiments) that deserve a two-carrier analysis. The set of Equations (5.6) must then be replaced by

$$\mu_H = \frac{p\mu_p\mu_{Hp} - n\mu_n\mu_{Hn}}{p\mu_p + n\mu_n} \qquad \frac{\rho_B}{\rho_0} = \frac{p\mu_{0p} + n\mu_{0n}}{p\mu_p + n\mu_n} \times \frac{1}{1 + \mu_H^2 B^2} \tag{5.7}$$

where the absolute values of electron and hole drift mobilities $\mu_{n,p}$ under magnetic field are related to zero-field mobilities: $\mu_{n,p} \simeq \mu_{0n,p}/(1 + \mu_{0n,p}^2 B^2)$. The ambipolar Hall coefficient is still given by $R_H = \mu_H \rho_B$. At a high magnetic field, the Hall coefficient is no longer dependent on magnetic field, scattering process, or anisotropy: $R_H = 1/q(n-p)$.

There is nothing special about the Hall-effect setup and theory in good-quality SOI material. The Hall voltage is very small and can be affected by parasitic factors: misalignment of lateral contacts, nonohmic conduction, etc. For accuracy, the measurement must be repeated for several positive and negative values of current and magnetic field.

One might be concerned about possible in-depth inhomogeneities of the film. This is primarily suspected when poor values of Hall mobility are measured. More precise information is delivered by the Hall-effect profiling techniques described in Section 9.1. Also, in early SOS films, lateral strain effects were invoked to explain unusual features: degraded electron mobility, improved hole mobility and increased longitudinal magnetoresistance as compared to bulk Si. A detailed analysis proceeds from the Boltzmann equation and concludes on additional anisotropy effects.[4] Experimental results, obtained in various types of SOI samples, will be discussed in the following section.

At very high magnetic field and low temperature, quantization occurs and the magnetoresistance shows periodic oscillations in reciprocal magnetic field ($1/B$). According to the Shubnikov–de Haas model, peaks in carrier scattering and hence in magnetoresistance occur when each Landau level empties. The period of the oscillations gives the band anisotropy and the precise carrier concentration (without the uncertainty related to the Hall factor r_H). The carrier effective mass and mobility are deduced from the peak amplitudes.

5.4 Van der Pauw Measurements

Van der Pauw has extended the Hall-effect theory to samples of arbitrary shapes (Fig. 5.3d).[7] The surface is assumed to be single connected, which means that it does not contain isolated holes. The sample thickness t_{si} is uniform, and there are four contacts located on the sample edge. The current I_{ab} is fed through two contacts, and the voltage drop V_{cd} is detected between the two other terminals. This measurement defines a *pseudoresistance* $R_{ab,cd} = V_{cd}/I_{ab}$. Several combinations of the four contacts are used to determine the resistivity and the Hall coefficient.

RESISTIVITY

The current is injected between adjacent contacts, and the pseudoresistances $R_{12,34}$ and $R_{23,41}$ are evaluated. The two measurements are performed with the roles of the contacts being shifted by a quarter turn. The merit of this particular combination of pseudoresistances is to eliminate the unknown distance between contacts:

$$\exp(-\pi t_{si} R_{12,34}/\rho_0) + \exp(-\pi t_{si} R_{23,41}/\rho_0) = 1 \tag{5.8}$$

This relation was first demonstrated for semi-infinite samples where the contacts are aligned.[7] Conformal mapping allows the generalization of Equation (5.8) by showing that any single connected domain can be transformed in a half-plane.

Instead of extracting the resistivity ρ_0 from Equation (5.8), it is more convenient to use the equivalent expression

$$\rho_0 = \frac{\pi t_{si}}{\ln 2} \times \frac{R_{12,34} + R_{23,41}}{2} \times f \tag{5.9}$$

where f is a *configuration* coefficient related to the ratio between the two pseudoresistances

$$2 \exp\left(-\frac{\ln 2}{f}\right) \times \cosh\left(\frac{R_{12,34}/R_{23,41} - 1}{R_{12,34}/R_{23,41} + 1} \times \frac{\ln 2}{f}\right) = 1 \tag{5.10}$$

The value of f is determined from Figure 5.4b or from more precise tables. For perfectly symmetrical samples, where $R_{12,34} = R_{23,41}$, we have $f = 1$.

HALL COEFFICIENT

The current is now injected through nonsuccessive (diagonally opposite) contacts, and $R_{13,24}$ is measured. The application of a magnetic field B, perpendicular to the surface, modifies the "transversal" voltage but does not change the image of current lines. The Hall coefficient is derived from the variation of the pseudoresistance under the magnetic field[7]

$$R_H = \frac{t_{si}}{B} \times [R_{13,24}(B) - R_{13,24}(0)] \tag{5.11}$$

Again, the Hall mobility is given by $\mu_H = R_H/\rho_B$.

In fact, Equation (5.11) is a reasonable approximation only for weak magnetic fields. When the magnetoresistance is not negligible ($\rho_B/\rho_0 \neq 1$), we have[8]

$$V_{24}(B) = \frac{\rho_B}{\rho_0} \times V_{24}(0) + \frac{R_H B I_{13}}{t_{si}} \tag{5.12}$$

It is clear that the original van de Pauw equation (5.11) does not hold anymore and should be replaced by

$$R_H = \frac{t_{si}}{B} \times \frac{R_{13,24}(+B) - R_{13,24}(-B)}{2} \tag{5.13}$$

The idea is that in Equation (5.12), the magnetoresistance is an even function of B and can be eliminated by comparing two measurements performed for both directions of the magnetic field. The comparison to zero field data, used in Equation (5.11), is no longer justified. The averaging for $\pm B$, made with Equation (5.13), is essential because the "parasitic" offset $V_{24}(0)$ may highly exceed the pure Hall voltage, so that even small magnetoresistance values lead in Equation (5.11) to fatal errors in R_H. Under no circumstances can the current inversion $\pm I_{13}$ and averaging be used as a substitute for the magnetic field inversion made in Equation (5.13). Nevertheless, once the averaging with $\pm B$ values has been performed, further averaging for $\pm I_{13}$ is valuable as a supplementary precaution.

From Equation (5.12), one can also derive the magnetoresistance coefficient

$$\frac{\rho_B}{\rho_0} = \frac{R_{13,24}(+B) + R_{13,24}(-B)}{2R_{13,24}(0)} \tag{5.14}$$

Alternatively, the magnetoresistance is obtained by repeating the procedure of Equation (5.10) under the magnetic field. Lateral inhomogeneities and anisotropy effects should be suspected if the results vary when the roles of the contacts are interchanged.

A major hypothesis of van der Pauw's theory is the presence of ideal contacts: *point* contacts, exclusively located at the sample circumference, over the whole thickness. If the size of the contacts and/or their distance to the sample edge is not negligible compared with the sample perimeter, corrections factors are needed. Basically, the Hall voltage correction is proportional to the ratio between the largest dimension of the contact and the perimeter, whereas the resistivity correction is related to the square of this ratio and therefore is far less relevant.

Corrections factors have duly been calculated for a number of contact shapes,[1] very precisely defined and also difficult to fabricate. We feel that these sophisticated attempts at improvement in fact work against the marvelous flexibility of the original van der Pauw theory. In practice, it is much easier to enlarge the sample perimeter using "scissored" shapes (Figs. 5.3g,h). This still preserves the small device area, which is an appreciated feature for on-chip test structures. Since contacts with finite sizes suffer from local geometrical effects, it is suitable to have the contacts placed at regular distances (Fig. 5.3e–h) and to avoid samples with sharp angles.

The target of Hall-effect measurements is to monitor the optimization of SOI materials, which is achieved by increasing the carrier mobility and suppressing contamination sources. This implies a deep understanding of the physical mechanisms involved in carrier transport. Selected situations for SOI are illustrated below.

Figure 5.5a shows the relation between mobility and doping observed in early SOS films.[9, 10] The surprising increase of the hole mobility with doping level (for $N_A < 10^{17}\,\mathrm{cm}^{-3}$) was explained by the suppression of small depletion regions that exist around crystal imperfections. The probability of space-charge-scattering mechanisms is lowered; thereby, mobility is enhanced.[10] A mobility degradation in thinner layers was systematically reported and attributed to numerous defects that subsist in the bottom of the film.

Low-temperature measurements, $\mu_H(T)$, are very useful for the investigation of dominant scattering mechanisms. In Figure 5.5b, Hall mobility curves are compared for three families of SIMOX samples. Sample 1 gives evidence for the efficiency of high-temperature (1300 °C) annealing

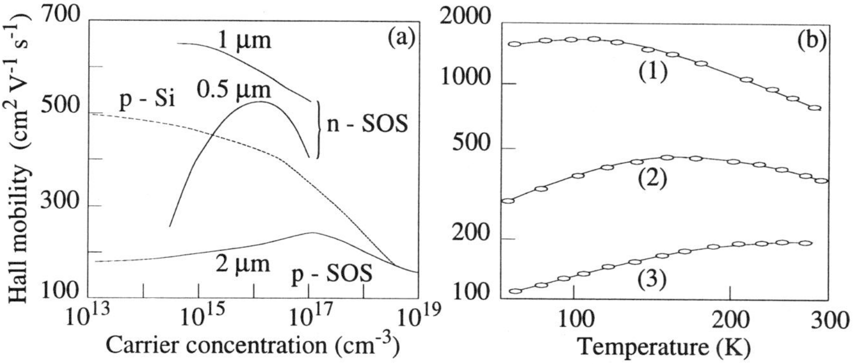

Figure 5.5 Hall-effect mobility (a) versus doping in SOS films[9, 10] and (b) versus temperature in SIMOX films.[11]

in argon ambient: the mobility is high, presents a peak at about 100 K, and is dominated above by acoustic phonon scattering ($\mu_H \sim T^{-1.5}$). This behavior compares well with the case of bulk Si.

Annealing in nitrogen (sample 2) is responsible for n-type contamination. The mobility is degraded, and its peak is shifted to a higher temperature. Sample 3, annealed at low temperature (1150 °C), has a very poor mobility that monotonously increases with temperature.[11] This behavior is unusual for bulk Si but has also been reported for SOS. It is notorious for being induced by the presence of a highly defective and doped region at the bottom of the film, which allows Coulombic scattering to dominate even at 300 K.

Most Hall measurements tend to reveal that doping overcompensation takes place during the synthesis of SIMOX structures: the film has n-type conduction although the starting wafer was p-type. The concentration and mobility of oxygen donors, which might be responsible for such a compensation, were analyzed on SIMOX samples annealed at various temperatures between 400 and 900 °C.[11] Figure 5.6a shows a generation of thermal donors at 550 °C and, in particular, a very substantial activation of *new donors* at 750 °C. The mobility–doping correlation fits the empirical law used in bulk Si: $\mu = \mu_0/(1 + N_D/N^\star)$, where $N^\star \simeq 10^{17}\,\mathrm{cm}^{-3}$. After deactivation of thermal donors at 650 °C, the sample exhibits excellent electron mobility (1250 cm^2/Vs) and a background doping as low as $2 \times 10^{15}\,\mathrm{cm}^{-3}$.

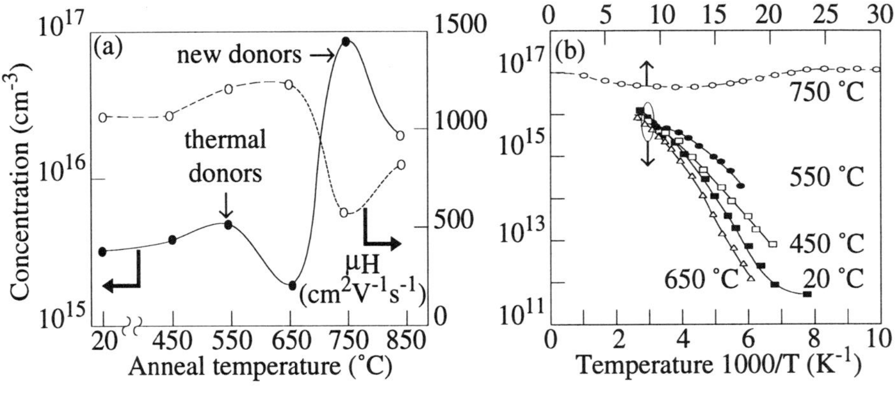

Figure 5.6 Oxygen donor activation in SIMOX samples annealed at 1300 °C and reannealed at several temperatures: (a) carrier concentration and Hall mobility variations;[11] (b) Arrhenius curves of electron concentration versus reciprocal temperature.[12]

From Arrhenius curves such as those in Figure 5.6b, it is inferred that thermal donors are divalent oxygen donors with a shallow state at 40–80 meV and a deep state at 220 eV.[12] In contrast, the concentration and mobility of "new" donors formed at 750 °C are almost independent of temperature below 50 K. The film appears to be degenerated, although the carrier concentration is one order of magnitude smaller than the theoretical value expected to cause degeneracy. It is speculated that the formation of new oxygen donors is inhomogeneous, being concentrated in a very narrow region (10–15 nm) close to the interface between the film and the buried oxide.

A degenerated behavior was also observed for the transition layer, which originates from the high density of fixed charges existing at the Si–Al_2O_3 interface in SOS films: again, the carrier mobility and concentration are constant below 25 K, as shown in Figure 5.7.[13]

The temperature variation of the carrier concentration[2] identifies the dopants by revealing their activation energy. The following relations hold

[2]When the film thickness is unknown or in-depth inhomogeneities are presumed, the Hall coefficient of Equations (5.3) or (5.13) is resolved in terms of sheet carrier concentration: $t_{si}/R_H \sim n t_{si}$.

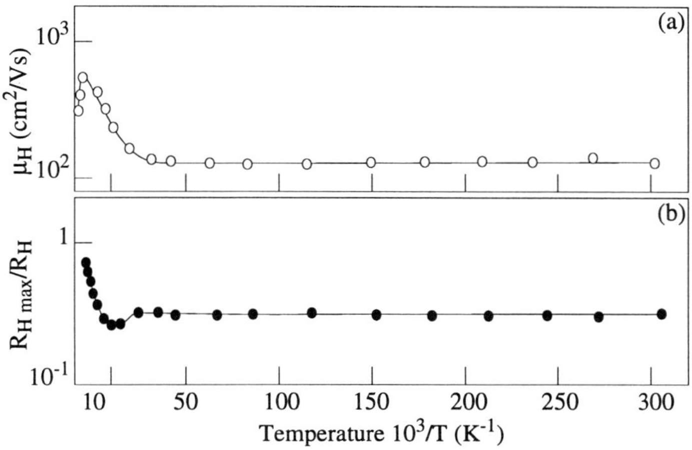

Figure 5.7 Temperature dependence of (a) Hall mobility and (b) Hall coefficient in n-type SOS showing a degenerated behavior.[13]

for a single donor level and for a compensated n-type film, respectively:

$$n \simeq \sqrt{\frac{N_D N_C}{g_D}} \exp\left(-\frac{E_D}{2kT}\right)$$
$$n \simeq \frac{N_D - N_A}{g_D N_A} N_C \exp\left(-\frac{E_D}{kT}\right) \tag{5.15}$$

where $g_D = 2$ is the degeneracy factor, E_D is the donor activation energy, and $N_C \simeq 4 \times 10^{15} T^{1.5}$.

The experimental data is used to draw the Arrhenius plot, $\ln nT^{-1.5}$ vs. $1/T$. The activation energy is determined from the slope and the doping concentration from the plateau observed at high temperature (Fig. 5.8a). Ignoring the multiplying coefficient $T^{-1.5}$ on the vertical axis may be a source of inaccuracy. In principle, regions with several plateaus and slopes should indicate the coexistence of intentional and unintentional dopants. Unfortunately, the concentrations and activation energies of dopant species are not separated enough, so Arrhenius curves become confusing.

Hall effect "spectroscopy" may be used to enhance the resolution. It consists in taking the first- or second-order derivative of the carrier concentration with respect to the reciprocal temperature.[14] The methods

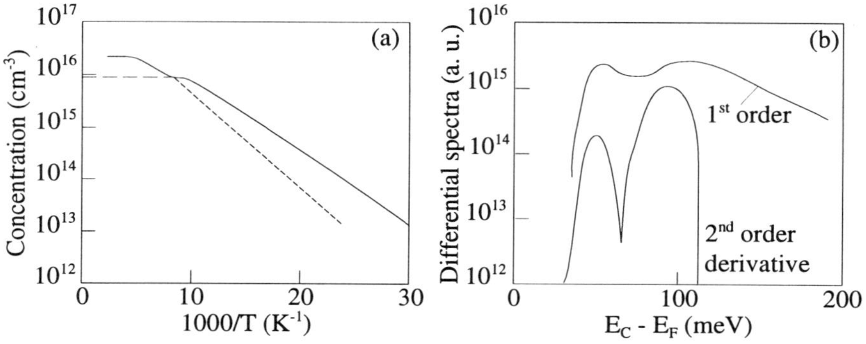

Figure 5.8 (a) Arrhenius curves and (b) spectroscopic methods based on the first- and second-order derivatives of the carrier concentration versus temperature in SIMOX films containing two donor dopants (after Kleveland et al.[14]).

are compared in Figure 5.8 for two coexisting dopants. The ambiguous structure affecting the linearity of the Arrhenius plot (Fig. 5.8a) is clearly converted in spectroscopic peaks (Fig. 5.8b). The positions of the peaks yield the activation energies, the accuracy being improved with the second-order derivative. The first derivative shows better correlation between peak magnitude and doping concentration.

To achieve noise-free curves, a good experimental resolution is required: equidistant points in the reciprocal temperature range (for Arrhenius plots) or in the temperature range (steps of $\Delta T \simeq 2$–$10\,\mathrm{K}$, for spectroscopic methods). This implies the use of computer-controlled van der Pauw measurements.

5.5 Photoconductivity

A variety of characterization methods are using light as an additional experimental parameter. The measurement setup is more complex, but it is justified if particular features are expected. We focus on photoconductivity, which is the simplest and most frequently utilized method in SOI materials.

The illumination wavelength is selected such that the photon energy exceeds the bandgap ($\lambda < hc/E_G$) and causes carrier generation.

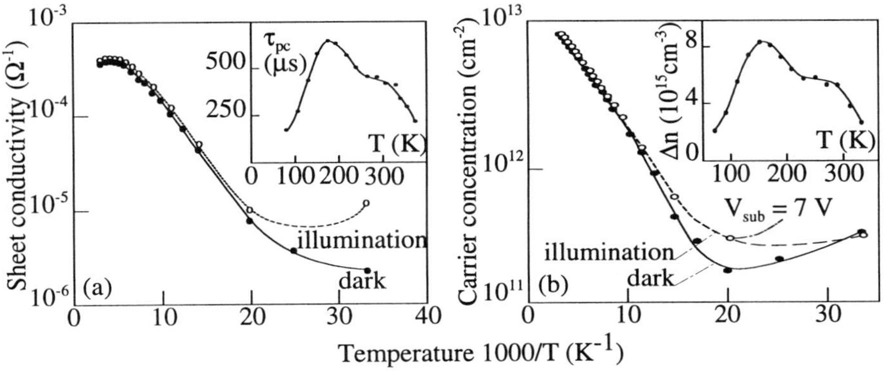

Figure 5.9 (a) Sheet conductivity and (b) sheet carrier concentration, under illumination and darkness, versus temperature in SIMOX. The insets show the photoconductivity lifetime and excess concentration of photogenerated carriers (after Papaioannou et al.[15]).

Electron-hole pairs contribute to the conductivity

$$\sigma_{pc} = \frac{q}{t_{si}} \int_0^{t_{si}} \left[n\mu_n + \alpha\phi_0 e^{-\alpha z} (\tau_n\mu_n + \tau_p\mu_p) \right] dz \tag{5.16}$$

where $\tau_{n,p}$ are *recombination* lifetimes for electrons and holes in a n-type film and ϕ_0 is the photon flux. In general, the absorption coefficient α is very large compared to the reciprocal thickness of SOI films ($\alpha \gg 1/t_{si}$) and the sample is uniformly illuminated.

The excess photoconductivity $\Delta\sigma_{pc}$ is expressed as

$$\Delta\sigma_{pc} \simeq \frac{q\alpha\phi_0}{t_{si}} \int_0^{t_{si}} (\tau_n\mu_n + \tau_p\mu_p)\, dz \tag{5.17}$$

Typical temperature-dependent conductivity curves are shown in Figure 5.9a and explained as follows. Only a small fraction of photons are absorbed. For a photon flux of $10^{17}\,\text{cm}^{-2}\text{s}^{-1}$ and $\alpha \simeq 150\,\text{cm}^{-1}$, the number of photogenerated pairs (10^{14}–$10^{15}\,\text{cm}^{-3}$) is far below the doping level (10^{16}–$10^{17}\,\text{cm}^{-3}$). Illumination is therefore unable to affect the conductivity behavior at room temperature. Only at very low temperature, where thermally generated carriers freeze out, does photoconductivity play a significant role. A global *photoconductivity lifetime* τ_{pc} is defined

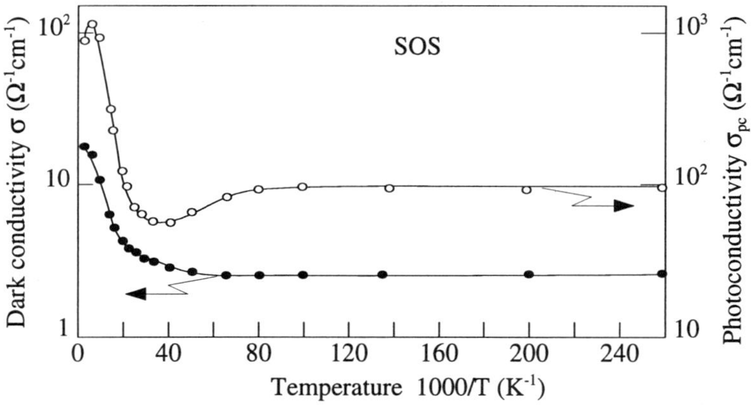

Figure 5.10 Dark conductivity and photoconductivity against temperature in n-type SOS films (after Robert et al.[13]).

to overcome the difficulty raised by the independent variations of $\mu_{n,p}$ and $\tau_{n,p}$ with temperature:

$$\tau_{pc} = \frac{1}{q\alpha\phi_0} \times \frac{\Delta\sigma_{pc}}{\mu_H} \tag{5.18}$$

where the temperature dependence of the Hall mobility is determined from van der Pauw measurements under darkness. Although photoconductivity yields an overestimated recombination lifetime, the values shown in the inset of Figure 5.9a confirm the excellent quality of SIMOX films. Above 150 K, τ_{pc} decreases with increasing temperature, as in bulk Si. The maximum value corresponds to the crossover between Fermi level and impurity level. A detailed analysis of τ_{pc} should also take into consideration the contributions of interface recombination velocities.

The carrier concentrations are directly obtained from Hall-effect measurements (Fig. 5.9b). Caution is needed to mask the contacts of the sample in order to avoid a parasitic photovoltaic effect. Under illumination, the two-carrier equation (5.7) yields

$$R_{Hi} = \frac{r_H}{q(n+\Delta n)}\left(1 - \frac{\mu_p^2 \Delta n}{\mu_n^2 (n+\Delta n)}\right)\left(1 + \frac{\mu_p \Delta n}{\mu_n (n+\Delta n)}\right)^{-2} \tag{5.19}$$

In bulk silicon, the ratio between electron and hole mobilities is $\mu_n/\mu_p \simeq 3$. The electron concentration is known from the Hall coefficient, R_{Hd}, measured in darkness. The excess carrier concentration is obtained by solving Equation (5.19) for Δn. For weak photoexcitation, the brackets of Equation (5.19) can be neglected:

$$\Delta n \simeq \frac{1}{q}\left(\frac{1}{R_{Hi}} - \frac{1}{R_{Hd}}\right) \tag{5.20}$$

The excess carrier concentration is proportional to the electron recombination lifetime: $\Delta n = \tau_n \alpha \phi_0$. Indeed, the temperature variations of Δn (inset of Fig. 5.9b) and τ_{pc} (inset of Fig. 5.9a) are fully consistent.

A hopping conduction mechanism has been invoked to explain the low-temperature behavior of carrier concentration shown in Figure 5.9b. Taking into account hopping electrons (h) with extremely small mobility and conduction band electrons (c), we have from the equivalent of Equation (5.7)[15]

$$R_{Hd} = \frac{\mu_c \sigma_c + \mu_h \sigma_h}{(\sigma_c + \sigma_h)^2} \simeq \frac{\mu_c \sigma_c}{(\sigma_c + \sigma_h)^2} \tag{5.21}$$

This relation predicts, for $\sigma_c = \sigma_h$, a maximum value of R_{Hd}, i.e., a minimum carrier concentration, which is experimentally visible for 50 K in Figure 5.9b. Above 50 K, $\sigma_c \gg \sigma_h$ and $n \sim \exp(-E_D/kT)$. At lower temperatures, σ_h prevails and $n \sim \exp[(E_D - 2E_h)/kT]$. Since $E_h \ll E_D$, the carrier concentration apparently increases below 50 K. At low temperature, photogenerated carriers represent a substantial contribution to the concentration of conduction-band electrons. This opposes the transition to hopping-dominated conduction. The model is confirmed by applying a positive bias on the substrate. Electron accumulation in the more defective interface region accentuates the influence of the hopping mechanism.

A second example of photoconductivity measurement and interpretation is shown in Figure 5.10. SOS samples were illuminated with a high-intensity source ($\Delta n \gg n$). Under darkness, the conductivity exhibits a degenerated behavior below 25 K, already discussed in relation with Figure 5.7. The photoconductivity is constant at very low temperature and thermally activated above 40 K, but presents a surprising minimum around 30 K.

The proposed scenario invokes the presence of long-range potential fluctuations resulting from a random spatial distribution of the band energy. According to the Fermi-level position, some carriers can get localized in the potential wells. In the degenerated situation, E_F is situated above the potential well, and all photocarriers are free. At higher temperature, E_F crosses the wells, localization occurs, and only a fraction of the total number of electrons contribute to the current. As the temperature increases further, E_F comes below the conduction band edge, and thereby the fraction of free electrons increases rapidly towards 100%. The average depth of potential fluctuations was found from the fitting of the photoconductivity minimum.[13]

There are many other types of experiments based on photoconductivity principles:

- Photoconductivity decay after exposure to a short laser pulse is monitored to determine the recombination lifetime.

- By varying the incident photon energy, photoconductivity peaks may be observed that reflect the existence of deep traps.

- The time-of-flight technique, referred to as the Haynes–Shockley experiment, consists of applying a narrow light impulse near the terminal of a biased rectangular sample. The distribution of the photogenerated minority carriers is sensed by a p–n junction situated at the opposite terminal. The time-of-flight (delay between carrier injection and collection) gives the mobility of minority carriers. The widened shape of the collected pulse is a measure of the diffusion coefficient, whereas the reduction of the pulse area accounts for the recombination lifetime.

- The photomagnetoelectric (PME) effect combines light and magnetic field actions. The sample surface is illuminated, and a transversal magnetic field is applied in the plane of the film. The vertical diffusion current, resulting from the photocarriers gradient, is "rotated" by the magnetic field. This gives rise to a longitudinal short-circuit current or open-circuit voltage, from which the carrier generation–recombination properties can be calculated. A more advanced version is the field-assisted PME effect: a longitudinal electric field is applied to reinforce the vertical carrier gradient by magnetoconcentration.[17] According to the direction and intensity of the Lorentz force, more or less carriers are concentrated at the film–oxide interface. The resulting nonlinearity of the longitudinal $I(V)$ characteristics is exploited for parameter extraction.

5.6 PICTS

The identification and suppression of deep traps is primarily important for the optimization of SOI wafers. Deep traps are detectable by a number of techniques, including DLTS, that are feasible on MOS capacitors, diodes, and transistors. The idea behind *Photo-Induced Current Transient Spectroscopy (PICTS)* is to analyze the deep traps before any device processing.

The test vehicle is nothing but the as-grown SOI film, illuminated by a low-intensity light. If the number of photogenerated pairs is small enough, the concentration of the majority carriers and their quasi-Fermi level are only slightly affected. In an n-type film, the change in electron occupancy of bulk and interface traps can be safely ignored. In contrast, the hole population increases sharply, their quasi-Fermi level shifts considerably towards the valence band, and their trap occupancy is modified. Only photogenerated holes are therefore candidates for being trapped. To this purpose, the illumination is maintained until a quasi-stationary state of excitation is reached.

The neutrality condition for excess carriers is

$$\Delta n = \Delta p + \sum_i p_i \tag{5.22}$$

where p_i accounts for trapped holes. The summation is restricted to those traps that are actually swept during the shift of the quasi-Fermi level of holes.

After interruption of the illumination, the photoconductivity decay is characterized by distinct time constants corresponding successively to a very fast recombination of free carriers ($< 100\,\mu$s), a slower emission from interface traps, and a much slower emission from bulk traps. At the end of the initial transient, when one-to-one recombination of excess electrons and holes is completed, the instantaneous concentration of trapped minority carriers is balanced by the remaining excess majority carriers:

$$\Delta n(t) = \sum_i p_i = \sum_i p_i(0)\exp(-e_i t) + \int D_{it}(E)\exp(-et)dE \tag{5.23}$$

Here, a distinction was made between bulk hole traps and interface traps. The energy levels of hole traps are discrete, and each trap is characterized by an emission rate e_i. Similarly, $e(E)$ is the emission rate of the

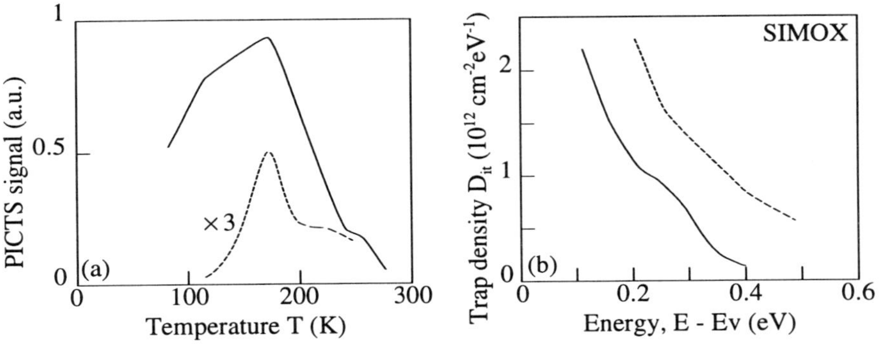

Figure 5.11 (a) PICTS spectra and (b) extracted profiles of interface traps in a SIMOX film (——) and in a SIMOX P^+PP^+ transistor (- - -) (after Papaioannou et al.[18, 19]).

continuum of interface traps $D_{it}(E)$. The integral is considered for the energy range defined by the quasi-Fermi levels of electrons and holes.

The key feature of PICTS is that the emission rate of trapped minority carriers is mirrored by the decay rate of excess majority carriers. It is easy to determine the excess electron concentration by applying a low voltage across the sample and monitoring the transient photocurrent

$$\Delta I(t) = (\alpha q \mu_n V) \times \Delta n(t) \tag{5.24}$$

where prefactor α accounts for the geometrical parameters of the sample.

The trap spectroscopy makes use of a temperature ramp to scan various traps. The transient current is processed with the rate window method developed for DLTS and described in Section 7.4. Same samples are used for PICTS and Hall-effect measurements. The excess photoconductivity is limited to a few percent, which corresponds to an excess carrier concentration of 10^{14}–10^{15} cm^{-3}. A conventional DLTS system is used to analyze the PICTS signal. The temperature is scanned with a slow ramp (40 mK/s) from 90 to 300 K. The rate window is kept constant in the range 10–1000 s^{-1} and the voltage (0.1–0.2 V) is selected in the ohmic region.

Typical PICTS spectra are reproduced in Figure 5.11a. A smooth distribution versus temperature (plain curve) illustrates the case of respectable semiconductors,[3] where interface traps prevail over bulk traps. The energy profile, derived from Equations (5.23) and (5.24), is[18]

$$D_{it}(E) = \frac{\Delta I}{0.56\alpha q\mu_n kTV} \tag{5.25}$$

and the energy E is defined with respect to the valence band:

$$E = E_V + kT\ln(0.48v_{th}\sigma N_V/f) \tag{5.26}$$

The coefficients 0.56 and 0.48 are specific to the DLTS correlator, σ is the capture cross section, and other notations are as usual. The $D_{it}(E)$ profiles of Figure 5.11b were calculated by taking $\sigma = 10^{-16}\,\text{cm}^2$. This assumption is not crucial, since for a larger value of σ the curves would just be shifted towards slightly deeper energy levels.

The trap distribution presumably reflects the quality of the interface between film and buried oxide. The structure observed around 0.3 eV is attributed to residual bulk traps with low density. The peak shown by the dotted curve of Figure 5.11a represents a more obvious signature of localized traps. The contribution of interface traps is still visible as a continuous background signal. From this double signal, the interface trap profile (Fig. 5.11b) and the concentration of bulk traps is resolved. The activation energy of the trap is deduced from the Arrhenius plot.

PICTS stands as a highly specialized and tedious method that, in turn, gives a unique opportunity for investigating the trapping mechanisms in SOI films. PICTS has also been successfully applied to SOI transistors.[19] It can therefore be expected to work in Ψ–MOSFETs (Section 4.6) as well, thereby extending the number of parameters available by quick evaluation of as-grown SOI wafers.

We have reviewed, in this chapter, the potential and limitations of standard methods applied to the characterization of the transport properties in thin SOI films. A general problem is the large value of the sheet resistance, which demands the use of very high impedance detection systems. In point-contact measurements, a trade-off is necessary for achieving good ohmic contacts while limiting the probe penetration into the film. The depletion regions induced by interfacial charges or electrical contacts may extend over the whole film and prevent current injection. The success of the experiment may, however, be ensured by

[3] Including SOI.

biasing the substrate, rather than by intentionally doping the film. The substrate biasing is a pure SOI effect that makes available a new family of experiments (in-depth profiling, PICTS, MOS-like transport, etc.).

We have focused only on techniques that have been used extensively in SOI. Of course, many other characterization methods, such as thermoelectric power, photomagnetoelectric effect, time-of-flight, etc., have been elaborated and deserve to be applied on SOI.

Bibliography

[1] D.K. Schroder, *Semiconductor Material and Device Characterization,* J. Wiley & Sons, New York, 1990.

[2] S. Cristoloveanu, J. Pumfrey, E. Scheid, P.L.F. Hemment, and R.P. Arrowsmith, "Thermal donor and new donor generation in SOI material formed by oxygen implantation," *Electron. Lett.*, vol. 21, p. 802, 1985.

[3] P. Normand, D. Tsoukalas, N. Guillemot, and P. Chenevier, "A pile-up phenomenon during arsenic diffusion in silicon-on-insulator structures formed by oxygen implantation," *J. Appl. Phys.*, vol. 66, p. 3585, 1989.

[4] S. Cristoloveanu, G. Ghibaudo, and G. Kamarinos, "Le matériau silicium sur saphir en France. Revue des propriétés physico-chimiques et électriques," *Rev. Phys. Appl.*, vol. 19, p. 161, 1984.

[5] S. Cristoloveanu, "Silicon films on sapphire," *Rep. Prog. Phys.*, vol. 50, p. 327, 1987.

[6] A.C. Berr, *Galvanomagnetic Effects in Semiconductors,* Academic Press, New York, 1963.

[7] Van der Pauw, "A method of measuring specific resistivity and Hall effect of discs of arbitrary shapes," *Philips Res. Repts.*, vol. 13, p. 1, 1958.

[8] D. Marion, S. Cristoloveanu, and A. Chovet, "High-magnetic-field Van der Pauw method: magnetoresistance measurement and applications," *IEE Proc.*, vol. 129, p. 125, 1982.

[9] D.J. Dumin and P.H. Robinson, "Carrier transport in thin silicon films," *J. Appl. Phys.*, vol. 39, p. 2759, 1968.

[10] H. Schlötterer, "Mechanical and electrical properties of epitaxial silicon films on spinel," *Solid-State Electron.*, vol. 11, p. 947, 1968.

[11] S. Cristoloveanu, S. Gardner, C. Jaussaud, J. Margail, A.J. Auberton–Hervé, and M. Bruel, "Silicon on insulator material formed by oxygen implantation and high-temperature annealing: carrier transport, oxygen activity, and interface properties," *J. Appl. Phys.*, vol. 62, p. 2793, 1987.

[12] F. Vettese, J. Sicart, J.L. Robert, S. Cristoloveanu, and M. Bruel, "Electrical properties of oxygen thermal donors in silicon films synthesized by oxygen implantation," *J. Appl. Phys.*, vol. 65, p. 1208, 1989.

[13] J.L. Robert, J.M. Dusseau, P. Girard, and J. Sicart, "A model of conduction in inhomogeneous degenerate semiconductors: application to silicon on sapphire films," *J. Appl. Phys.*, vol. 54, p. 1903, 1983.

[14] B. Kleveland, S. Cristoloveanu, and J. Sicart, "Novel Hall effect spectroscopy of impurity levels in semiconductors," *Solid-State Electron.*, vol. 33, p. 743, 1990.

[15] G. Papaioannou, S. Cristoloveanu, and P. Hemment, "Low temperature properties and phototransport in silicon-on-insulator films synthesized by oxygen implantation," *J. Appl. Phys.*, vol. 63, p. 4575, 1988.

[16] S. Cristoloveanu, "Oxygen-related activity and other specific electrical properties of SIMOX," *Vacuum*, vol. 42, p. 371, 1991.

[17] K.N. Kang and S. Cristoloveanu, "Theory of the field-assisted photomagnetoelectric effect in thin films; experiment on n-type GaAs and applications," *Thin Solid Films*, vol. 114, p. 335, 1984.

[18] G. Papaioannou, V. Ioannou–Sougleridis, S. Cristoloveanu, and C. Jaussaud, "Photoinduced current transient spectroscopy in silicon-on-insulator films formed by oxygen implantation," *J. Appl. Phys.*, vol. 65, p. 3725, 1989.

[19] G. Papaioannou, V. Ioannou–Sougleridis, and S. Cristoloveanu, "Characterization of SIMOX–MOS transistors by photo-induced current transient spectroscopy," *Semicond. Sci. Technol.*, vol. 6, p. 937, 1991.

Chapter 6

SIS Capacitor-Based Characterization Techniques

In this chapter, we describe several capacitance characterization techniques for analyzing the interface and bulk defects at the top and bottom interfaces and in the buried oxide of the SOI material by using its inherent silicon–insulator–silicon (SIS) capacitor structure. The techniques include small-signal a.c. capacitance and conductance as well as transient capacitance methods.[1, 2] The capacitance and conductance methods consist of low-frequency, high-frequency, and combined high–low-frequency measurements. Two variants of the transient capacitance techniques are examined: the bias-scan and temperature-scan deep-level-transient spectroscopy (DLTS).

The detailed theoretical and experimental aspects of these techniques will be discussed and used for extracting the basic interface and bulk parameters of the SOI materials. The two interfaces under investigation, namely, the film–buried-oxide (F/O) and buried-oxide–substrate (O/S), will be indicated by the subscripts 2 and 3, respectively. This notation was made for the sake of simplicity and consistency, because the front MOS interface (i.e., gate oxide-film), for which the subscript 1 is normally reserved, does not exist in an SIS capacitor.

The advantages for using the SIS capacitor structure instead of the standard MOS capacitor structure are that (1) it requires only a few processing steps for fabricating such a test structure and hence provides a quick turnaround time for the SOI material characterization, (2) the analysis of fully processed MOS capacitors on SOI substrates is much more difficult, since there are three different interfaces involved and a number of parameters that have to be controlled, and (3) besides the interface coupling problem, the accuracy of MOS capacitance measurements is questionable due to the existence of large series resistances and parasitic capacitances. To illustrate our points, we also present, in Section 6.5, conventional C–V measurements on the MOS capacitor structure formed on thick films or directly on the SOI substrate. It is worth noting that the SIS capacitor structure is particularly attractive and suitable for the quick characterization of interface and defect properties in the starting SOI wafers prior to the fabrication of complex MOS integrated circuits.

Section 6.1 describes the basic principles and experimental details of capacitance and conductance techniques applied to an SIS capacitor structure in SOI materials. The bias-scan and temperature-scan DLTS techniques, depicted in Sections 6.2 and 6.3, are dedicated to the extraction of interface trap parameters. We shall show how the capacitance C–V, conductance $G(\omega)$, and DLTS techniques can be extended from the MOS capacitors to the SIS capacitors and what their inherent limitations are. The determination of generation lifetimes in SIS capacitors using the Zerbst method is described in Section 6.4. The standard capacitance technique applied to an MOS–SOI capacitor is discussed in Section 6.5.

6.1 Capacitance and Conductance Techniques

6.1.1 Principle and Experimental Setup

Capacitance and conductance techniques are very frequently utilized for extracting the parameters of the Si–SiO_2 interface in bulk-silicon MOS systems. Although the method becomes more complex, it can still be applied to the SIS capacitor structure formed on an SOI substrate. The guideline is the modification of conventional MOS capacitor theory to include the top and bottom Si/SiO_2 interfaces of the SOI buried oxide. By introducing a coupling factor, which accounts for the interaction of the two surface potentials, a small-signal equivalent circuit can be defined for the SIS capacitor, and then reduced to an MOS-like equivalent circuit. Based on this model, the expressions for the surface potential relating to the gate bias voltage can be derived.

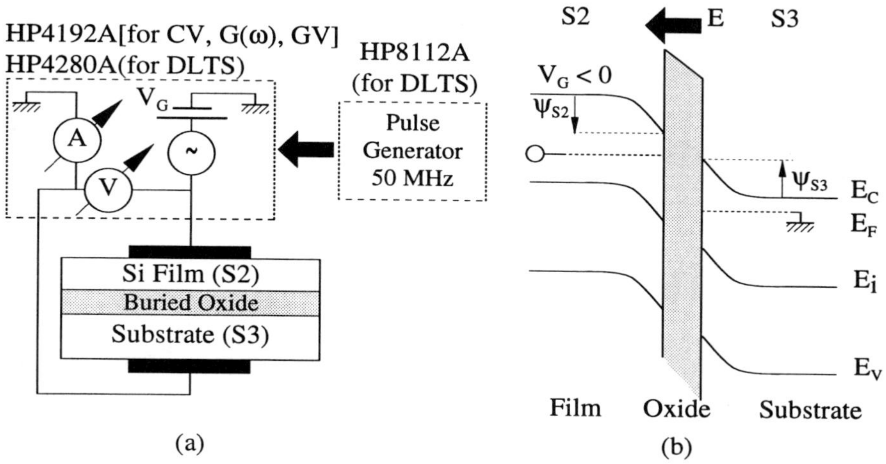

Figure 6.1 (a) Experimental setup for the capacitance, conductance, and DLTS measurements on an SIS capacitor, and (b) energy band diagram of an n–i–n SIS capacitor

The device parameters such as doping concentration, oxide thickness, fixed oxide charge, and interface-trap density profile can be determined from the small-signal a.c. capacitance (C–V) and conductance (G–V) measurements on the SIS capacitor. Examples for extracting the interface parameters in a SIMOX SIS capacitor are presented. It is shown that by carefully accounting for the charge coupling between the top and bottom interfaces of an SIS capacitor, it is possible to obtain an accurate determination of trap densities at both interfaces.

An SIS (n–i–n or p–i–p) capacitor structure (see Fig. 6.1a) can be readily formed on a starting SOI wafer with minimum processing steps. For a n-type SOI wafer, good ohmic contacts are prepared by depositing a N^+ epilayer on the top Si film of the SOI substrate. Ohmic contacts to the front and back surfaces are made by spinning phosphorus-doped glass on both sides of the wafer, followed by a 950 °C, 15-minute drive-in diffusion.

Trenches were etched in the Si film down to the buried oxide, forming isolated silicon islands on the buried oxide. The typical active area of the mesa-etched SIS capacitor structure is $200 \times 200\,\mu m^2$. Although the plasma etch used to isolate the islands may cause some sidewall

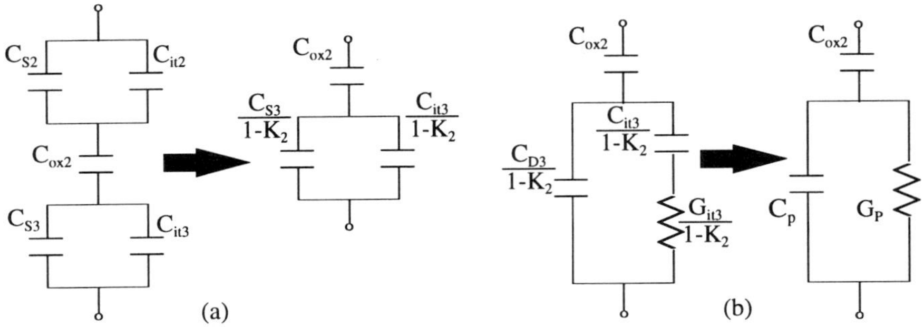

Figure 6.2 (a) Simplified small-signal equivalent circuit for an SIS capacitor and the MOS-like equivalent circuit; (b) simplified small-signal equivalent circuit for an SIS capacitor in depletion for a single level interface trap and its MOS-like equivalent circuit.

damage, the sidewall area is only about 1 % of the capacitor area, making its contribution to the capacitance measurements negligible. The C–V and G–V measurements are performed with an HP–4192A impedance analyzer. Figure 6.1a shows a schematic drawing of the SIS capacitor and the experimental setup for the capacitance, conductance, and DLTS measurements. The energy band diagram for an n–i–n SIS capacitor is reproduced in Figure 6.1b.

The data acquisitions and parameter extraction for the experimental setup shown in Figure 6.1a can be carried out by using a computer with a built-in software program for analysis of the C–V, G–V, and DLTS data. The practical examples for the C–V and G–V measurements and related characterization have been obtained on SIMOX SIS capacitors.

6.1.2 Model and Equivalent Circuit

The small-signal equivalent circuit presented here is derived for an n–Si/SiO_2/n–Si (n–i–n) SIS capacitor structure and is shown in Figure 6.2a together with the MOS-like equivalent circuit. Figure 6.2b presents the simplified small-signal equivalent circuit of an SIS capacitor in depletion, for a single-level interface trap, and the MOS-like equivalent circuit.

The simplified equivalent circuit of Figure 6.2a is also applicable to p–i–p SIS capacitors by changing the direction of voltage scan, and can be adapted to p–i–n capacitors. However, such an asymmetric structure is less desirable, since two depletion regions can be formed simultaneously, and hence it becomes more difficult to distinguish the roles of each interface. For simplicity, the top Si film (F) is referred to as S2 and the substrate (S) as S3.

Basic assumptions used in the analysis include the following: (1) doping densities in S2 and S3 are uniform, (2) gate bias is applied to S2 with reference to S3 (see Fig. 6.1), (3) metal–semiconductor work functions are identical for the ohmic contacts at S2 and S3, (4) surface potential versus d.c. gate bias is independent of the frequency of the small-signal gate bias, and (5) the buried oxide is a perfect insulator with negligible leakage current, and no mobile ions, internal space charge, or polarized molecules.

As shown in Figure 6.1, if a gate bias V_G is applied to the film and the substrate is grounded, then a set of equations can be derived from the global equilibrium and Gauss' law:[3, 4]

$$V_G = \psi_{s_3} - \psi_{s_2} + V_{ox_2} + W_{23} \tag{6.1}$$

$$Q_T = Q_{s_2} + Q_{it_2} + Q_{f_2} = -Q_{s_3} - Q_{it_3} - Q_{f_3} \tag{6.2}$$

$$V_{ox_2} = Q_T / C_{ox_2} \tag{6.3}$$

$$W_{23} = \frac{kT}{q} \ln(\frac{N_{D_3}}{N_{D_2}}) \tag{6.4}$$

where ψ_s is the surface potential, Q_T is the total charge in the buried-oxide capacitor, V_{ox_2} is the potential drop across the buried oxide, W_{23} is the Fermi potential difference between S2 and S3; N_D, Q_s, Q_{it}, and Q_f denote the doping density, surface charge (including contributions of inversion, accumulation, or depletion layers), interface trapped charge, and fixed oxide charge, respectively.

The total capacitance C_T of the SIS capacitor can now be expressed by[3, 4]

$$\frac{1}{C_T} = \left(\frac{dQ_T}{dV_G}\right)^{-1} = \frac{1}{C_{s_2} + C_{it_2}} + \frac{1}{C_{ox_2}} + \frac{1}{C_{s_3} + C_{it_3}} \tag{6.5}$$

where $C_s = -dQ_s/d\psi_s$ is the surface capacitance, $C_{it} = -dQ_{it}/d\psi_s$ is the static capacitance of the interface trap, and C_{ox_2} is the buried-oxide capacitance.

In the following analysis, we consider the case of a negative gate bias that tends to accumulate S2 and deplete S3. A symmetrical analysis can be readily made for a positive bias that tends to deplete S2 and accumulate S3.

By varying the gate-bias voltage, the surface potential at both interfaces will change. The coupling of the surface potentials not only depends on silicon doping but also on the interface properties such as interface traps and fixed oxide charges. Therefore, it is appropriate here to define a coupling factor $K_2(\psi_{s_3})$ and to express it in terms of C_s and C_{it} by differentiating Equation (6.2) with respect to ψ_s:

$$K_2(\psi_{s_3}) \equiv \frac{d\psi_{s_2}}{d\psi_{s_3}} = -\frac{C_{s_3} + C_{it_3}}{C_{s_2} + C_{it_2}} \tag{6.6}$$

In analogy to MOS theory,[5] $d\psi_{s_3}/dV_G$ is obtained as a function of the surface capacitance and interface trap capacitance by differentiating Equation (6.1) with respect to surface potential,ψ_{s_3}, which yields

$$\frac{d\psi_{s_3}}{dV_G} = -\frac{C_{ox_2}}{C_{s_3} + C_{it_3} + C_{ox_2}(1 - K_2)} \tag{6.7}$$

Using Equation (6.7), the total capacitance of an SIS capacitor given by Equation (6.5) is rewritten as

$$\frac{1}{C_T} = \frac{1}{C_{ox_2}} + \frac{1 - K_2}{C_{s_3} + C_{it_3}} \tag{6.8}$$

According to Equation (6.8), the equivalent circuit of an SIS capacitor reduces to an MOS-like equivalent circuit, as shown in Figure 6.2. For $K_2 = 0$, there is no charge coupling, and the equivalent circuit for an SIS capacitor is identical to that of a conventional MOS capacitor.

For $|K_2| < 1$, the total capacitance decreases with increasing negative gate bias and is governed by the buried-oxide thickness, depletion capacitance, and interface capacitance of S3. Although Equation (6.8) still holds for positive gate bias, a more suggestive image is given by the symmetrical expression obtained by interchanging the subscripts 2 and 3. For $|K_2| < 1$, the roles of the depletion and interface capacitances of S2 become dominant, and C_T is now used to characterize the upper interface of the buried oxide.

The interface-trap density as a function of gate bias or surface potential can be determined from the measured admittance of an SIS capacitor by using: (1) the high-frequency capacitance, (2) the low-frequency capacitance, (3) the combined high–low-frequency capacitance, and (4) the conductance measurements.

CAPACITANCE METHOD

Following the high-frequency MOS capacitance method developed by Terman[6], the interface trap capacitance is derived from Equations (6.1) through (6.4):

$$C_{it_3}(\psi_{s3}) = C_{ox_2}\left[\left(\frac{d\psi_{s3}}{dV_G}\right)^{-1} - (1-K_2)\right] - C_{s_3}(\psi_{s3}) \qquad (6.9)$$

It is noted that if K_2 and $\psi_{s_3}(V_G)$ are known, then C_{it_3} is easily extracted.

When the signal frequency is low enough, then the interface traps follow the a.c. bias, and the C–V characteristics may be represented by the equivalent circuit shown in Figure 6.2a. Using this figure, C_{it_3} is determined from the measured low-frequency capacitance $C_{LF}(V_G)$,[7, 8] which implies that $C_{s_3}(V_G)$ and $K_2(V_G)$ are known. Now, solving Equation (6.8) for C_{it_3} yields

$$C_{it_3}(V_G) = (1-K_2)\left(\frac{1}{C_{LF}} - \frac{1}{C_{ox_2}}\right)^{-1} - C_{s_3}(V_G) \qquad (6.10)$$

The value of $C_{s_3}(V_G)$ is found by first calculating C_{s_3} as a function of ψ_{s_3} using the conventional MOS theory for the depletion region and then relating $C_{s_3}(\psi_{s_3})$ to $C_{s_3}(V_G)$, provided that ψ_{s_3} versus V_G is known.

We can also apply the combined high–low-frequency capacitance method of the MOS theory to an SIS capacitor.[7, 8] Since the interface traps are no longer able to follow the signal, the high-frequency capacitance C_{HF} is obtained from Equation (6.8) by setting $C_{it_3}(\omega) = 0$, which yields

$$\frac{1}{C_{HF}} = \frac{1}{C_{ox_2}} + \frac{1-K_2}{C_{s_3}} \qquad (6.11)$$

Solving Equation (6.11) for C_{s_3} and then substituting it into Equation (6.10) gives

$$C_{it_3}(V_G) = (1-K_2)\left[\left(\frac{1}{C_{LF}} - \frac{1}{C_{ox_2}}\right)^{-1} - \left(\frac{1}{C_{HF}} - \frac{1}{C_{ox_2}}\right)^{-1}\right] \qquad (6.12)$$

The advantage of Equation (6.12) is to obtain C_{it_3} directly from the C–V measurements. This method eliminates the uncertainties introduced by using the calculated C_{s_3}, which requires the knowledge of correct band bending.

CONDUCTANCE METHOD

The conductance method[5, 9] can also be employed for determining the interface-trap density in an SIS capacitor operated in the depletion region. The a.c. current density, $i_T(t)$, obtained by differentiating Equation (6.2) with respect to time, is given by

$$i_T(t) = \frac{dQ_T}{dt} = (j\omega C_{d_3} + Y_{it_3})d\psi_{s3} \tag{6.13}$$

where C_{d_3} is the depletion-layer capacitance; $Y_{it_3} = G_{it_3} + i\omega C_{it_3}$ is the admittance of a series RC network and depends on surface potential. By differentiating Equation (6.1) with respect to time, one obtains

$$\begin{aligned} dV_G &= d\psi_{s_3} - d\psi_{s_2} + \frac{i_T(t)}{j\omega C_{ox_2}} \\ &= \left[Z_{s_3}(1 - K_3) + \frac{1}{j\omega C_{ox_2}}\right] i_T(t) \end{aligned} \tag{6.14}$$

where

$$Z_{s_3} = (j\omega C_{d_3} + Y_{it_3})^{-1} \tag{6.15}$$

The term inside the bracket in Equation (6.14) is the impedance of the equivalent circuit in which C_{ox_2} is in series with $Z_{s_3}(1-K_3)$. The equivalent circuit for an SIS capacitor derived from Equations (6.13) and (6.14) is shown in Figure 6.2b, where C_{it_3} and G_{it_3} are, respectively, the capacitance and conductance associated with the interface traps at S3.

The interface state time constant τ_{it}, defined by C_{it_3}/G_{it_3}, predicts the frequency behavior of the interface traps. The parallel branch of the equivalent circuit of Figure 6.2b can be converted into a frequency-dependent capacitance C_p and conductance G_p, which are given by

$$C_p = (1 - K_2)^{-1}\left(C_{d_3} + \frac{C_{it_3}}{1 + \omega^2\tau_{it}^2}\right) \tag{6.16}$$

$$\frac{G_p}{\omega} = (1 - K_2)^{-1}\left(\frac{\omega\tau_{it}C_{it_3}}{1 + \omega^2\tau_{it}^2}\right) \tag{6.17}$$

The above expressions are for a single-level interface trap. However, the continuous distribution of traps throughout the bandgap can be taken into account by assuming that $(1 - K_2)$ is a slow-varying function of surface potential in a range of kT/q around the Fermi level; one obtains:

$$C_p = (1 - K_2)^{-1} \left[C_{d_3} + \frac{C_{it_3}}{\omega\tau_{it}} \arctan(\omega\tau_{it}) \right] \tag{6.18}$$

$$\frac{G_p}{\omega} = (1 - K_2)^{-1} \left[\frac{C_{it_3}}{2\omega\tau_{it}} \ln(1 + \omega^2\tau_{it}^2) \right] \tag{6.19}$$

A comparison of the expressions for the capacitance and conductance methods in MOS and SIS capacitors reveals that Equations (6.16) through (6.19) differ from the MOS capacitor theory only by a factor of $(1 - K_2)$. Therefore, the conventional MOS characterization methods for interface-trap analysis can be extended to an SIS capacitor by simply multiplying the measured interface-trap density by $(1 - K_2)$ in order to account for the charge coupling between the two interfaces.

6.1.3 Coupling of Surface Potentials

To understand the SIS capacitance and conductance methods described above, it is important to examine the physical aspects of the coupling factor, which contains all the information about the film, buried-oxide layer, and substrate. For a given set of interface parameters (the fixed oxide charge and interface-trap densities), ψ_{s_2} and ψ_{s_3} are determined by using Equations (6.1) through (6.4) for each gate bias. Therefore, $C_{s_2}, C_{s_3}, C_{it_2}$, and C_{it_3} as a function of the surface potential can be calculated, and the coupling factor is deduced from Equation (6.6).

Figure 6.3a shows the coupling factor K_2 as a function of surface potential ψ_{s_3} in a symmetrical (i.e., $N_{D_2} = N_{D_3}$) n–i–n SIS capacitor with doping density as a parameter. The relationship between ψ_2 and ψ_3 is illustrated in Figure 6.3b for the case of negligible defect densities and different doping levels in the film and in the substrate. The apparent complex behavior of the coupling factor may be explained by focusing our attention on the flat-band, accumulation, depletion, and strong-inversion regions to illustrate the underlying basic concepts.

For identical doping ($N_{D_2} = N_{D_3}$), both interfaces are at the flat-band condition for $V_G = 0$ and $K_2 = -1$. As the gate bias becomes more positive, S2 is first depleted and then driven from depletion into weak inversion, whereas S3 is still in accumulation (region II). As a result, C_{s_2} decreases and C_{s_3} increases progressively, causing an increase of $|K_2|$. At the onset of strong inversion in S2 (region I), the maximum depletion width is reached and C_{s_2} saturates under high-frequency operation, causing a rapid increase of $|K_2|$. In contrast to low-frequency

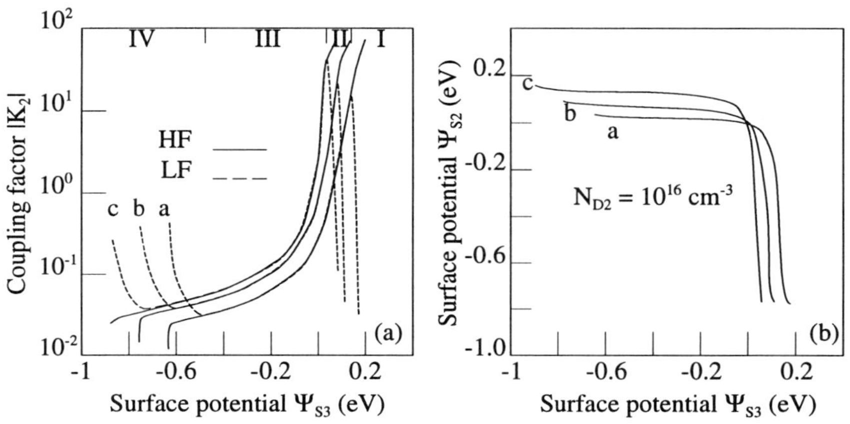

Figure 6.3 (a) Coupling factor K_2 versus surface potential ψ_{s_3} in an n–i–n SIS capacitor with doping density as a parameter (assuming $N_{D_2} = N_{D_3}$). (b) Surface potential ψ_{s_2} versus ψ_{s_3} with doping density of S3 as a parameter ($t_{ox_2} = 350\,\text{nm}$ and $N_{D_2} = 10^{16}\,\text{cm}^{-3}$, with N_{D_3} in curves a,b, and c equal to 10^{15}, 10^{16}, and $10^{17}\,\text{cm}^{-3}$, respectively).

operation, C_{s_2} also increases exponentially. The low-frequency surface capacitances in S2 (strong inversion) and S3 (strong accumulation) and coupling factor are given by[5]

$$C_{s_2} = \frac{C_{FB}}{\sqrt{2}}\frac{n_i}{N_{D_2}}\exp\left(-\frac{q\psi_{s2}}{2kT}\right) \tag{6.20}$$

$$C_{s3} = \frac{C_{FB}}{\sqrt{2}}\exp\left(\frac{q\psi_{s3}}{2kT}\right) \tag{6.21}$$

and

$$K_2 = -\frac{N_{D2}}{n_i}\exp\left[\frac{q(\psi_{s_2}+\psi_{s3})}{2kT}\right] \tag{6.22}$$

where C_{FB} is the flat-band surface capacitance. For $\psi_{s_2} < 0$, $\psi_{s_3} > 0$, and $\psi_{s_2} > \psi_{s_3}$, a decrease of $|K_2|$ is expected. As the gate bias becomes negative, S2 is in accumulation and S3 in depletion (region III). Similar to the case of positive gate bias, a decrease of $|K_2|$ is obtained. At the onset of strong inversion in S3 (region IV), $|K_2|$ decreases under high-frequency operation and increases under low-frequency operation. The influence of the substrate doping concentration on coefficient K_2 is

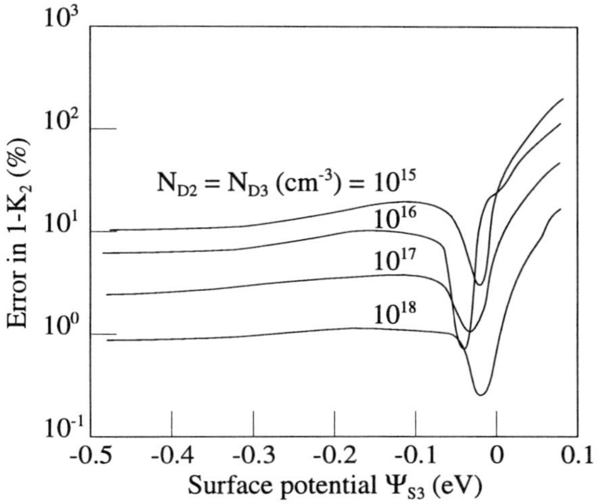

Figure 6.4 The error between $(1 - K_2)$ and $(1 - K_2')$ as a function of surface potential in SIS structures ($t_{ox_2} = 350\,\text{nm}$, $D_{it_2} = D_{it_3} = 5 \times 10^{11}\,\text{cm}^{-2}\text{eV}^{-1}$ and $N_{D_2} = N_{D_3} = 10^{15}$, 10^{16}, 10^{17}, and $10^{18}\,\text{cm}^{-3}$, respectively, for the four curves). Fixed oxide charges are assumed to be zero at the S2 and S3 interfaces.

illustrated in Figure 6.3b, which shows that for a given surface potential under depletion or weak inversion, $|K_2|$ increases with increasing N_{D_3}.

For convenience, the coupling factor given in Equation (6.6) can be approximated by

$$K_2' = -\frac{C_{s_3}}{C_{s_2}} \tag{6.23}$$

The error introduced by Equation (6.23) is defined as

$$\text{Error} = \left|\frac{(K_2' - K_2)}{(1 - K_2)}\right| = \left|\frac{C_{s_2}C_{it_3} - C_{s_3}C_{it_2}}{C_{s_2}(C_{s_2} + C_{s_3} + C_{it_2} + C_{it_3})}\right| \tag{6.24}$$

Figure 6.4 shows a plot of error (in percent) as a function of the surface potential for four different donor concentrations. The error of using K_2' in the depletion region is less than 10 % for $N_D > 10^{16}\,\text{cm}^{-3}$. Even for $N_D = 10^{15}\,\text{cm}^{-3}$, the maximum error does not exceed 18 %. The exact correction factor can be obtained by using an approximated coupling factor as a first-trial value to obtain the corresponding interface-trap densities, D_{it_2} and D_{it_3}, from the experiment. A new coupling factor is then determined from Equation (6.6). Finally, an iteration procedure will yield rapidly the accurate coupling factor.

6.1.4 Surface Potential Versus Gate Bias

In a bulk-silicon MOS capacitor, the band bending as a function of gate bias is deduced directly from the high–low-frequency capacitance method.[5] However, due to the two interfaces associated with an SIS capacitor, the low-frequency capacitance method can only be used to determine the surface potential difference ($\Delta\psi_s \equiv \psi_{s3} - \psi_{s2}$) versus the gate bias. In analogy to an MOS capacitor, this is given by[8]

$$\frac{d(\psi_{s3} - \psi_{s2})}{dV_G} = 1 - \frac{C_{LF}}{C_{ox_2}} \tag{6.25}$$

Integrating Equation (6.25), one obtains

$$\Delta\psi_s(V_G) = \Delta\psi_s(V_{Go}) + \int_{V_{G_0}}^{V_G} dV_G \left[1 - \frac{C_{LF}(V_G)}{C_{ox_2}}\right] \tag{6.26}$$

In Equation (6.26), V_{G_0} is chosen to minimize the uncertainty in $\Delta\psi_s$ vs. V_{Go}. For example, V_{Go} can be taken in accumulation or in inversion, where band bending is only a weak function of gate bias. Once $\Delta\psi_s$ is found, the surface potential ψ_{s3} is calculated from the charge-conservation relation given in Equations (6.1) through (6.4), by solving

$$Q_{s3}(\psi_{s3}) = C_{ox_2}[\Delta\psi_s - V_G + W_{23}] - [Q_{f3} + Q_{it3}] \tag{6.27}$$

The parameters on the right-hand side of Equation (6.27) can be extracted experimentally. For example, the fixed oxide charge is determined by measuring the voltage shift of the C–V curve, the surface potential difference by integrating the low-frequency C–V curve, and the interface trapped charge by integrating the interface trap density obtained from the high–low-frequency capacitance measurement. Thus, for a given gate bias, values of ψ_{s2} and ψ_{s3} are calculated.

The band bending may be determined by using the high-frequency capacitance method.[6] From Equations (6.1) through (6.4), the theoretical high-frequency $C(\psi_{s2})$ and $C(\psi_{s3})$ curves are generated by taking the fixed oxide charge into account. The results are then compared with the measured high-frequency C–V curves to obtain ψ_{s2} and ψ_{s3} versus gate bias V_G. It is noteworthy to point out that this procedure is valid

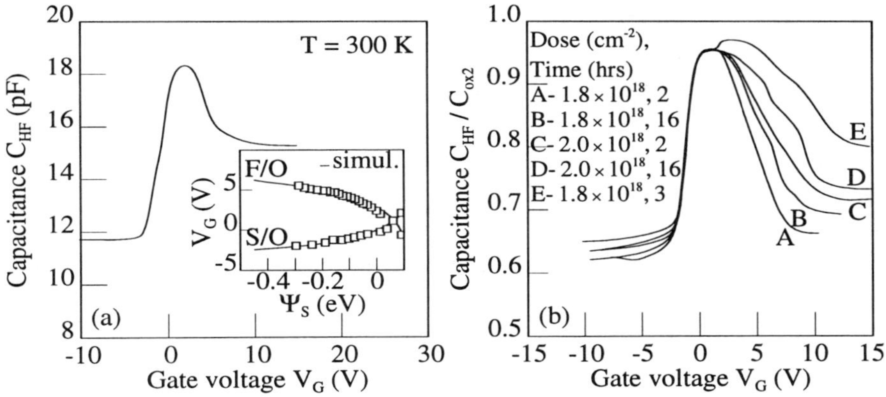

Figure 6.5 (a) High-frequency C–V curve for a typical SIMOX SIS capacitor; insert shows the gate bias versus surface potential determined by using the modified Terman method.[6] (b) High-frequency C–V curves for various oxygen implant doses, anneal temperatures, and times.

only for $Q_{s_2} + Q_{f2} \gg Q_{it_2}$, an easily satisfied condition because the fixed oxide charge density is generally higher than the integrated interface-trap charge.[5] Furthermore, as Q_{s_2} increases exponentially with surface potential in accumulation, a small deviation from the flat band can meet the above requirement.

6.1.5 Parameter Extraction

Prior to extracting the interface properties, we need to know the SIS capacitor parameters, such as buried-oxide thickness, doping concentration, and fixed oxide charge. The buried-oxide thickness t_{ox_2} can be determined by three different methods. First, if the doping concentration in either S2 or S3 is known, and the semiconductor layer is partially depleted, then the buried-oxide thickness is found from the measured minimum high-frequency capacitance C_{HF}^{min} as shown in Figure 6.5a. Second, the buried-oxide thickness can be deduced from the measured maximum low-frequency capacitance C_{LF}^{max}, which is equal to the buried-oxide capacitance. Third, if the conductivity types of S2 and S3 are not the same, then the oxide thickness is calculated from the measured maximum high-frequency capacitance C_{HF}^{max}.[6]

The fixed oxide charge densities (Q_{f_2} and Q_{f_3}) are determined from the voltage shift of the measured high-frequency C–V curve by comparing it with the ideal C–V curve. Applying a gate bias to accumulate one interface gives the fixed oxide charge at the other interface. Figure 6.5b shows the high-frequency C–V curves for SIMOX SIS capacitors synthesized with different implanted oxygen doses and annealing times. The stretch-out and shift in the C–V curves were observed at positive gate biases and were attributed to an increase of interface-trap density and fixed oxide charge density at the upper oxide interface. Once the buried-oxide thickness is known, the doping concentrations in S2 and S3 are calculated from the minimum high-frequency capacitance regardless of the interface-trap density.[6]

However, this method is less accurate if the thickness of S2 is too small, so that the film becomes fully depleted before reaching the minimum capacitance. In this case, the buried-oxide thickness is no longer accessible from the high-frequency minimum capacitance value. In analogy to the MOS theory,[5] an independent characterization method is needed to determine the doping concentration before the layer is fully depleted. For example, the substrate doping density N_{D_3} can be deduced from

$$N_{D_3}(w_3) = -2\left\{q\epsilon_{si}\frac{d}{dV_G}\left[\frac{1}{(1-K_2)C_{HF}^2}\right]\right\}^{-1} \tag{6.28}$$

where w_3 is the distance from the substrate/buried-oxide interface. Since the precise value of $(1 - K_2)$ requires the prior evaluation of the doping concentrations in S2 and S3, an iteration procedure becomes necessary. A rapid convergence is expected, since $(1 - K_2)$ is relatively insensitive to the doping concentration, especially near the midgap.

The corrected and apparent $1/C^2$ vs. V_G curves, shown in Figure 6.6 for a SIMOX SIS capacitor, have been obtained with and without taking into account the coupling factor. By neglecting the voltage stretch-out caused by interface traps, doping concentrations in S2 and S3 are extracted from the slope of the $1/(1-K_2)C_{HF}^2$ plot, which gives $N_{D_2} = 5 \times 10^{15}\,\text{cm}^{-3}$ and $N_{D_3} = 1.8 \times 10^{15}\,\text{cm}^{-3}$. As shown in Figure 6.6, ignoring the coupling factor could lead to an erroneous result.

It is interesting to note that the doping concentration in S3 determined from the minimum high-frequency capacitance shown in Figure 6.5a gives the same value as that determined from the slope of the $1/C^2$ vs. V_G plot of Figure 6.6. However, it overestimates by a factor of two the

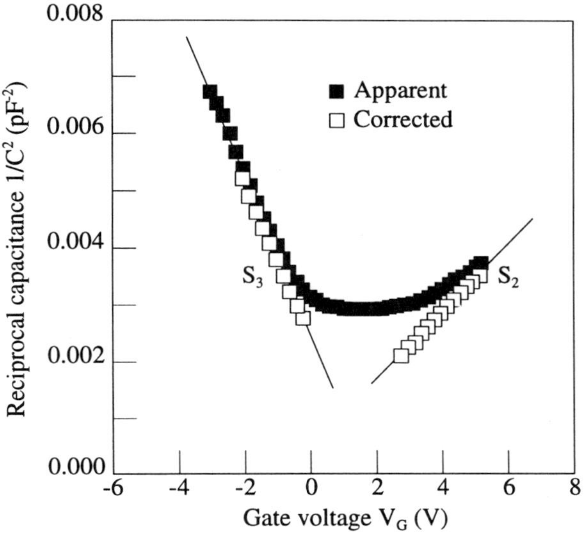

Figure 6.6 Apparent ($K_2 = 0$) and corrected $1/C^2$ versus gate bias curves deduced from Figure 6.5a. An approximated coupling factor K_2' is used in the correction.

doping level in S2, indicating that the film is fully depleted at $V_G \simeq 5\,\mathrm{V}$. The buried-oxide-layer thickness is found from the measured minimum high-frequency capacitance in S3, which yields $t_{ox_2} = 362\,\mathrm{nm}$.

Fixed oxide charge densities are determined by comparing the voltage shift of the C–V curve in weak inversion with the ideal C–V curve, and the result yields values of $Q_{f_2}/q = 1.9 \times 10^{11}\,\mathrm{cm}^{-2}$ and $Q_{f_3}/q = 1.2 \times 10^{10}\,\mathrm{cm}^{-2}$, respectively.

The high–low-frequency capacitance method has been widely used in analyzing the interface properties of an MOS capacitor. A problem in SIMOX is the leakage current, which may exceed that of a thermally oxidized MOS capacitor. Therefore, for low-frequency C–V measurements, losses in the buried-oxide measured at room temperature cannot be neglected. It is usually tedious and difficult to account for these losses. Instead, one can use the high-frequency quasi-static C–V method, in which the measurement is performed under external excitation (heat or light) in order to reduce the minority carrier response time. Thus, the quasi-static condition can be easily satisfied at a relatively high measure-

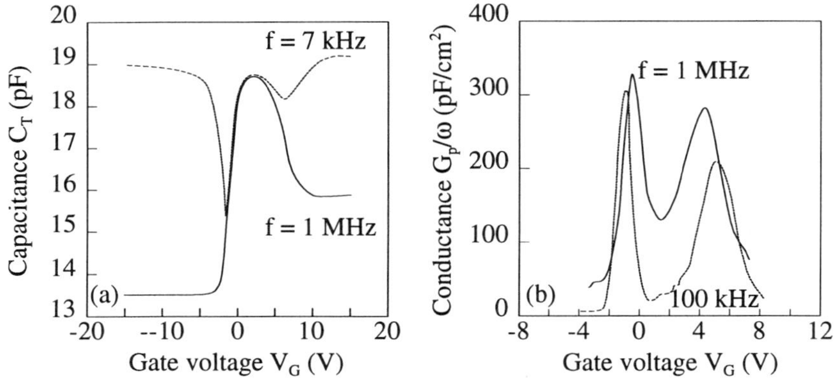

Figure 6.7 (a) The quasi-static (7 kHz) and high-frequency (1 MHz) C–V curves measured in an SIS SIMOX capacitor at 425 K. (b) The parallel conductance versus gate voltage (G/ω–V) curves taken at $f = 100$ kHz and 1 MHz.

ment frequency.

Figure 6.7a shows the quasi-static and high-frequency C–V curves measured at 425 K. In the quasi-static C–V curve, the two valleys correspond to depletion in S2 and S3. The thickness of the buried-oxide ($t_{ox_2} = 366$ nm), obtained from the maximum low-frequency capacitance, is within 2 % of that obtained from the minimum high-frequency capacitance data. The interface-trap densities are determined from Equation (6.12) by comparing the quasi-static and high-frequency C–V curves measured at 425 K, and the surface potential versus gate bias is found by solving Equation (6.26).

Figure 6.7b shows the conductance–voltage G/ω vs. V_G curves measured at $f = 100$ kHz and 1 MHz. The two conductance peaks correspond to depletion in S2 and S3. A maximum value in the conductance versus gate-bias curve reveals that the loss is primarily caused by the interface traps rather than by the bulk traps. As the frequency increases, both peaks shift towards the flat-band voltage and tend to merge into a single peak. This implies that, at high-frequencies, the interface-trap losses from S2 and S3 will interact with one another. Consequently, the extraction of the components G_p and C_p of the interface-trap parallel admittance will be erroneous in this region if the coupling factor is not taken into account.

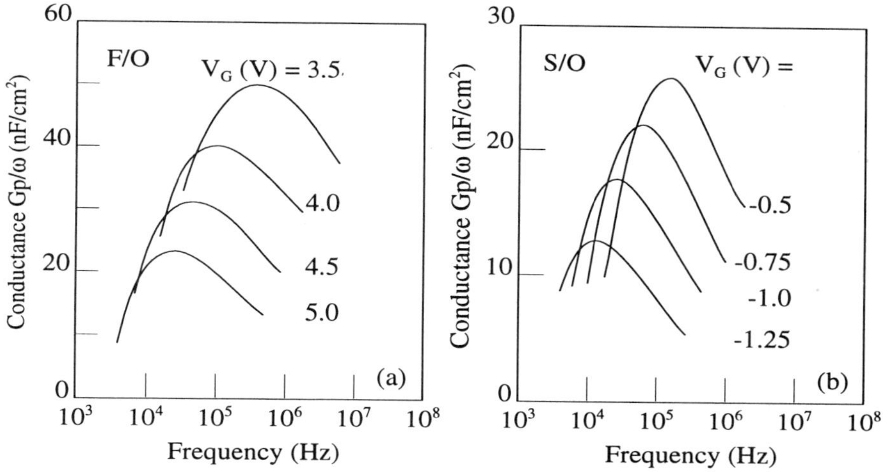

Figure 6.8 The parallel conductance G_p/ω of interface traps versus frequency measured at different gate voltages for (a) the film/oxide interface and (b) the substrate/oxide interface of a SIMOX SIS capacitor ($T = 300\,\mathrm{K}$).

Figures 6.8a and 6.8b show the conductance G_p versus frequency in S2 and S3 with gate bias V_G as a parameter. Using the continuum interface-state model, the interface-trap density D_{it} and the capture cross section σ_n are determined from the magnitude and position of the G_p peaks in these curves. The interface-trap density profile $D_{it}(E)$ is constructed by using the interface-state energies calculating from the surface potential versus gate-voltage curves shown in Figure 6.5a.

Figure 6.9a summarizes the interface-trap density profiles determined by the capacitance and conductance methods for the F/O and O/S interfaces of a SIMOX SIS capacitor. The discrepancy between the interface-trap densities resulting from the capacitance and conductance methods can be reduced by using the statistical model of Nicollian and Goetzberger, which accounts for surface potential fluctuations.[9] Figure 6.9b shows the capture cross section of the interface traps versus energy in the upper-half bandgap of silicon, determined by the conductance method for the upper and lower interfaces of the buried-oxide.

From Figure 6.9a, it is noted that the trap densities at both interfaces increase rapidly toward the band edge with a minimum occurring

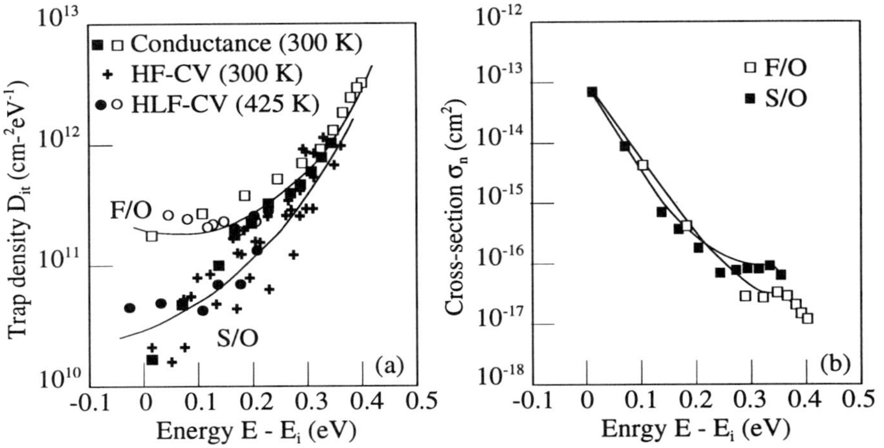

Figure 6.9 (a) Interface state profiles determined by high-frequency C–V (HFCV), high–low-frequency C–V (HLF–CV), and conductance methods for (a) the film/oxide (F/O) and (b) the substrate/oxide (S/O) interfaces. (b) The capture cross sections of the interface traps in the upper-half bandgap for the film/oxide and substrate/oxide interfaces.

near midgap, similar to that observed in thermally grown oxides. The results show that the midgap interface-trap density at the F/O interface is approximately an order of magnitude higher than that observed at the S/O interface. This is fully consistent with the amounts of implantation-induced damage on both sides of the BOX.

In summary, a model for electrical characterization of SIS capacitors by the capacitance and conductance methods has been derived. By introducing a coupling factor, an MOS-like equivalent circuit is obtained for the SIS capacitor. Therefore, the existing measurement techniques and interface parameter extraction methods for the MOS capacitor can be directly applied to an SIS capacitor. Determination of the surface potential as a function of gate bias requires a combination of theoretical calculations, and low- or high-frequency capacitance measurements. The results for SIMOX samples revealed that the midgap interface-trap density at the upper F/O interface is substantially higher than that of the S/O interface.

6.2 Bias-Scan DLTS Technique

The Deep-Level Transient Spectroscopy (DLTS) technique has been used extensively in the study of defect densities, energy levels, and capture cross sections of bulk and interface traps in p–n junctions, Schottky diodes, and MOS capacitors fabricated on various semiconductors. However, in thin SOI structures, the conventional capacitance DLTS method was unable to provide satisfactory results concerning bulk traps and interface states in thin silicon films due to high series resistance.[11] This obstacle has been overcome by performing conductance or current DLTS measurements in SOI MOSFETs.[12]

Since the interface traps are continuously distributed in the energy bandgap, the conventional DLTS technique encounters problems. This is due to the fact that DLTS signals are contributed by the continuous interface trap states with different capture cross sections and different activation energies. Furthermore, bulk traps and generation–recombination centers are expected to be numerous in the SOI film, making it difficult to distinguish the different sources of transient signals.

To solve this problem, Johnson and Wang proposed an energy-resolved DLTS technique.[13, 14] Katsube and Kakimato[15] have discretized the distributed interface states by applying small trap-filling pulses leading to improved energy resolution. Yamasaki et al.[16] have successfully separated bulk traps and interface states by applying different pulse voltages. Kumar and Iyer[17] have reported a bias-scan current DLTS technique for interface-state characterization. This experiment is performed at a fixed temperature instead of the commonly used temperature-scan DLTS at a fixed bias.

6.2.1 Physical Principle and Experimental Setup

Described in this section is a bias-scan DLTS technique for determining the distribution profile of interface-state density in an SIS capacitor. The theoretical aspects of the bias-scan DLTS will be discussed in Section 6.2.3. In this technique, a small trap-filling pulse, superimposed on the quiescent bias, is scanned at a fixed temperature and at a fixed rate window. As the Fermi level moves across the bandgap, a peak in the DLTS signal is obtained when the emission time constant of the surface states in a small energy interval of the forbidden gap is equal to the rate window.

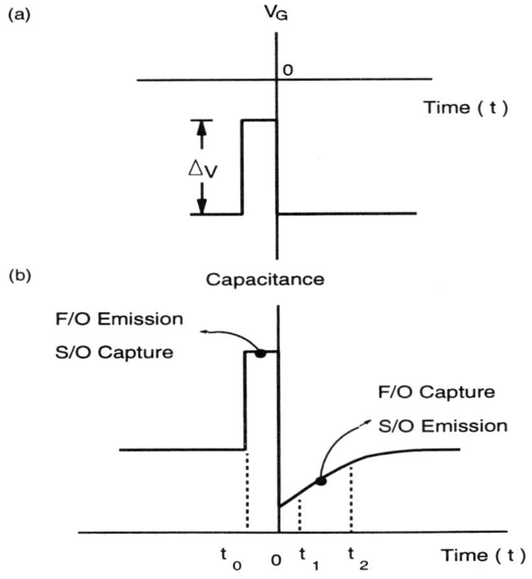

Figure 6.10 (a) Sequences of a trap-filling pulse and (b) capacitance transient observed in an n–i–n SIS capacitor.

The magnitude of the peak gives D_{it}, whereas the surface potential corresponding to the bias at the DLTS signal peak defines the energy of the interface states. The variation of the rate window or device temperature would allow profiling the interface-state distribution in the bandgap of the SOI material.

The capture cross sections can be obtained by using the detailed balance relation between the emission rate and capture rate, while the energies of interface states are determined independently from the surface potential. Therefore, one needs to determine the surface potential as a function of the gate bias (see inset of Figure 6.5a). The C–V measurement discussed in the previous section represents a good solution since it also provides the buried-oxide capacitance and doping density.

The bias-scan DLTS measurements are performed at selected temperatures and rate windows using a HP–4280A capacitance meter. The test sample is placed in a Joule–Thomson cryostat, where the temperature can be varied between 77 and 400 K. The experimental setup for the bias-scan DLTS measurements is shown in Figure 6.1a.

6.2.2 Carrier Dynamics

Figure 6.10 shows the sequences of a trap-filling pulse (Fig. 6.10a) and the capacitance change observed in an n–i–n SIS capacitor with a quiescent gate bias V_G applied to top silicon film (Fig. 6.10b). Since the substrate is grounded, $V_G > 0$ depletes the F/O interface and $V_G < 0$ depletes the S/O interface. In the case of a moderate fixed oxide charge density, it is generally true that if one interface is in depletion, the other interface will be in accumulation. In the following analysis, we consider the case of a "positive" filling pulse superimposed on a negative quiescent bias, which allows the characterization of the S/O interface. A similar analysis applies to a "negative" filling pulse superimposed on a positive gate bias for the characterization of the F/O interface. Applying a positive filling pulse to an SIS capacitor fills the states below the Fermi level at the S/O interface and empties the states above the Fermi level at the F/O interface ($t_0 < t < 0$). At the end of the pulse ($t = 0$), the S/O interface starts to emit the captured carriers and the F/O interface tends to fill the empty interface states below the Fermi level.

The transient capacitance signals consist of both carrier capture and emission at the F/O and S/O interfaces. The capture process at the F/O interface is assumed to be an exponential function of time with a characteristic filling time constant given by

$$\frac{1}{\tau_c} = \sigma_n v_{th} n + e_n + e_p \tag{6.29}$$

where $e_{n,p}$ are the emission rates for electrons and holes, σ_n is the electron capture-cross section, v_{th} is the thermal velocity, and n is the electron concentration. The filling time constant can be easily extracted from Equation (6.29). For example, substituting values of $\sigma_n = 10^{-15}\,\text{cm}^2$, $v_{th} = 1.2 \times 10^7$ cm/s, and $n = 10^{15}$ cm^{-3} into Equation (6.29), where $\sigma_n v_{th} n \gg e_{n,p}$, one obtains $\tau_c = 83\,\text{ns}$. The key point is that the filling time constant τ_c is much smaller than the capacitance sampling times t_1 and t_2 of the selected rate windows. Therefore, the measured capacitance transients are not affected by the presence of F/O interface traps due to their rapid capture of electrons under the prevailing accumulation condition. This implies that the accumulated F/O interface has no influence on the slow capacitance transient $\Delta C(t)$, which is fully dominated by carrier emission from the S/O interface.

In the case when a small quiescent bias is applied to the SIS capacitor, the F/O interface will be driven into depletion instead of accumulation.

Therefore, the carrier capture transient will be very slow due to the lack of free carriers at the F/O interface. The carrier capture time may vary from nanoseconds to microseconds, depending on the applied gate-bias voltages. When the sampling times t_1 and t_2 are comparable to the carrier capture time, the DLTS signal is a combination of the two sources, namely, carrier emission from the S/O interface and carrier capture at the F/O interface. Under this condition, no clear, useful information concerning trap parameters may be obtained. It is easy to identify this case, since capture of the majority carriers gives a positive capacitance transient (i.e., $\Delta C > 0$), whereas emission of the majority carriers produces a negative capacitance transient (i.e., $\Delta C < 0$).

6.2.3 Theory and Parameter Extraction

In the bias-scan DLTS experiment, the S/O interface is always kept in depletion or weak accumulation to avoid minority carrier generation or slow-carrier capture at the F/O interface. A small positive pulse ΔV_G superimposed on the quiescent bias voltage is then applied to the film to deplete the S/O interface, while the F/O interface is in accumulation during the filling pulse. By quickly returning the bias voltage to its quiescent value after the S/O interface states located below the Fermi level are filled, the electrons from the filled traps above the new position of the Fermi level are emitted, with a characteristic emission rate depending exponentially on temperature.

The electron transition between the interface states and the conduction band produces a capacitance transient that is sampled at time t_1 and t_2. Since variation of the film capacitance at the F/O interface will not be detected due to the rapid-carrier capture, the total capacitance transient simply reflects the change of the depletion-layer width in the substrate. Assuming that $\Delta C = C_T(t_1) - C_T(t_2)$ is much smaller than the total capacitance C_T, one obtains the approximated expression[17]

$$\Delta C \simeq -\frac{C_{so}^3}{q\epsilon_{si}N_{D3}C_{ox2}}\int_{E_v}^{E_c} S(e_n)\cdot D_{it3}(E)\cdot(f_2 - f_3)\cdot dE \tag{6.30}$$

where

$$S(e_n) = \exp[-e_n(E)t_1] - \exp[(-e_n(E)t_2] \tag{6.31}$$

and

$$\frac{1}{C_{so}} = \frac{1}{C_{ox2}} + \frac{1}{C_{s3}} \tag{6.32}$$

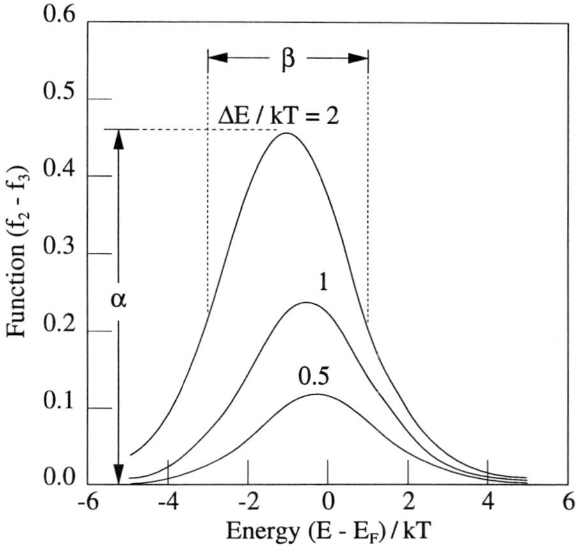

Figure 6.11 The function $(f_2 - f_3)$ versus energy $(E - E_F)/kT$. The parameter $\Delta E/kT$ is related to the trap filling pulse ΔV; α denotes the height and β is the full width at half maximum.

C_{ox2} is the buried-oxide capacitance, C_{s3} is the substrate capacitance, N_{D3} is the donor density, and f_2 and f_3 are the electron distribution functions of the surface states at the quiescent bias and during the filling pulse, respectively. Figure 6.11 shows the calculated $(f_2 - f_3)$ versus energy $(E - E_F)/kT$ with the change of surface potential ΔE (caused by a small trap-filling pulse ΔV) as a parameter.

The integral in Equation (6.30) is derived for the states with energy level E_T, and the result yields[16]

$$\Delta C \simeq -\frac{C_{so}^3}{q\epsilon_{si}N_{D3}C_{ox2}} D_{it3}(E_T)S[e_n(E_T)]\alpha(\Delta E)\beta(\Delta E) \qquad (6.33)$$

where α is the height and β is the full width at half maximum of the $(f_2 - f_3)$ function for a given filling pulse height.

The energy E_T of the interface traps may be obtained from the surface potential corresponding to the peak of the bias-scan DLTS signal, which is determined by using the high-frequency C–V measurements as described in the previous section. The capture cross section is found from the detailed balance relation

$$e_n = \sigma_n v_{th} N_C \exp\left(-\frac{E_C - E_T}{kT}\right) \tag{6.34}$$

where N_C is the effective density of conduction band states.

The emission rate at the DLTS signal peak is determined by the rate window, $e_n = \ln(t_2/t_1)/(t_2 - t_1)$. Indeed, the DLTS signal reaches a maximum value when the emission rate matches the rate window. As the rate window or the temperature are varied, the peaks occur at different positions of the Fermi level in the bandgap. Thus, a direct profiling of interface state distribution is obtained.

In high-frequency operation, the interface traps cannot follow the a.c. signals. Thus, $C_{it_3} = 0$, and C_{so} is expressed by rearranging the total high-frequency capacitance C_T, given by Equation (6.11), in terms of measurable quantities

$$C_{so} = (1 - K_2)\left(\frac{1}{C_T} - \frac{K_2}{C_{ox_2}}\right)^{-1} \tag{6.35}$$

The interface-state density can be readily determined from the measured capacitance using Equation (6.33). The doping concentrations and buried-oxide thickness are evaluated from the high-frequency C–V curves, whereas the surface potential versus gate bias is calculated using the modified Terman method.[6]

The bias-scan DLTS experiments are carried out at selected temperatures and rate windows. Figure 6.12 shows the DLTS signals as a function of bias voltage for different sample temperatures (Fig. 6.12a) and rate windows (Fig. 6.12b) at the F/O and S/O interfaces, respectively. A small filling-pulse height of 100 mV and a pulse width of 1 ms were used in these DLTS measurements.

A slow d.c. bias ramp rate is used to prevent any hysteresis between the forward and reverse ramp directions. As the temperature decreases, the emission rates for all the interface traps are reduced. Therefore, to find the states having a particular emission rate, one needs to explore the states closer to the conduction band edge. On the other hand, if a faster rate window is chosen for a fixed temperature measurement, then to explore the states matching the rate window requires smaller quiescent bias. Figure 6.13 shows the interface-state density as a function of energy in the forbidden gap for the F/O and S/O interfaces in a SIMOX SIS

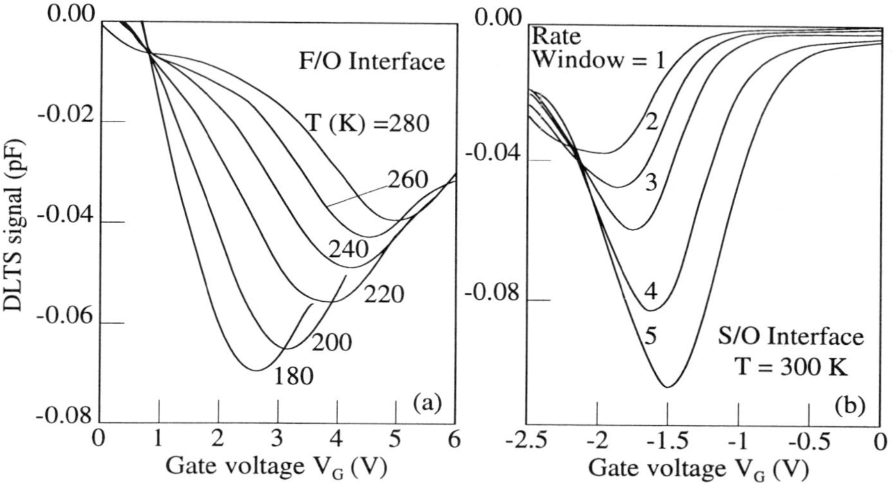

Figure 6.12 Bias-scan DLTS signals for (a) the F/O interface for different temperatures and at a fixed rate window $t_2/t_1 = 100/20\,\mu s$, and (b) the S/O interface at $T = 300\,K$ and for different rate windows $t_2/t_1 =$ (1) 1000/200, (2) 500/100, (3) 250/50, (4) 100/20, and (5) $50/10\,\mu s$.

capacitor as determined by the conductance (solid line) and bias-scan DLTS (square dots) techniques.

In general, the transient capacitance signals in an SIS capacitor are contributed by the bulk traps, the interface-state traps, and minority carrier generation. Because of high densities of bulk defects and interface states in some SOI wafers, the minority carrier generation is significant within the observed transient time in the conventional DLTS measurements on SIS capacitors. In contrast, in the bias-scan DLTS technique, the minority carrier generation can be neglected since the Fermi level is always kept above the midgap. Furthermore, the minority carrier generation process can also be suppressed by performing the experiment at lower temperatures. This is clearly illustrated in Figure 6.15a, which shows that the minority carrier generation dominates the DLTS signals for $T \geq 300\,K$, and is greatly reduced as temperature decreases from 300 to 220 K.

The bulk traps in SIMOX wafers generally consist of discrete energy levels.[18] Since the bias-scan DLTS measurements are performed at

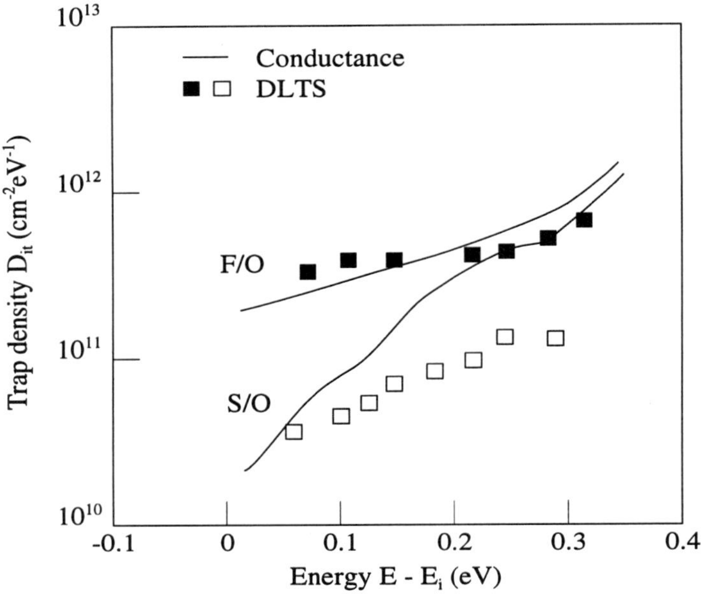

Figure 6.13 Interface-state density distribution determined by the conductance (solid lines) and the bias-scan DLTS (squares) measurements at the F/O and S/O interfaces.

a fixed temperature, the emission rate for the discrete level is fixed, resulting in a constant shift of the baseline in the DLTS signals. This can be easily distinguished from the contribution of the interface states. If the discrete level is spatially nonuniform and dominant over the interface states, the bias-scan DLTS is similar to the techniques used in bulk-defect profiling.[19]

Since variation of the bias voltage is proportional to the change of depletion edge, the peak position of the DLTS signal caused by the bulk trap should be independent of the rate window or temperature used in the measurement, allowing distinction between bulk traps and the temperature- or rate-window-dependent interface-state signal.

In Figures 6.12a and 6.12b, the peak position changes with quiescent bias as the interface states of different energies are being sampled. It is also noted that the baselines of the bias-scan DLTS signals shown in Figures 6.12a and 6.12b are much smaller than the transient signals from the interface states at both the F/O and S/O interfaces, indicating that the transient signals from the bulk traps are not significant for the rate windows and temperatures selected in these measurements. This result is consistent with the conductance measurement data discussed earlier.

The interface-trap distributions at the F/O and S/O interfaces of a SIMOX SIS capacitor determined by the bias-scan DLTS and the conductance methods are compared in Figure 6.13. The results reveal that the density of midgap states of the F/O interface is about an order of magnitude higher than that of the S/O interface.

In general, the interface-trap density determined from the bias-scan DLTS is slightly smaller than that deduced by the conductance method. This difference increases when the interface-trap density is nonuniformly distributed across the band gap. In the conductance method, the curves G_p/ω versus frequency (with bias as a parameter) will spread over a frequency range determined only by the time-constant dispersion.[9]

By contrast, in the bias-scan DLTS, ΔC versus bias (with rate window or temperature as a parameter) will spread over a bias range determined by the time-constant dispersion and the interface-states profile. Since the interface states are more uniformly distributed across the bandgap at the F/O interface, the agreement between these two methods is much better at the F/O interface than that at the S/O interface. To eliminate the spreading caused by the nonuniform profile of interface states, the correlation DLTS technique[20] may be used, where ΔC is measured as a function of the time-constant of the exponential decay weighting function with bias as a parameter.

In conclusion, we have described a bias-scan DLTS technique for profiling the interface-state density at the F/O and S/O interfaces of an SIS capacitor. Interface-state densities resolved by the bias-scan DLTS and conductance methods are found to be in reasonable agreement. The capture cross sections at the F/O and S/O interfaces, determined from the detailed balance equation for the SIMOX sample, are found to be in the 10^{-14}–10^{-16} cm^2 range.

6.3 Temperature-Scan DLTS Technique

Described in this section is the conventional temperature-scan DLTS technique, which may be employed to determine the interface-state densities at the F/O and S/O interfaces of an SIS capacitor. The C–V and

DLTS measurements are performed using the same experimental setup shown in Figure 6.1a.

6.3.1 Decoupling the F/O and S/O Interfaces

The total capacitance C_T of an SIS capacitor was described in Equation (6.8) by including a coupling factor K_2. Figure 6.14 shows the normalized capacitances of $C_{s_2}, C_{s_3}, C_{ox_2}$, and C_T versus gate-bias voltage for an SIS capacitor, where the concentrations of interface traps and fixed oxide charges are assumed to be zero at the F/O and S/O interfaces.

When the substrate is biased into strong inversion, the film is in accumulation with $C_{s_2} \gg C_{s_3}$, and the total capacitance is simply a series combination of oxide capacitance C_{ox_2} and interface capacitance C_{s_3} at the S/O interface, as in a conventional MOS capacitor (i.e., $K_2 = 0$). The measured capacitance transients will not be affected by the presence of F/O interface-states, due to their rapid capture of electrons. This condition can be easily satisfied by pulsing the S/O interface into deep depletion, causing accumulation at the F/O interface. Reciprocally, F/O interface traps may be detected by reversing the polarity of the gate pulse. Thus, the conventional temperature-scan DLTS method may be applied to an SIS capacitor.

6.3.2 Theory and Parameter Extraction

For electron emission from a continuous distribution of interface states, the DLTS signal ΔC is given by[17]

$$\Delta C = [C_T(t_1) - C_T(t_2)]_{max} \simeq -\frac{kTC_T^3}{q\epsilon_{si}N_D C_{ox_2}} D_{it}(E) \cdot \ln\left(\frac{t_2}{t_1}\right) \qquad (6.36)$$

It is noted that if the electron-capture cross section is not strongly dependent on energy, then the DLTS signal will peak at an emission rate given by $e_n = \ln(t_2/t_1)/(t_2 - t_1)$. The corresponding energy, relative to midgap, is

$$E = \frac{E_g}{2} - kT \cdot \ln\left[\frac{\sigma_n v_{th} N_C (t_2 - t_1)}{ln(t_2/t_1)}\right] \qquad (6.37)$$

Therefore, the temperature scan of the DLTS signals is directly related to the energy distribution of the interface state density: from Equations (6.36) and (6.37), $\Delta C/T$ and E are proportional to D_{it} and T, respectively.

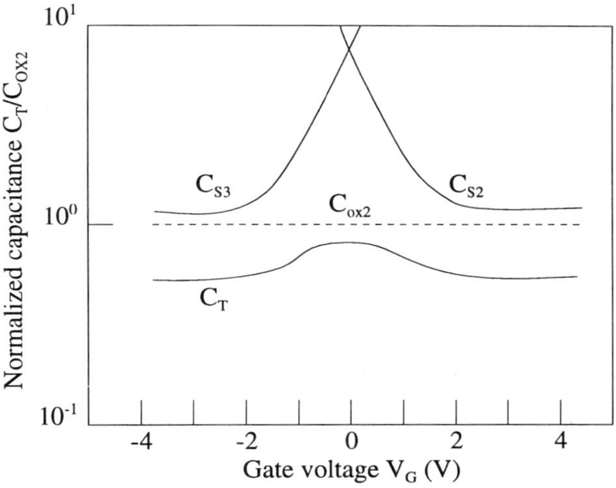

Figure 6.14 Normalized values of the total capacitance (C_T), surface capacitances (C_{s_2}, C_{s_3}), and buried-oxide capacitance (C_{ox_2}) versus gate voltage for an SIS capacitor, where $N_{D_2} = N_{D_3} = 10^{15}\,\text{cm}^{-3}$ and $t_{ox_2} = 350\,\text{nm}$.

Figure 6.15a shows the DLTS scans of the F/O and S/O interface traps obtained by biasing the SIS capacitors into deep depletion and periodically pulsing the gate voltage into accumulation in order to populate both the interface states and deep-level traps in the depletion layer. The two SIMOX SIS capacitors probed in Figure 6.15 were processed identically except for the postimplantation anneal temperature. The DLTS signals at low temperatures correspond to states near the conduction band edge, while the signal at higher temperatures measures the states near the midgap. The rapid drop in the absolute values of the DLTS signals at temperatures below 120 K may arise from a dependence of the capture cross section on energy near the conduction band edge. The increase in DLTS signals above 250 K is ascribed to the onset of minority carrier generation. The DLTS analysis is applicable over the region of the spectrum where the signal is dominated by electron emission, and the capture cross section is assumed to be a constant. The interface-state distribution at midgap cannot be determined from the conventional DLTS measurements due to the onset of minority carrier generation at high temperatures.

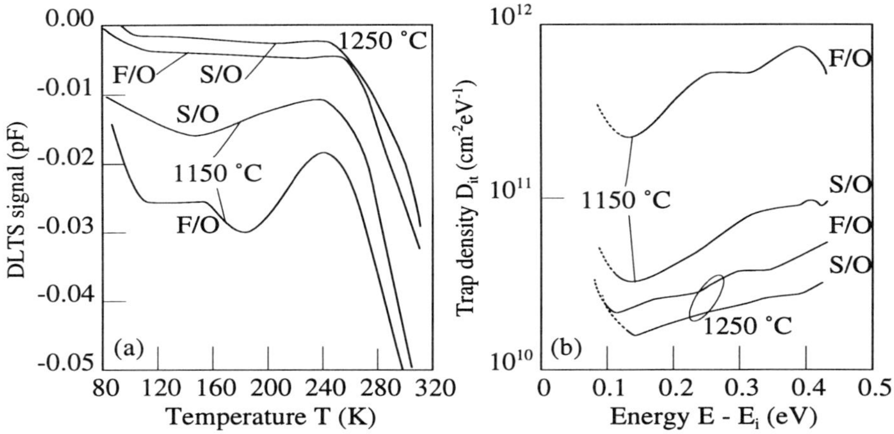

Figure 6.15 (a) DLTS scans for SIS capacitors annealed at 1150 °C for three hours and 1250 °C for two hours. (b) Interface state profiles obtained from (a), by assuming a capture cross section of 10^{-15} cm^2.

Figure 6.15b shows the trap profiles of the F/O and S/O interfaces obtained from the temperature-scan DLTS data of Figure 15a, by assuming a constant capture cross section of 10^{-15} cm^2. As is frequently the case for SIMOX materials, D_{it} increases towards the conduction band edge and is higher at the F/O interface than at the S/O interface, especially for the low-temperature annealed sample. It is noted that a decrease of σ_n by one order of magnitude causes a shift of D_{it} curves towards the conduction band edge by only 2.3 kT.

In conclusion, we have described in this section a conventional temperature-scan DLTS technique for determining the interface-state density profile at the F/O and S/O interfaces of a SIS capacitor. Pulsing the SIS capacitor into deep depletion eliminates the charge coupling of film and substrate, and hence separates the fast transient signals at one interface from the slow carrier emission transients at the other interface.

6.4 Zerbst Method and Generation Lifetime

The generation lifetime is an important parameter for determining the quality of SOI materials and devices. The Zerbst method can be applied to determine the average generation lifetimes in SOI materials using the

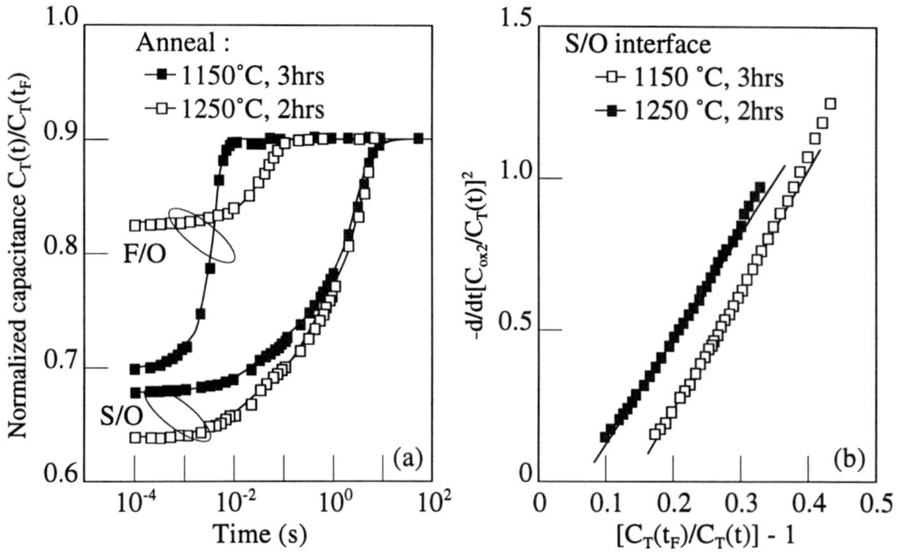

Figure 6.16 (a) C–t curves for the film and substrate of SIS capacitors fabricated on SIMOX annealed at 1150 °C and 1250 °C. (b) Corresponding Zerbst plots for the substrate transients.

same SIS capacitor structure described in the previous sections. The Zerbst (C–t) analysis is based on the measurements of capacitance transient caused by pulsing one interface, either F/O or S/O interface in SIS capacitors, from accumulation into deep depletion. This technique eliminates transients arising from the other interface of the SIS capacitor and ensures that the charge coupling at both interfaces is negligible.

Figure 6.16a shows the C–t curves measured in the film and the substrate of a SIMOX SIS capacitor treated with different postimplant annealings. The recovery time required for the capacitance to change from its initial deep depletion value $C_T(0)$ to the final value $C_T(t_F)$ is designated as t_F.

The C–t transients of Figure 6.16a reveal that t_F is about 8 seconds for the substrate and less than 0.1 second for the film, implying that the relaxation time in the substrate is two to three orders of magnitude longer than in the film for this particular sample. Furthermore, the comparison of the C–t data for the two wafers (Fig. 6.16a) indicates that the recovery time, and hence the carrier generation lifetime, in the film is a factor of 10 higher for sample annealed at 1250 °C over the 1150 °C annealed SIMOX sample. However, no significant improvement in the generation lifetime by higher temperature annealing was observed in the substrate.

The so-called Zerbst plots are constructed from the transient C–t curves:[21]

$$-\frac{d}{dt}\left[\frac{C_{ox2}}{C_T(t)}\right]^2 = \frac{2n_i}{\tau_g N_D}\frac{C_{ox2}}{C_T(t_F)}\left[\frac{C_T(t_F)}{C_T(t)} - 1\right] \tag{6.38}$$

The generation lifetime τ_g is deduced from the slope of Zerbst plot shown in Figure 16b. It is noted that the existence of a linear region in the Zerbst plots for the substrate (Fig. 16b) implies that the generation of minority carriers is proportional to the deep depletion-layer width of the SIS capacitor. Similar Zerbst plots are obtained for the film of this SIS capacitor. The generation lifetime ($\simeq 0.2\,\mu$s) in the film of SIMOX material, annealed at 1250 °C for two hours, is comparable to that reported for recrystallized SOI–ZMR wafers.[22]

The pulsed C–t method described above is one of the most successful and widely used techniques for determining the generation lifetimes in bulk-silicon MOS capacitors in which only one interface is involved. In this section, we have demonstrated that this same technique may be applied to determine the generation lifetimes in both the film and the substrate of SOI materials using an SIS capacitor structure. The detailed theory of this technique, exposed in the classic Zerbst paper,[21] will be reviewed and adapted to SOI in Section 9.5. It is noteworthy that the standard Zerbst method can only measure the average generation lifetimes in the film and the substrate of SOI materials. To determine the generation lifetime profiles, a more complicated procedure such as the differential current/capacitance technique[23] should be used. This will be discussed in details in Chapter 7.

6.5 MOS Capacitance Method

In addition to the SIS capacitor, it is also possible to use the conventional MOS C–V and G–ω techniques to characterize the interface states at the Si/SiO_2 interfaces in thick, partially depleted film with body contact. A metal-gate oxide structure is formed on top of the SOI film. However, if the silicon overlay is very thin (fully depleted case), then it is rather difficult to interpret the C–V data. Indeed, the direct measurement of the capacitance of an SOI-MOS capacitor structure can yield a complicated C–V curve in which capacitances may be contributed from the accumulation, depletion, and inversion states at the front and back interfaces of the silicon film and at the top of the silicon substrate.[24, 25] Although analytical model of the SOI capacitor is available,[24] it is usually easier to simulate the SOI–MOS capacitor structure numerically and to compare the simulated results with the measured data for the charges in the oxide and at the Si–SiO_2 interfaces.[25, 26]

Lee and Cristoloveanu[24] have proposed a simple technique to determine accurately the C–V curves, free from the influence of parasitic capacitances inherent in an SOI–MOS structure. The method is based on the capacitance measurements between the various terminals of an SOI–MOS device and the interpretation of C–V data according to an appropriate equivalent circuit.

Figures 6.17a and 6.17b show the schematic diagram of a p-type capacitor structure and the equivalent circuit. R_s is the series resistance of the silicon island, whose complex behavior represents a major source of errors. By using a specially designed interdigitated capacitor, the series resistance R_s of the p-silicon island can be eliminated. The resultant simplified equivalent circuit is shown in Figure 6.17c.

Another approach makes use of the high–low-frequency method on an SOI–MOS capacitor to investigate the layer thickness, interface-trap distributions, and doping.[26] In this method, the high-frequency, small-signal sinusoidal steady-state response of the two-terminal SOI–MOS capacitor is solved numerically using perturbation analysis of the basic semiconductor equations. The perturbation analysis is equivalent to that presented for the MOS capacitor with the addition of two Si–SiO_2 interfaces. The model includes the effects of a floating semiconductor region, interface traps, and nonuniform doping. Simulations show a difference between the high-frequency and quasi-static solutions for accumulation, which is attributed to the dipole layer in the silicon film. Applying the two-terminal model to the quasi-static and high-frequency C–V measurements, the interface-trap distributions at the three interfaces of the SOI–MOS capacitor can be determined.[25, 26]

For a thick SOI film ($t_{si} \geq 1$–$2\,\mu$m) with body contact, the capacitance (C–V) and conductance (G–ω) measurements follow the conventional guidelines of bulk-silicon MOS capacitors. Although the conductance technique is more accurate, the capacitance method can give

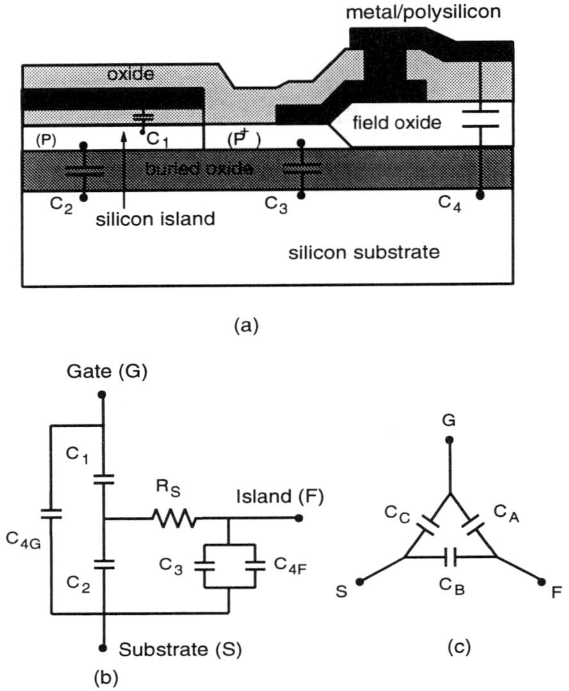

Figure 6.17 (a) Schematic diagram of an SOI–MOS capacitor (C_1 is the active gate capacitance, and C_2, C_3, and C_4 are parasitic capacitances), (b) equivalent circuit of an SOI–MOS capacitor with series resistance R_s, and (c) simplified equivalent circuit for interdigitated capacitor (after Lee and Cristoloveanu[24]).

rapid evaluation of flat-band shift and the total interface trapped charge. Figure 6.18 illustrates a typical low-frequency C–V stretch-out curve (dashed line) due to interface trapped charges in a silicon MOS capacitor with thermally grown oxide. At high frequencies (i.e., $\omega\tau \gg 1$), the interface-traps cannot follow the a.c. signal, so the total capacitance of the MOS system is the series combination of the oxide capacitance ($C_{ox} = \epsilon_{ox}/t_{ox_2}$) and the semiconductor depletion-layer capacitance C_d:

$$C = \frac{C_{ox}C_d}{C_{ox} + C_d} \tag{6.39}$$

For a given oxide thickness t_{ox_2}, the value of C_{ox} is constant and corresponds to the maximum capacitance of the system. The depletion capacitance C_d can be expressed by[5]

$$C_d = \frac{\partial Q_s}{\partial \psi_s} = \frac{\epsilon_{si}}{\sqrt{2}L_D} \frac{[1 - e^{-\beta\psi_s} + (n_{po}/p_{po})(e^{\beta\psi_s} - 1)]}{F(\beta\psi_s, n_{po}, p_{po})} \tag{6.40}$$

where

$$F(\beta\psi_s, n_{po}, p_{po}) = \left[(e^{-\beta\psi_s} + \beta\psi_s - 1) + (n_{po}/p_{po})(e^{-\beta\psi_s} - \beta\psi_s - 1)\right]^{1/2} \geq 0 \tag{6.41}$$

$\beta = q/kT$, ψ_s is the surface potential, and $L_D^2 = kT\epsilon_{si}/p_{po}q^2$ is the extrinsic Debye length for holes. Equation (6.40) shows that C_d depends on the applied gate voltage V_G ($V_G = \psi_s + V_{ox}$, V_{ox} being the voltage drop across the oxide).

The solid line, shown in Figure 6.18, is the ideal C–V curve at high-frequency, which excludes the capacitance of interface-trap charges. The effect of interface traps is to cause a variable shift of the ideal MOS C–V curve along the voltage axis due to the gradual charging of interface traps. More charges on the metal are needed to create a given surface potential. The interface-trap density can be expressed by[5]

$$D_{it} = \frac{C_{ox}}{q}[(d\psi_s/dV_G)^{-1} - 1] - \frac{C_d}{q} \tag{6.42}$$

where $d\psi_s/dV_G$ is the slope of a ψ_s-versus-V_G curve, and C_d is calculated from the known doping density with Equation (6.40). The integration method is used to evaluate ψ_s vs. V_G using the low-frequency C–V measurements.[9] If the oxide trapped charges and mobile ionic charges are neglected, then $V_G = V_{ox} + \psi_s$, $dQ_{ox} = C_{ox}dV_{ox} = CdV_G$, and

$$\frac{d\psi_s}{dV_G} = 1 - \frac{C}{C_{ox}} \tag{6.43}$$

or

$$\frac{d\psi_s}{dV_{ox}} = \frac{C_{ox}}{C} - 1 \tag{6.44}$$

Integration of Equation (6.43) from V_1 to V_2 yields

$$\psi_s(V_1) - \psi_s(V_2) = \int_{V_2}^{V_1} \left[1 - \frac{C}{C_{ox}}\right] dV_G \tag{6.45}$$

Equation (6.45) shows that the surface potential can be determined at any given bias voltage by integrating a curve of $(1 - C/C_{ox})$. Note

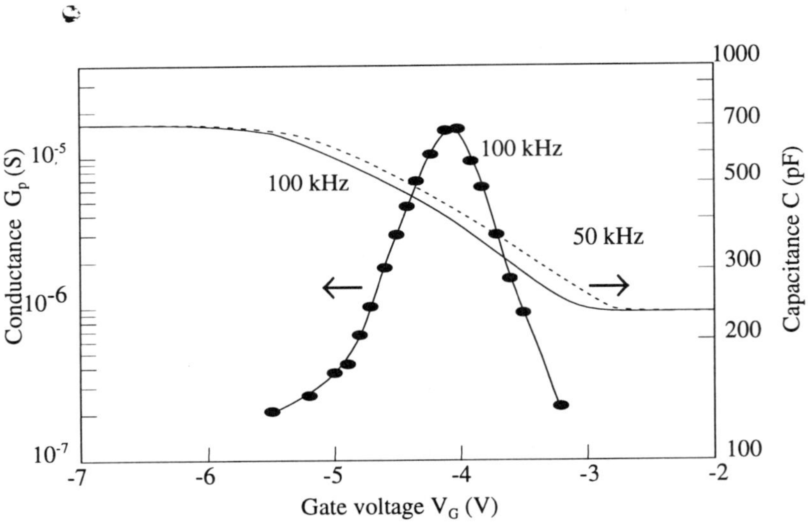

Figure 6.18 A comparison of capacitance and conductance measurement in an MOS capacitor at two frequencies. The stretch-out of the C–V curve at low frequency is due to the interface trapped charges.

that Equation (6.45) is valid only when the interface trapped charges are in equilibrium during the C–V measurements, which implies that the capacitance measurement frequency must be low enough so that all interface traps can follow both the gate bias and the a.c. signal. From the above results and the requirement of charge neutrality in the MOS system, the following relation is reached:

$$\frac{d\psi_s}{dV_{ox}} = \frac{\epsilon_{ox}/d_{ox}}{(dQ_s/d\psi_s) + qD_{it}(q\psi_s)} \tag{6.46}$$

With the help of Equations (6.44) and (6.45), a curve of $d\psi_s/dV_{ox}$ versus ψ_s can be obtained directly using the low-frequency capacitance measurement on an MOS capacitor. If the doping density and the temperature are known, this curve can be compared with the one given by Equation (6.46) to determine accurately the interface-state density D_{it} in the MOS capacitor.

Besides the above capacitance method, the conductance technique discussed in Sections 6.1 and 9.4 also allows investigation of the interface trapped charges in an MOS capacitor.[9] Although both capacitance and conductance versus voltage curves contain identical information about the interface traps, greater inaccuracies arise in extracting this information from the capacitance measurement because the measured capacitance consists of oxide capacitance, depletion-layer capacitance, and interface-trap capacitance. On the other hand, the conductance method yields more accurate and reliable information about interface traps because the measured conductance is directly proportional to the interface traps. This is particularly so when the interface-trap density is low as in the thermally oxidized SiO_2–Si interface.

The conventional capacitance and conductance techniques discussed above can be applied to study the interface properties of the top interface of an SOI–MOS capacitor with partially depleted thick film and body contact. In this case, a gate metal and thermal oxide layer are deposited on the top of the silicon film to form an MOS capacitor on SOI substrate. If the silicon film is removed from the SOI substrate and a gate metal is deposited on the buried oxide, then the properties of the buried oxide and BOX-substrate interface can be inspected by C–V or G–V methods. Alternatively, the silicon overlay may be overdoped in order to act as a gate that controls the depletion region in the bulk-Si substrate.

In conclusion, we have demonstrated in this chapter that several electrical characterization techniques, including the high–low-frequency capacitance and conductance, the bias-scan and temperature-scan DLTS, and the pulsed capacitance methods, can be applied in SIS capacitor structures on SOI wafers to determine key physical parameters such as bulk defects and interface-state densities at both interfaces of film/buried-oxide/substrate, and the generation lifetimes in the SOI material. Based on an MOS-like equivalent circuit, the expressions for extracting the interface-state densities and doping density from the capacitance and conductance measurements were derived. From bias-scan and temperature-scan DLTS measurements, the bulk defects and interface-state densities at both interfaces of SIMOX wafers are determined. The generation lifetimes in the film and substrate of the SOI wafer are deduced from the pulsed C–t measurements and Zerbst plot.

There are relevant advantages of the SIS capacitor over the conventional MOS capacitor structure for electrical characterization of SOI materials: (1) it is an inherent structure for the SOI material, (2) it requires very few processing steps (except for the top and bottom ohmic contacts and mesa-etch, no additional layers are needed to form an SIS capacitor structure), and (3) a quick-turnaround C–V measurements can be performed on an SIS capacitor for parameter extraction. Most importantly, the SIS capacitor measurements are valuable for the SOI wafer screening prior to full processing of VLSI circuits.

The conventional MOS capacitance and conductance measurement and the modeling of a two-terminal MOS capacitor is directly transposable from bulk silicon to partially depleted SOI. It can also be applied for characterization of the buried-oxide, underlying interface, and silicon substrate.

Bibliography

[1] H.S. Chen and S.S. Li, "A model for analyzing the interface properties of a semiconductor–insulator–semiconductor structure—I: Capacitance and conductance techniques," *IEEE Trans. Electron Devices*, vol. 39, p. 1740, 1992.

[2] H.S. Chen and S.S. Li, "A model for analyzing the interface properties of a semiconductor–insulator–semiconductor structure—II: Transient capacitance technique," *IEEE Trans. Electron Devices*, vol. 39, p. 1747, 1992.

[3] F.T. Brady, S.S. Li, D.E. Burk, and W.A. Krull, "Determination of the fixed oxide charge and interface trap densities for buried oxide layers formed by oxygen implantation," *Appl. Phys. Lett.*, vol. 14, p. 886, 1988.

[4] K. Nagai, T. Sekigawa, and Y. Hayashi, "Capacitance-voltage characteristics of semiconductor–insulator–semiconductor (SIS) structure," *Solid-State Electron.*, vol. 28, p. 789, 1985.

[5] E.H. Nicollian and J.R. Brews, *MOS Physics and Technology*, Wiley, New York, 1982.

[6] L.M. Terman, "An investigation of surface states at a silicon/silicon oxide interface employing metal–oxide–silicon diode," *Solid-State Electron.*, vol. 5, p. 285, 1962.

[7] M. Kuhn, "A quasi-static technique for MOS C–V and surface state measurements," *Solid-State Electron.*, vol. 13, p. 873, 1970.

[8] R. Castangé and A. Vapaille, "Description of the SiO_2–Si interface properties by means of very low frequency MOS capacitance measurements," *Surface Sci.*, vol. 28, p. 157, 1971.

[9] E.H. Nicollian and A. Goetzberger, "The Si–SiO_2 interface-electrical properties as determined by the MIS conductance technique," *Bell Syst. Technol. J.*, vol. 46, p. 1055, 1967.

[10] C.N. Berglund, "Surface states at steam-grown silicon–silicon dioxide interfaces," *IEEE Trans. Electron Devices*, vol. 13, p. 701, 1966.

[11] P.K. McLarty, J.W. Cole, K.F. Galloway, D.E. Ioannou, and S.E. Bernacki, "Deep level transient spectroscopy studies of epitaxial silicon layers on silicon-on-insulator substrates formed by oxygen implantation," *Appl. Phys. Lett.*, vol. 51, p. 1078, 1987.

[12] P.K. McLarty, D.E. Ioannou, and H.L. Hughes, "Deep states in silicon-on-insulator substrates prepared by oxygen implantation using current deep level transient spectroscopy," *Appl. Phys. Lett.*, vol. 53, p. 871, 1988.

[13] N.M. Johnson, "Energy-resolved DLTS measurement of interface states in MIS structures," *Appl. Phys. Lett.*, vol. 34, p. 802, 1979.

[14] K.L. Wang, "A determination of interface state energy during the capture of electrons and holes using DLTS," *IEEE Trans. Electron Devices*, vol. 26, p. 819, 1979.

[15] T. Katsube and K. Kakimoto, "Temperature and energy dependencies of capture cross sections at surface states in Si metal–oxide–semiconductor diodes measured by deep level transient spectroscopy," *J. Appl. Phys.*, vol. 52, p. 3504, 1981.

[16] K. Yamasaki, M. Yoshida, and T. Sugano, "Deep level transient spectroscopy of bulk traps and interface states in Si MOS diodes," *Jpn. J. Appl. Phys.*, vol. 18, p. 113, 1979.

[17] V. Kumar and S.B. Iyer, "Characterization of surface states in MOS capacitors by a modified DLTS technique," *Phys. Stat. Solidi (a)*, vol. 76, p. 637, 1983.

[18] P.K. McLarty, D.E. Ioannou, and J.P. Colinge, "Bulk traps in ultrathin SIMOX MOSFET's by current DLTS," *IEEE Electron Device Lett.*, vol. 9, p. 545, 1988.

[19] Y. Zohta and M.O. Watanabe, "On the determination of the spatial distribution of deep centers in semiconducting thin films from capacitance transient spectroscopy," *J. Appl. Phys.*, vol. 53, p. 1809, 1982.

[20] G.L. Miller, D.V. Lang, and K.C. Kimerling, "Capacitance transient spectroscopy," *Ann. Rev. Material Soc.*, p. 377, 1977.

[21] M. Zerbst, "Relaxation effects at semiconductor–insulator interfaces", *Z. Angew. Phys.*, vol. 22, p. 30, 1966.

[22] D.P. Vu, A. Chantre, H. Mingam, and G. Vincent, "Electrical properties of halogen lamp recrystallized silicon films on SiO_2," *J. Appl. Phys.*, vol. 56, p. 1682, 1984.

[23] H.S. Chen, F.T. Brady, S.S. Li, and W.A. Krull, "A new technique for determining the generation lifetime profile in thin semiconductor films with application to silicon-on-insulator (SOI) substrate," *IEEE Electron Device Lett.*, vol. 10, p. 496, 1989.

[24] J.H. Lee and S. Cristoloveanu, "Accurate technique for C–V measurements on SOI structures excluding parasitic capacitance effects," *IEEE Electron Device Lett.*, vol. 7, p. 537, 1986.

[25] D. Flandre and F. Van De Wiele, "A new analytical model for the two-terminal MOS capacitor on SOI substrate," *IEEE Electron Device Lett.*, vol. 9, p. 296, 1988.

[26] M. Gaitan and P. Roitman, "Small signal model of the MOSOS capacitor," *IEEE SOS/SOI Technol. Conf. Proc.*, p. 48, 1989.

[27] S.M. Sze, *Physics of Semiconductor Devices*, Wiley, New York, 1981.

Chapter 7

Diode Measurements

7.1 Current–Voltage Measurements in a P–N Diode

In this section, we discuss the characterization of SOI material properties by using the current–voltage (I–V) characteristics of p–n diodes formed on a SOI substrate. SOI diodes have either vertical or lateral configuration (Fig. 7.1). Vertical diodes are fabricated on thick epitaxial films and behave like bulk-silicon diodes except that the material properties may be quite different. Thin lateral diodes are more typical SOI devices. The forward- and reverse-biased I–V characteristics of a p–n diode make it possible to determine the quality of silicon film, buried-oxide, and interface properties of the SOI material.

The I–V curves of p–n diodes are readily recorded with an HP–4140B pA-meter by keeping the substrate connected to the ground. For a typical SOI p–n diode, the ideality factor is evaluated from the slope of the forward I–V curve for $0.1 < V_F < 0.5\,\mathrm{V}$. The reverse-bias leakage current can be measured at $V_R = -5\,\mathrm{V}$.

The deviation of ideality factor n from unity at low forward-bias voltage is attributed to carrier recombination at the junction space-charge region or surface. In a *short* diode, free from series-resistance effects, values of n vary between 1 and 2. The possible generation–recombination (G–R) processes in SOI p–n diodes include bulk G–R in the junction space-charge region, silicon surface recombination, silicon-film/buried-oxide interface recombination, and silicon/lateral-isolation interface recombination. For Si planar p^+–n junctions, the generation current in the depletion region is usually much larger than the other G–R processes.

Under reverse bias and for $p_{no} \gg n_{po}$ and $|V_R| > 3kT$, the total reverse current density J_R (in A/cm^2) can be approximated by the sum of the diffusion component in the quasi-neutral region and the generation current in the depletion region w_d, which reads[1]

$$J_R = q\sqrt{\frac{D_p}{\tau_p}}\frac{n_i^2}{N_D} + \frac{qn_iw_d}{\tau_e} \tag{7.1}$$

with

$$\tau_e = \frac{\sigma_p\sigma_n v_{th}N_t}{\sigma_n \exp(\frac{E_t-E_i}{kT}) + \sigma_p \exp(\frac{E_i-E_t}{kT})} \tag{7.2}$$

where τ_e is the electron lifetime and other symbols have their usual meaning. For silicon film with small n_i, the generation current governs the reverse leakage current. At sufficiently high temperature, the diffusion current will gradually become the prevailing component.

In a high-quality SOI p–n diode, the generation and/or diffusion currents dominate under reverse-bias conditions; the reverse leakage current is very small, and its dependence on the reverse voltage is weak until the near-breakdown voltage region, where avalanche multiplication takes over.

In poor materials, the surface leakage current may prevail; then the reverse current will increase rapidly with increasing reverse bias, and a soft breakdown may occur at the junction. A particular feature of SOI diodes is that the contribution of the back film/buried-oxide interface to the leakage current can be enhanced or attenuated according to the voltage applied to the silicon substrate (back gate).

Studies of junction breakdown characteristics on SOI p–n diodes provide information concerning the breakdown mechanisms in various SOI devices. Chen et al. use an analytical model and PISCES II simulation to characterize junction breakdown in SOI diodes.[2] The model depicts the influence of the back-gate bias and buried-oxide properties on the parameters of a fully depleted SOI diode. Breakdown in a thin, fully depleted SOI diode is quite different from that observed in a thicker, partially depleted diode, because the junction curvature effect is reduced. This yields a higher breakdown voltage than in bulk silicon or thick SOI film diodes. The breakdown voltage is limited by the finite silicon film thickness. High densities of fixed oxide charges and interface states tend to reduce breakdown voltage of SOI diodes. This model was supported by the breakdown-voltage measurements performed on SIMOX diodes.[2]

Under forward-bias conditions, the current conduction in SOI p–n diodes may be dominated by two main components: the recombination current at low-bias voltage and the diffusion current at higher voltage. The total current density is approximated by the sum of recombination and the diffusion currents[1]

$$J_F = q\sqrt{\frac{D_p}{\tau_p}}\frac{n_i^2}{N_D}\exp(qV/kT) + \frac{qw_d}{2}\sigma v_{th} N_t n_i \exp(qV/2kT) \qquad (7.3)$$

The measured I–V curve is represented by an empirical formula

$$J_F \sim \exp(qV/nkT) \qquad (7.4)$$

where the diode ideality factor n approaches 2 when the recombination current becomes dominant, and is equal to 1 in the ideal case of a prevailing diffusion current. When both currents are comparable, n has a value varying between 1 and 2.

A systematic study of an SOI p–n diode implies measuring the I–V characteristics under forward- and reverse-bias conditions, and for various back-gate biases. Based on the above discussion, it is clear that the breakdown voltage, the ideality coefficient, and the forward and reverse currents can elucidate the various generation–recombination mechanisms, provide the carrier lifetime, and yield information on the quality of the silicon film, buried-oxide, and interfaces.

7.2 Differential Current/Capacitance Method

The generation current in a reverse-biased diode can be evaluated by using the Sah–Noyce–Shockley model.[1] Since the junction capacitance C is inversely proportional to the depletion-layer width w_d, the product of bulk generation current and junction capacitance should be independent of the applied reverse-bias voltage. In practical situations, the discrepancy has been attributed to the surface leakage[2] and the field-dependent generation lifetime.[3] Calzolar and Graffi[4] have incriminated the considerable difference existing between the depletion-layer width w_d and the generation-layer width w_i, under reverse-bias conditions. In fact, it has been shown that the generation rate drops rapidly to negligible values outside w_i.

In this section, we describe a differential current/capacitance technique for determining the generation lifetime profile in reverse-biased

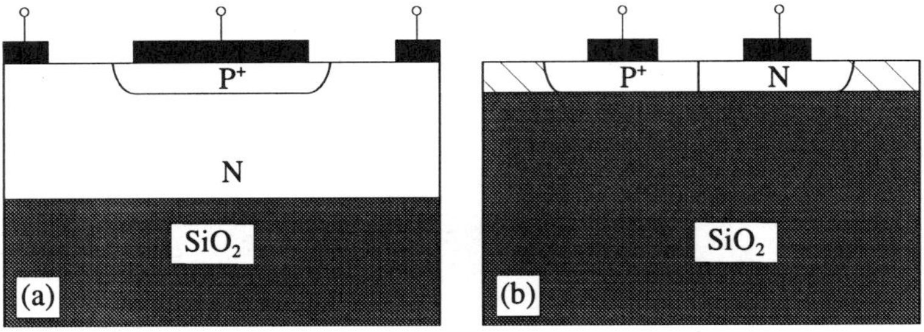

Figure 7.1 Schematic of p^+–n diodes with (a) vertical and (b) lateral configurations formed on an SOI substrate.

p^+–n diodes formed on SOI substrates. The results for several SIMOX samples are compared with the average generation lifetimes derived from the conventional Zerbst method.

7.2.1 Principle and Experimental Setup

The differential current–voltage (I–V) and capacitance–voltage (C–V) technique provides the generation lifetime profile in the superficial layer of the SOI wafer. The method involves a combined I–V and C–V measurement on a reverse-biased p^+–n diode formed on a SOI substrate.

The test structure shown in Figure 7.1a consists of a thin n-type SOI film, with a p^+–n junction or a Schottky diode fabricated on top of it, for I–V and C–V measurements. This test structure limits further expansion of the depletion region when the Si film is fully depleted, and hence provides an easy way to account for the field-dependent generation current. The experimental setup for the I–V and C–V measurements is similar to that shown in Figure 6.1a. The measurements are performed with an HP–4140 pA meter (I–V) and HP–4280A capacitance-meter (C–V) by keeping the substrate grounded.

7.2.2 Model

The current components in a reverse-biased p–n diode consist of the diffusion current I_{dif}, bulk generation current I_{gen}^b, surface generation current I_{gen}^s, and the field-enhanced current I_{fe}. The last component is the field-dependent current, which is not included in the first three current components. The generation current resulting from the lateral extension of the depletion region is also implicitly included in this term. For a p$^+$–n diode, if the n-layer is partially depleted under an applied reverse-bias V_R, then the total reverse current is given by[5]

$$I_R(V_R) = A_s q n_i S_1 + A_b q \int_0^{w_i(V_R)} \frac{n_i}{\tau_g(x)} dx + I_{dif}(V_R) + I_{fe}(V_R) \quad (7.5)$$

where S_1 is the surface generation velocity, A_b is the area of the bulk space-charge region, and A_s is the surface area of the diode.

For a p$^+$–n diode formed on an SOI substrate, the diffusion current is negligible due to the small intrinsic carrier concentration and thin Si film. In addition, the field-enhanced reverse current is important only when V_R is large. In general, the generation layer width w_i remains much smaller than w_d until V_R is sufficiently high and renders I_{fe} no longer negligible. For a p$^+$–n diode, the generation layer width is given by[4]

$$w_i = \left(\frac{2\epsilon_{si} kT}{q^2 N_D}\right)^{1/2} \left[\left(\ln\frac{N_D}{n_i} + \frac{qV_R}{kT}\right)^{1/2} - \left(\ln\frac{N_D}{n_i}\right)^{1/2}\right] \quad (7.6)$$

where the symbols used in Equations (7.5) and (7.6) have their usual meaning.

Since I_{gen}^s is independent of bias voltage, and I_{dif} and I_{fe} are negligible, we can rewrite Equation (7.5) in a differential form by using $dw_d/dV_R = (\epsilon_{si} A_b/C^2) dC/dV_R$. Therefore, from Equations (7.5) and (7.6), the generation lifetime can be expressed in terms of measurable quantities as

$$\tau_g(w_i) = \left(\frac{\epsilon_{si} q n_i A_b^2}{C^2}\right) \times \left(\frac{dw_i}{dw_d}\right) \times \left(\frac{dC/dV_R}{dI_R/dV_R}\right) \quad (7.7)$$

where

$$\frac{dw_i}{dw_d} = \left[\frac{V_R + (kT/q)\ln(N_A N_D/n_i^2)}{V_R + (kT/q)\ln(N_D/n_i)}\right]^{1/2} \quad (7.8)$$

is the correction factor, which takes into account the difference in the variations of w_i and w_d.

It is interesting to note that even though w_i may differ significantly from w_d, Equation (7.8) reveals that dw_i/dw_d is close to unity over a wide range of bias voltages V_R and doping concentrations N_D.

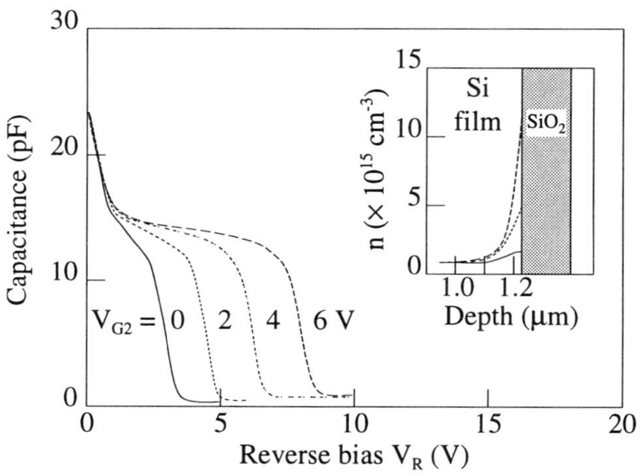

Figure 7.2 Capacitance-voltage curves for a p$^+$–n SOI diode with back-gate bias V_{G_2} as a parameter. The inset shows the corresponding free-carrier concentration profiles.

7.2.3 Parameter Extraction

To illustrate the differential current/capacitance technique,[6] the I–V and C–V measurements are made on several p$^+$–n diodes formed on SIMOX substrates with a junction depth of 0.4 $\pm$ 0.1 μm and a diode area of 450 $\times$ 450 μm^2. The doping densities and generation lifetime profiles for these samples are obtained with the help of Equation (7.7).

The measurements are performed on SIMOX wafers synthesized with different oxygen implant doses and postimplant annealing treatments. Carrier concentration, determined by C–V profiling, is generally uniform across the epitaxial layer except near the buried oxide, where the fixed oxide charge tends to attract free carriers toward the interface.

The effect of series resistance on C–V measurements was taken into account by considering both capacitive and resistive components of the diode.[7] Also, a positive back-gate (substrate) bias was applied to accumulate the film/oxide (F/O) interface, which in turn significantly reduced the series resistance and provided a very accurate doping concentration profile beyond a distance of 0.2 μm from the F/O interface.

The doping profile in this region was found to be uniform by the differential capacitance technique. The correction of carrier concentra-

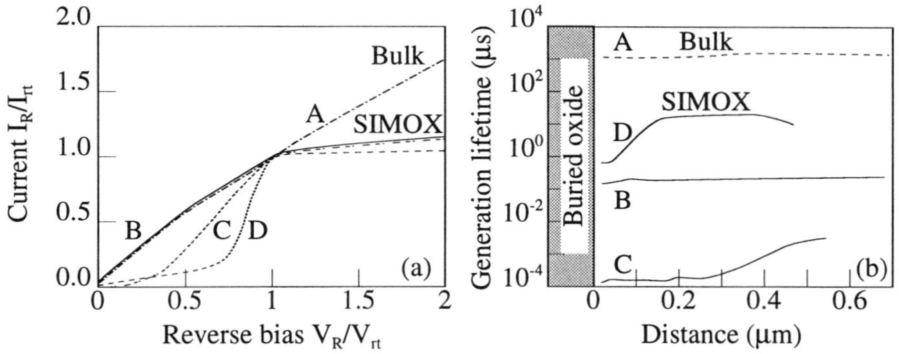

Figure 7.3 (a) Normalized reverse-bias I–V curves for bulk and SOI diodes. V_R is normalized to the reach-through voltage V_{rt}, and I_R is normalized to the corresponding reach-through current I_{rt}. (b) Generation lifetimes versus distance from the F/O interface, determined from (a).

tion profiles near reach-through is possible by computer simulation.[8] Figure 7.2 shows the capacitance–voltage characteristics of a p^+–n SOI diode, recorded for various back-gate biases V_{G_2}, and the corresponding free-carrier concentration profile. For simplicity, we assume uniform doping concentration across the entire Si film in the following analysis.

The I–V curves for several p–n diodes formed on the SIMOX wafers are shown in Figure 7.3a. Sample A is a bulk-Si diode and samples B, C, and D are different SIMOX diodes. The I–V curves of SIMOX diodes are normalized to the reach-through voltage V_{rt} (defined by the corner of the reverse-biased I–V characteristics) and current I_{rt}. For $V_R > V_{rt}$, I_R is nearly constant due to reach-through of Si film. A slight increase of I_R in this region arises from the contribution of I_{fe}.

The corresponding generation lifetime profile (Fig. 7.3b) is derived by using Equation (7.7) and assuming a uniform doping concentration. Sample A (bulk diode) has a constant τ_g across the n-layer, and the value of τ_g is close to that reported by Schroder.[9] Samples C and D show a significant increase of generation current as V_R approaches V_{rt}, and a large reduction of τ_g at the film/buried-oxide interface. Sample B has a similar I–V characteristic to that of bulk diode for $V_R < V_{rt}$, with a relative uniform τ_g compared to that of samples C and D. In sample D, the lifetime is much higher, due to a better control of the contamination

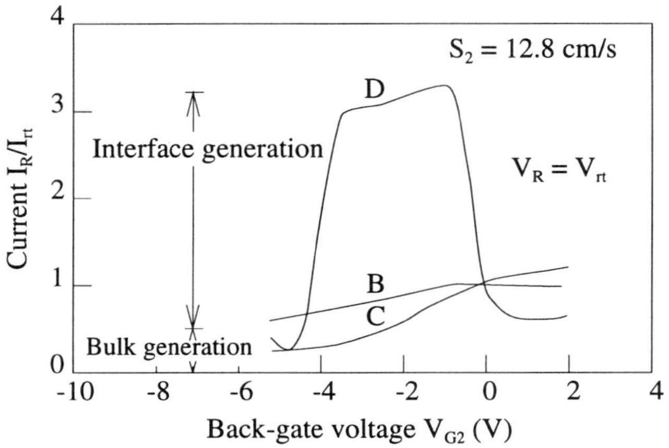

Figure 7.4 Normalized reverse current versus back-gate bias V_{G_2} for three SIMOX diodes with parameters given in Table 1.

level;[10] the dramatic reduction of τ_g within a Debye length ($\sim 0.1\,\mu$m) from the F/O interface is caused mainly by the increase of generation current from the F/O interface states. However, this phenomenon is not seen in sample B and C, since the bulk generation current dominates the F/O interface generation current in both samples.

The above results are confirmed by the back-gate-controlled diode measurements shown in Figure 7.4 (see also Section 7.3). The measurements were made by changing the back-gate bias V_{G_2} while keeping V_R constant. When the F/O interface is accumulated or inverted, only those centers that are within the film contribute to the generation current. When the F/O interface is depleted, interface generation provides yet another contribution to the total generation current. This contribution may result in a peak in the I_R versus V_{G_2} curve, as shown in Figure 7.4. It is interesting to note that this peak can only be observed in sample D rather than samples B and C.

For $V_R > V_{rt}$, since the depletion-layer width is mainly limited by the buried-oxide-layer thickness, any increase in I_R would come from I_{fe}. Therefore, an estimation of error may be obtained from this part of the I–V characteristics. As shown in Figure 7.3a, the increase of I_{fe} for $V_{rt} < V_R < 2V_{rt}$ is less than 11 % of the total current, which suggests an error of about the same magnitude (if I_{fe} is negligible for $V_R < V_{rt}$). It

Table 7.1 Summary of processing parameters and Zerbst lifetimes* for the bulk (A) and SIMOX (B, C, and D) diodes.

Sample	Anneal (°C, hrs.)	Dose (cm^{-2})	N_D (cm^{-3})	V_{rt} (V)	I_{rt} (A)	τ_g (μs)
A	–	–	2.0×10^{15}	3.4	2.8×10^{-13}	
B	1250, 2	$1.8{\times}10^{18}$	1.6×10^{15}	3.4	2.3×10^{-9}	0.12
C	1250, 2	$2.0{\times}10^{18}$	3.0×10^{15}	1.8	7.3×10^{-7}	0.02
D	1285, 2	$1.8{\times}10^{18}$	1.6×10^{15}	1.0	5.8×10^{-11}	4.4

* Zerbst generation lifetimes (τ_g) are calculated at $0.4 \pm 0.1\,\mu$m from the film/buried-oxide interface.

is noted that values of the generation lifetimes differ by more than two orders of magnitude for samples C and D. The error caused by I_{fe} is essentially negligible in this case.

Table 7.1 summarizes the device parameters and values of the *average* generation lifetimes calculated from Zerbst analysis of capacitance transients on the buried-oxide capacitors.[11] The Zerbst lifetimes and the generation lifetimes determined by the differential current/capacitance technique are in reasonable agreement for the corresponding samples inspected. It is noteworthy that the layer width in which Zerbst analysis is applicable is only $\sim 0.1\,\mu$m for these devices. The discrepancy of τ_g, determined by Zerbst method and by the present technique, in sample C is believed to be caused by film nonuniformity, while the discrepancy in sample D is attributed to the predominance of the F/O interface generation over the bulk generation. Furthermore, the short generation lifetime in sample C makes it more difficult to find a linear region to extract generation lifetime from the Zerbst plot. For sample B, which has a uniform generation lifetime, the agreement between the Zerbst method and the differential technique is greatly improved.

In conclusion, we have demonstrated a differential current/capacitance technique for determining the generation lifetime profile in a p–n diode formed on the SOI substrate. The method requires the use of I–V and C–V measurements on a reversed-biased p–n diode. It also provides a simple procedure for identifying any error caused by the field-enhanced current. The spatial resolution of this technique is identical to that of C–V profiling technique, which is about one Debye length.

7.3 Gated-Diode Measurements

Gated diodes are very appropriate devices for film and interface characterization in partially and fully depleted SOI structures, since many different measurements can be performed. Gate-controlled diodes with enclosed geometry, where the parasitic influence of the Si island edges is eliminated, are the best-suited test vehicles. In this section, we present the analysis and measurements of leakage current in gated diodes.

7.3.1 Principle and Experimental Setup

The leakage current I_R of a reversed-biased gate-controlled p^+–p–n^+ diode as shown in Figure 7.5a is basically induced by the carrier generation mechanisms. The different regions of the structure where carriers are generated may be identified by recording the $I_R(V_G)$ curve illustrated in Figure 7.5b. For $V_G < V_{FB}$, the interface is accumulated and the leakage current I_1 originates from the junction–controlled depletion region only. For $V_{FB} < V_G < V_T$, the carrier generation at the *depleted* interface prevails. Interface inversion, for $V_G > V_T$, causes the interfacial generation rate to vanish.[12]

The *Gated–Diode Leakage* (GDL) method is attractive for the simplicity and static mode of operation. The measurement setup consists of a picoammeter and a hot chuck, whose temperature may be regulated. The basic limitation is the relatively low sensitivity, the measured currents being in the pA range. This difficulty can be overcome by increasing the temperature, the probing surface (large area devices or parallel interconnection of several diodes), or the reverse bias.

7.3.2 Model

The rise and fall edges of the characteristic of Figure 7.5b correspond to the flat-band voltage V_{FB} and the threshold voltage V_T, respectively. The *interface–induced* excess current $(I_3 - I_4)$ is proportional to the generation velocity S:[12]

$$I_3 - I_4 = \frac{qn_iS}{2}(LW) \tag{7.9}$$

where q is the electron charge, n_i is the intrinsic concentration, and L is the *effective* length of the p^- base: $L = L_G - \Delta L$.

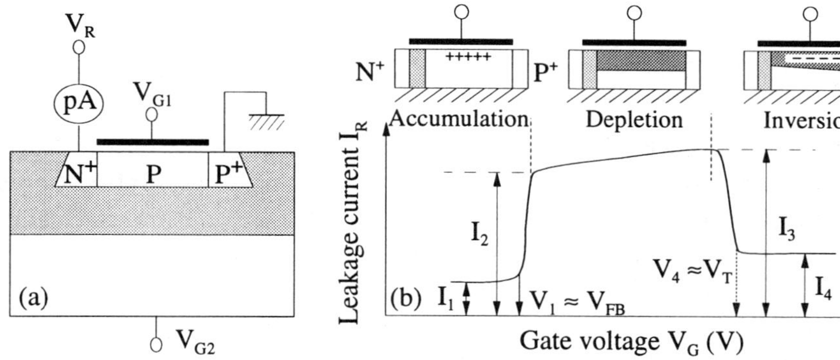

Figure 7.5 (a) Schematics of gated diode measurements performed in an MOS p^+–p–n^+ diode on SOI and (b) typical variation of the leakage current versus front-gate bias.

The slope of the "plateau" ($I_3 > I_2$) illustrates the expansion of the gate–controlled depletion region

$$I_3 - I_2 = \frac{qn_i}{2}\frac{x_d}{\tau_g}(LW) \tag{7.10}$$

where τ_g is the average generation lifetime in the depletion region of thickness x_d. When the parasitic diffusion currents are negligible, i.e., for $L \le L_s$, where $L_s \simeq 3\,\mu$m is a critical length,[13] it is normally found that $I_4 - I_1 = I_3 - I_2$.

The description given here may, however, fail in state-of-the-art devices. For instance, in short–channel diodes, an "abnormal" relation $I_1 > I_4$ may be induced by the *inversion* of the substrate doping type close to the drain (due to the lateral diffusion of the n^+ drain doping under the gate[14] or to the expansion of the depletion region in the overlapped LDD region).[15]) Also, the inhomogeneous distribution of the surface potential along the channel from the n^+ contact to the p^+ contact (for $L > L_s$) may lead to $I_3 < I_2$ as V_G increases; the interface region close to the p^+ contact becomes gradually masked by the onset of strong inversion, whereas the region near the n^+ terminal is still depleted.[13]

A structure may appear in the region $V_G < V_T$ of $I_R(V_G)$ curves, which reflects the superposition of various contributions (accumulated p-substrate and inverted n-substrate zones, inhomogeneous profile of defects along the channel, etc.).[15, 16] The spatial defect-distribution can be resolved by coupling GDL current measurements with two-dimensional simulations.

7.3.3 Parameter Extraction

The GDL method can be directly adapted to thin-film SOI diodes, in order to determine, from $I_R(V_{G_1})$ curves, the properties of the *front* interface (Si film–thermal oxide). Reversed measurements, performed by varying the back-gate bias V_{G_2}, are used to explore the properties of the *back* interface (film–implanted oxide).[17]

The parameter extraction methods are very similar to those used for *charge pumping* and will be described in more detail in Section 9.2. For instance, drawing the surface leakage current $(I_3 - I_4)$ versus gate length L_G results in a straight line whose intercept with the horizontal axis gives ΔL. This extrapolation must therefore be restricted to $L \leq 3\,\mu$m, for Pierret's effect to be avoided.[1]

The slopes of $(I_3 - I_4)$ vs. L lines are used to qualify the front interface in terms of generation velocity, which is related to the density of interface traps:[12]

$$S_1 = \sigma_{n_1} v_{th} (\pi kT) D_{it1} \qquad (7.11)$$

where v_{th} is the carrier thermal velocity and σ_{n_1} is the capture cross section, obtained by combining GDL with charge-pumping experiments. In device-grade SIMOX wafers, S_1 is very low (7–35 cm/s), confirming the excellent quality of the front interface.[17]

The values of V_{T_1} and V_{FB_1} are estimated from the edges of GDL curves. Numerical simulations have shown that the voltages V_1 and V_4 defined in Figure 7.5b in fact correspond to majority and minority carrier concentrations, respectively, exceeding by 20 times the intrinsic concentration.[18] When the back interface is *accumulated* ($\psi_{s2} = \psi_a$, Fig. 7.6a), the left edge is not modified by V_R, whereas the right edge, related to V_{T_1}, is shifted substantially towards a higher V_{G_1} bias. The average doping level $\overline{N_A}$ is determined from the variation of $V_{T_1}^{acc}$ with V_R[14]

$$V_{T_1}^{acc} = V_{FB_1} + \left(1 + \frac{C_{si}}{C_{ox_1}}\right)(2\phi_{F_1} + V_R) - \frac{C_{si}}{C_{ox_1}}\psi_a - \frac{q\overline{N_A}}{C_{ox_1}} t_{si} \qquad (7.12)$$

[1]In long diodes ($L > L_s$), the depletion region no longer covers the whole interface, and the surface generation velocity is underestimated.[13]

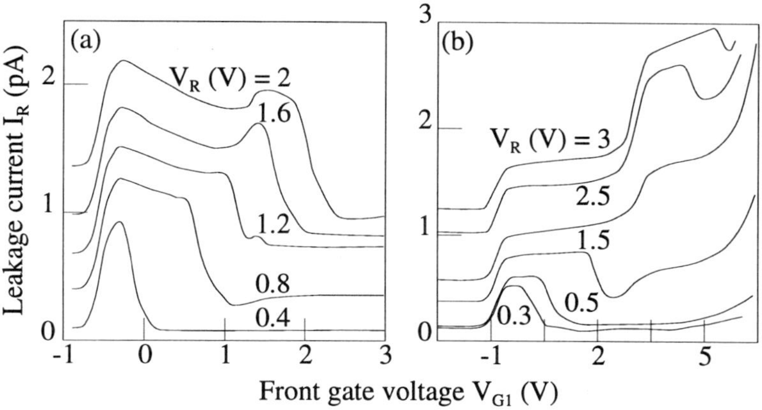

Figure 7.6 Front-gate leakage current for different reverse junction biases in an MOS p^+–p–n^+ diode on SIMOX: (a) back interface accumulated, (b) back interface grounded.

where ϕ_{F_1} is the Fermi potential near the top interface. For *inversion* at the back interface, the $I_R(V_{G_1})$ curves are shifted towards more negative V_{G_1} values due to the decrease of V_{T_1} and V_{FB_1}.

If the back interface is *depleted* or grounded and large reverse biases are used, $I_R(V_{G_1})$ curves become distorted (Fig. 7.6b). The increase of the junction leakage current I_1 with V_R is visible for $V_{G_1} < 0$, and merely reflects the extension of the junction–related depletion zone. This additional contribution is also responsible for the increase of the plateau level $I_{2,3}$. More interestingly, for $V_R \geq 1.5\,\text{V}$, a second hump emerges, whose onset is always defined by $V_{G_1} \simeq 2.5\,\text{V}$, whatever the bias V_R. This dramatic modification of the conventional bulk-Si curve is actually a typical feature of SOI, due to carriers being generated at the *back* interface as well. The second hump occurs when the Si film becomes *fully depleted*, i.e., for $V_{G_1} = V_{FD_1}$,

$$V_{FD_1} = V_{FB_1} + \frac{q\overline{N_A}}{2\epsilon_{si}} t_{si}^2 + \frac{q\overline{N_A}}{C_{ox_1}} t_{si} \tag{7.13}$$

Equation (7.13) shows, in agreement with Figure 7.6b, that V_{FD_1} does not depend on V_R and is fairly consistent with the parameters of the SOI structure.

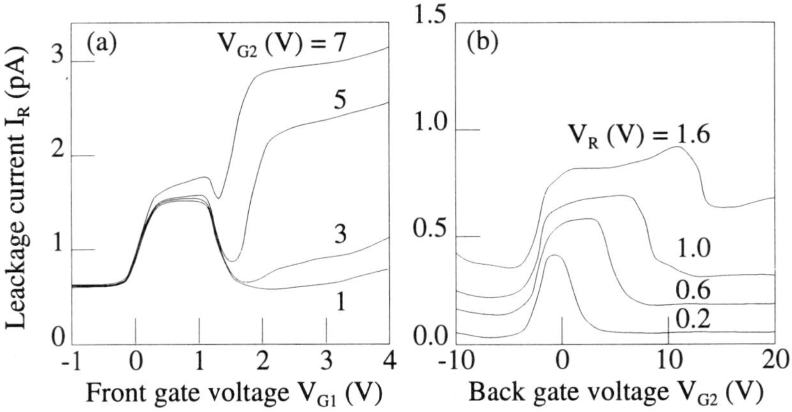

Figure 7.7 Leakage current versus (a) front-gate voltage for various back-gate biases, and versus (b) back-gate bias for various reverse voltages (SIMOX MOS–p^+–p–n^+ diode with $L = 4\,\mu$m).

Furthermore, it is clear that strong inversion can be achieved at the front interface for $V_{G_1} > V_{FD_1}$. Figure 7.6b shows that, for $V_R \simeq 2.5$ V, the V_{T_1}–related edge takes the form of a dip, whose position now is shifted by an amount equal to ΔV_R. The depth of the dip is smaller than the difference $(I_3 - I_4)$ measured for $V_R < 1$ V. The reason is that a parasitic tunneling current (assisted by the interface traps) normally emerges for large gate biases.[15] As a matter of fact, the leakage current increases rapidly for $V_{G_1} > 4$ V, even if V_R is small.

7.3.4 Interface Coupling Effects

The leakage current curves are plotted in Figure 7.7a for several substrate biases V_{G_2}. When V_{G_2} is increased and approaches the back-interface flat-band voltage V_{FB_2}, a second plateau appears on $I_R(V_{G_1})$ curves. This contribution is due to carrier generation at the back interface. Indeed, it is known that increasing V_{G_1} induces not only front-interface inversion but also a reduction of V_{T_2} and V_{FB_2} voltages, gradually driving the back interface into depletion.[59] Figure 7.7a shows that the activation of generation centers at the back interface happens at a lower V_{G_1} bias if V_{G_2} is increased.

Finally, the valleys observed in Figure 7.7a indicate the transition between the formation of a strong inversion layer at the front interface and a depletion region at the bottom of the film. The level of the second plateau reflects the degree of depletion (i.e., the bending of energy bands) achieved at the back interface, which is a function of both V_{G_1} and V_{G_2}.

An example of reciprocal measurements is given in Figure 7.7b, where the leakage current is plotted versus back-gate bias, while keeping the front gate in accumulation. The typical GDL features are still observed: the back-interface flat-band voltage is constant, whereas V_{T_2} increases with increasing the reverse bias V_R. The structure appearing for $V_R \geq 1.6\,\mathrm{V}$ is again explained by the contribution of the opposite (front) interface.

It is found from the plateau level that the back-interface generation velocity is 2 to 8 times larger than at the front interface. This difference in quality between the top and bottom regions of the film is consistent with the value of the generation lifetime (given by $I_3 - I_2$): $1\,\mu$s near the back interface and $6\,\mu$s near the front interface.

7.3.5 Complementary Aspects

The GDL method is very simple, and for standard conditions of operation, the interpretation is straightforward. It applies to short- and long-channel MOS–PIN diodes and is directly adaptable to the drain (or source) junction leakage currents in MOSFETs, where the lateral profile of hot-carrier-induced defects may be resolved using the influence of V_R.[15, 16] Large reverse biases are used to enhance the method sensitivity, but this may induce other phenomena: interface coupling, carrier tunneling, and strongly inhomogeneous surface potential along the channel, which causes a nonuniform surface generation rate.

GDL accesses rather involved "macroscopic" coefficients (carrier generation lifetime and surface generation velocities), which can be complemented by charge pumping. Of practical relevance for the technology optimization is the identification of the various sources of leakage current in MOS structures.

In SOI devices, probing the back interface does not imply any modification of the experimental setup, nor does it involve the parasitic contribution of the third interface situated beneath the buried oxide. The limited number of experimental variables in GDL makes hard the identification and deconvolution of various generation sources, since all of them are superposed upon a unique curve. $I_R(V_G)$ curves become very complicated when the contributions of the front and back interfaces are superimposed. An accurate fitting of the GDL characteristics would require numerical simulations because of the strong 2–D nature of the problem (vertical and lateral distributions of the electric field and depletion regions).

Preliminary *charge-pumping* experiments (see Chapter 9) are helpful to make easier the analysis of GDL curves. Although some results are overlapping, the two methods are complementary rather than mutually exclusive.

7.4 Deep-Level Transient Spectroscopy

Deep-level defects play an important role in determining the recombination and trapping mechanisms (and hence the minority carrier lifetimes) in a semiconductor material. Therefore, it is important to use a sensitive experimental tool for characterizing the deep-level defects in the SOI materials. Deep-level transient spectroscopy (DLTS)—a high-frequency transient capacitance technique—is the most sensitive method for defect characterization in a semiconductor.[20, 21] For example, by performing the DLTS measurement from 77 K to around 450 K, the emission spectrum of all the deep-level traps (both majority and minority carrier traps) in the forbidden gap of a semiconductor appears as positive or negative peaks on a flat baseline as a function of temperature.[20, 21]

The DLTS technique offers advantages such as high sensitivity, easy analysis of experimental data, and the capability of measuring traps over a wide energy range. By properly changing the experimental conditions, we can determine defect parameters, which include (1) minority and majority carrier traps, (2) the activation energy of deep-level traps, (3) the trap concentration and profile, and (4) electron and hole capture cross sections.

In addition, the electron and hole lifetimes can also be calculated from these measured defect parameters. Therefore, by carefully analyzing the DLTS data, most of the defect parameters associated with the deep-level defects in a SOI material can be revealed. We shall next discuss the theoretical and experimental aspects of the DLTS technique applied to a p^+–n or a n^+–p diode formed on SOI substrates.

7.4.1 Principle and Experimental Setup

The experimental setup for DLTS measurements is similar to that shown in Figure 6.1a and consists of an HP–4280 C–t meter (1 MHz) and an HP–8112A pulse generator (50 MHz). The test structure is either a p^+–n or an n^+–p diode formed on an SOI substrate as shown in Figure 7.1a. A PC is used to control the HP C–t meter for DLTS data acquisition and analysis. The DLTS technique is based on the transient capacitance change, associated with the thermal emission of charge carriers from a trap level to thermal equilibrium, after an initial nonequilibrium condition in the space-charge region of a p–n junction diode.

The polarity of the DLTS peak depends on the capacitance change after trapping of the minority or majority carriers. For example, an increase in the trapped minority carriers in the junction space-charge region (SCR) of a p^+–n diode would result in an increase in the junction capacitance of the diode. In this case, a minority carrier trap (i.e., hole trap) will produce a positive DLTS peak, whereas a majority carrier trap (i.e., electron trap) would display a negative DLTS peak.

For a p^+–n junction diode, the SCR extends mainly into the n-region, and the local charges are due to positively charged ionized donors. If a forward bias is applied, the minority carriers will be injected into this SCR region. Once the minority holes are trapped in a defect level, the net positive charge in the SCR will increase. This in turn will reduce the width of SCR, which will cause a positive capacitance change. In this case, the DLTS signal will have a positive peak. Similarly, if electrons are injected into the SCR and captured by the majority carrier traps, then the local charge density in the SCR is reduced and the depletion-layer width is widened, which in turn decreases the junction capacitance. Thus, majority carrier trapping will result in a negative DLTS peak.

7.4.2 Theory and Parameter Extraction

The peak height of a DLTS signal is directly related to the density of a trap level, which in turn is proportional to the change of junction capacitance $\Delta C(0)$ due to carrier emission from the trap level. Therefore, the defect density N_t can be calculated from the capacitance change $\Delta C(0)$ (or the DLTS peak height).

If $C(t)$ denotes the transient capacitance across the depletion layer of a Schottky barrier diode or a p^+–n junction diode, then using abrupt junction approximation, we can write

$$\begin{aligned} C(t) &= A\left[\frac{q\epsilon_{ox}\epsilon_{si}(N_D - N_t e^{-t/\tau})}{2(V_{bi} + V_R + kT/q)}\right]^{1/2} \\ &= C_o\left[1 - \left(\frac{N_t}{N_D}\right)e^{-t/\tau}\right]^{1/2} \end{aligned} \tag{7.14}$$

where τ is the thermal emission time constant; $C_o = C(V_R)$ is the junction capacitance measured at a quiescent reverse-bias voltage, V_R. If we use the binomial expansion in Equation (7.14) and assume that $N_t/N_D \ll 1$, then $C(t)$ can be simplified to

$$C(t) \simeq C_o\left[1 - \left(\frac{N_t}{2N_D}\right)e^{-t/\tau}\right] \tag{7.15}$$

For $t = 0$, we obtain

$$N_t \simeq 2N_D\frac{\Delta C(0)}{C_o} \tag{7.16}$$

where $\Delta C(0) = C_o - C(0)$ is the net capacitance change due to thermal emission of electrons from the trap level, and $C(0)$ is the capacitance measured at $t = 0$. $\Delta C(0)$ comes from the DLTS measurement, whereas the junction capacitance C_o and the background dopant density N_D are both extracted from the high-frequency C–V measurements. Therefore, the defect concentration N_t can be determined from Equation (7.16) by using DLTS and high-frequency (1 MHz) C–V measurements.

The decay time constant of the capacitance transient in the DLTS thermal scan is associated with a specific time constant that is equal to the reciprocal of the emission rate. For a given electron trap, the emission rate e_n is related to the capture cross section and the activation energy of the electron trap by

$$e_n = (\sigma_n v_{th} N_C/g)e^{(E_C - E_t)/kT} \tag{7.17}$$

where E_t is the activation energy of the electron trap, v_{th} is the average thermal velocity, N_C is the effective density of conduction band states, and g is the degeneracy factor. The electron-capture cross section σ_n, which depends on temperature, can be expressed by

$$\sigma_n = \sigma_o e^{-\Delta E_b/kT} \tag{7.18}$$

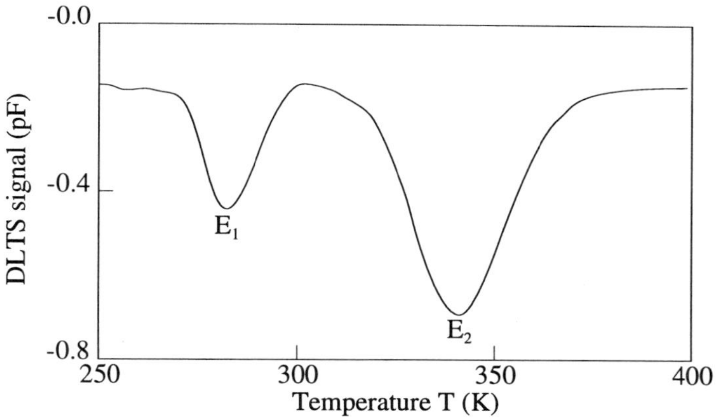

Figure 7.8 DLTS thermal scan of electron traps in a p$^+$–n diode formed on a SIMOX wafer ($E_1 = E_C - 0.41\,\text{eV}$, $E_2 = E_C - 0.58\,\text{eV}$).

where σ_o is the capture cross section when temperature approaches infinity, and ΔE_b is the activation energy of the capture cross section. Now substituting σ_n given by Equation (7.18) into Equation (7.17) and using the fact that N_C is proportional to $T^{3/2}$ and v_{th} is proportional to $T^{1/2}$, e_n reduces to

$$\begin{aligned} e_n &= BT^2 e^{(E_C - E_t - \Delta E_b)/kT} \\ &= BT^2 e^{(E_C - E_m)/kT} \end{aligned} \tag{7.19}$$

where B is a proportionality constant independent of temperature. From Equation (7.19), it is noted that the electron thermal emission rate e_n is a quasi-exponential function of the temperature. The change of capacitance transient can be derived from Equation (7.15), which yields

$$\Delta C(t) = C_o - C(t) \simeq C_o(N_t/2N_D)e^{-t/\tau} = \Delta C(0)e^{-t/\tau} \tag{7.20}$$

where τ is the reciprocal emission time constant.

The DLTS procedures for determining the activation energy of a deep-level trap in a semiconductor are described as follows. The first step of the experiment is to choose the rate windows t_1 and t_2 in a dual-gated integrator of a boxcar averager that is used in the DLTS system. Then, the capacitance change is measured at a preset t_1 and t_2 rate

Table 7.2 Energy levels and possible origins of electron and hole traps observed in SIMOX SOI films by DLTS measurements.

E_t (eV)	N_t/N_D	$\sigma(cm^2)$	Possible origin
$E_C - 0.12$	6.2E-2	5.7E-19	Mg, Fe
$E_C - 0.20$	1.0E-1	1.3E-18	V_3O, W
$E_C - 0.30$	1.2E-2	1.6E-18	V_2O
$E_C - 0.52$	4.0E-3	4.0E-14	Co
$E_V + 0.19$	3.7E-2	1.8E-18	Cu, Mn
$E_V + 0.45$	1.6E-2	2.9E-17	Fe
$E_V + 0.65$	3.6E-2	6.0E-16	Fe, O

window, which can be written as

$$\Delta C(t_1) = \Delta C(0)e^{-t_1/\tau} \tag{7.21}$$

$$\Delta C(t_2) = \Delta C(0)e^{-t_2/\tau} \tag{7.22}$$

The DLTS scan along the temperature axis is obtained by taking the difference of Equations (7.21) and (7.21), which produces a DLTS spectrum given by

$$S(\tau) = \Delta C(0)(e^{-t_1/\tau} - e^{-t_2/\tau}) \tag{7.23}$$

The maximum emission rate, τ_{max}^{-1}, can be obtained by differentiating $S(\tau)$ with respect to τ and setting $dS(\tau)/d\tau = 0$, which yields

$$\tau_{max} = \frac{(t_1 - t_2)}{\ln(t_1/t_2)} \tag{7.24}$$

Note that $S(\tau)$ reaches its maximum value at a characteristic temperature T_m corresponding to the maximum emission time constant τ_{max}. The emission rate is related to this value by $e_n = 1/\tau_{max}$, for each t_1 and t_2 rate-window setting. By changing the values of rate window t_1 and t_2 in the boxcar-gated integrator, a series of DLTS scans with different values of e_n and T_m can be obtained. From these DLTS thermal scans, an Arrhenius plot of e_nT^2 versus $1/T$ is drawn for a specific trap level, and the activation energy of the trap level can be calculated from the slope of this Arrhenius plot. A typical DLTS thermal scan in a SIMOX

p^+–n diode is shown in Figure 7.8, in which two electron traps with activation energies of E_C - E_t = 0.41 and 0.56 eV were identified.

It is established that oxygen implantation and subsequent fabrication processes introduce oxygen–vacancy complexes, dislocations, and metal impurities in SIMOX materials. Diffusion of implanted impurities requires consideration of vacancies, interstitials, vacancy–impurity species, and extended defects. Therefore, the generation of defects in SIMOX-based SOI devices is rather complicated. DLTS measurements on p–n junction diodes fabricated on earlier SIMOX wafers revealed various defect levels that can be divided into two groups: metallic impurities and oxygen-related defects. The defect energy levels and possible origins for both the electron and hole traps observed in the earlier SIMOX wafers are listed in Table 7.2.

In summary, we have demonstrated several techniques for characterizing the p–n and p–i–n diodes formed on SOI substrates. The static I–V measurements provide information concerning the recombination, generation, and breakdown mechanisms in the SOI diodes. The differential current/capacitance technique can be used to investigate the generation lifetime profile in thin SOI films. The gated-diode leakage method is applied to study the front and back interface properties of thin SOI diodes. For characterization of bulk defects in SOI materials, the DLTS technique is employed to determine the deep-level defects in SOI diodes. Several oxygen-related defects and metallic impurities have been detected in SOI wafers using the DLTS technique. In general, the I–V and C–V/C–t measurements on SOI diodes described in this chapter can be used to evaluate the quality of the thin Si overlay, front/back interfaces, and buried oxide of SOI materials.

Bibliography

[1] C.T. Sah, R.N. Noyce, and W. Schockley, "Carrier generation and recombination in p–n junction and p–n junction characteristics," *Proc. IRE*, vol. 45, p. 1228, 1957.

[2] H.S. Chen, S.S. Li, R.M. Fox, and W.A. Krull, "Increased junction breakdown voltages in silicon-on-insulator diodes," *IEEE Trans. Electron Devices*, vol. 36, p. 488, 1989.

[3] D.J. Sandiford, "Heat treatment centers and bulk currents in silicon p–n junctions," *J. Appl. Phys.*, vol. 30, p. 1981, 1959.

[4] P.U. Calzolar and M. Graffi, "A theoretical investigation on the generation current in silicon p–n junctions under reverse bias," *Solid-State Electron.*, vol. 15, p. 1003, 1972.

[5] S. M. Sze, *Physics of Semiconductor Devices*, Wiley, New York, 1981.

[6] D.P. Kennedy and R.R. O'Brien, "On the measurement of impurity atom distributions by the differential capacitance technique," *IBM J. Res. Dev.*, vol. 13, p. 212, 1969.

[7] A. Broniatowski, A Blosse, P.C. Srivastava, and J.C. Bourgoin, "Transient capacitance measurements on resistive samples," *J. Appl. Phys.*, vol. 54, p. 2907, 1983.

[8] J.D. Wiley and G.L. Miller, "Series resistance effects in semiconductor CV profiling," *IEEE Electron Devices*, vol. 22, p. 265, 1975.

[9] D.K. Schroder, "The concept of generation and recombination lifetimes in semiconductors," *IEEE Trans. Electron Devices*, vol. 29, p. 1336, 1982.

[10] L. Jastrzebski and A.C. Ipri, "The effect of 1300–1380 °C anneal temperatures and material contamination on the characteristics of CMOS/SIMOX devices," *IEEE Electron Device Lett.*, vol. 9, p. 151, 1988.

[11] F.T. Brady, S.S. Li, and W.A. Krull, " C–V and C–t analysis of buried oxide layers formed by high-dose oxygen implantation," *J. Electron. Mat.*, vol. 18, p. 385, 1989.

[12] A.S. Grove and D.J. Fitzgerald, "Surface effects on p–n junctions: characteristics of surface space-charge regions under non-equilibrium conditions," *Solid-State Electron.*, vol. 9, p. 783, 1966.

[13] R.F. Pierret, "The gate-controlled diode: measurement and steady-state lateral current flow in deeply depleted MOS structures," *Solid-State Electron.*, vol. 17, p. 1257, 1974.

[14] J. Oualid, J. Dugas, R. Jérisian, A. Bouassis, D. Labrunie, and J.M. Mirabel, "Interfacial generation of N–MOSFETs implanted or natural, with simple or double diffused drain," *J. Physique III*, vol. 1, p. 369, 1991.

[15] A. Acovic, M. Dutoit, and M. Ilegems, "Characterization of hot-electron-stressed MOSFETs by low-temperature measurements of the drain tunnel diode current," *IEEE Trans. Electron Devices*, vol. 37, p. 1467, 1990.

[16] P. Speckbacher, A. Asenov, M. Bollu, F. Koch, and W. Weber, "Hot-carrier-induced deep-level defects from gated-diode measurements on MOSFETs," *IEEE Electron Device Lett.*, vol. 11, p. 95, 1990.

[17] H. Seghir, S. Cristoloveanu, R. Jérisian, J. Oualid, and A.J. Auberton–Hervé, "Correlation of the leakage current and charge pumping in silicon on insulator gate–controlled diodes," *IEEE Trans. Electron Devices*, vol. 40, p. 1104, 1993.

[18] J.M. Moragues, E. Ciantar, R. Jérisian, J. Oualid, and S. Cristoloveanu, "Coupling between the front and back interfaces in gate–controlled P^+PN^+ diodes on silicon on insulator," *Microelectron. J.*, vol. 25, p. 307, 1994.

[19] H.K. Lim and J.C. Fossum, "Threshold voltage of thin-film silicon-on-insulator (SOI) MOSFETs," *IEEE Trans. Electron Devices*, vol. 30, p. 1244, 1983.

[20] D.V. Lang, "Deep-level transient spectroscopy: a new method to characterize traps in semiconductors," *J. Appl. Phys.*, vol. 45, p. 3023, 1974.

[21] D.V. Lang, "Fast capacitance transient apparatus: application to ZnO and O centers in GaP p–n junctions," *J. Appl. Phys.*, vol. 45, p. 3014, 1974.

[22] G.L. Miller, D.V. Lang, and L.C. Kimerling, "Capacitance transient spectroscopy," in *Annual Review of Material Science*, R.A. Huggins, R.H. Bube, and R.W. Roberts (eds.), Annual Reviews, Palo Alto, vol. 7, p. 377, 1977.

Chapter 8

MOS Transistor Characteristics

MOS transistor characteristics represent an invaluable and easy-to-access source of information on the intrinsic properties of SOI structures as well as on the process-induced defects. In this chapter, attention will be given to the most important phenomena and parameters of SOI–MOS transistors. We will focus on the ohmic region of operation. It is clear that the standard expressions existing for bulk-Si–MOSFETs apply, without any major modification, to SOI transistors, provided that they have a contact with the Si film (five-terminal devices) and are fabricated in non-fully depleted films (relatively thick or highly doped). Although the back gate-bias acts as an extra experimental parameter, its practical influence on the operation of partially depleted MOSFETs is rather limited.

In contrast, totally new relations had to be derived for fully depleted MOSFETs, whose behavior is more complex because it depends on both gate biases. SOI advocates keep saying that this represents a flexibility and richness in SOI devices, since the parameters of the SOI structure as well as the region of operation can be adjusted for ideal transistor performance. On the other hand, the detractors argue that the advantage of the back-gate bias for actual circuit operation is a *fully depleted* illusion. Also, they are more than reticent about the utilization of ad hoc transistor relationships that are totally absent from conventional semiconductor textbooks.[1]

---

[1]Will this argument still be acceptable once this book is published? Perhaps we should wait for a second edition!

At any rate, in case of interface coupling effects, a distinction will be made between the parameters of the front (subscript 1) and back (subscript 2) interfaces and gates. An absence of subscripts indicates that the relations are irrespective of the interface and can be applied equally for both channels.

8.1 Interface Coupling

The dual-gate configuration of SOI–MOS systems offers multiple options about how to bias, independently or simultaneously, the two interfaces. In thick, partially depleted films, triggering both channels at once results in the total current being simply the parallel combination of the two individual contributions. However, the *intrinsic* properties of the front channel are not modified at all by the back-gate bias.

The meaning of *interface coupling* is much beyond this simple context of superposed effects. It applies to fully depleted SOI devices, where the pure parameters (threshold voltage, transconductance, interface-trap response, etc.) of one channel are insidiously affected by the opposite gate voltage.

Most experimental results strikingly vary if the opposite interface is biased in accumulation or in inversion. However, the very subtle consequences of interface coupling only appear when at least one of the two interfaces is depleted. For instance, modifying the back-gate bias *directly* changes the back-surface potential and *indirectly* changes the apparent properties measured at the front interface; reciprocally, since the front interface potential was altered, this in turn will cause an additional shift of the back-channel potential. In this feedback mechanism, the surface potential is controlled not only by the adjacent gate but also by the opposite one. Interface coupling also means that front-gate measurements are all reminiscent of the bias and quality of the buried interface and oxide.

Weak interface coupling occurs in films thicker than one depletion region and thinner than the sum of the two depletion zones ($w_{d_{1,2}} < t_{si} < w_{d_1} + w_{d_2}$). More exciting is the case of *strong* coupling, where one gate is enough to deplete the whole film ($t_{si} < w_{d_{1,2}}$). How intense the coupling is depends on the doping and thickness of the Si overlay. For those applications in which the interface coupling effects are undesirable, the solution is to overimplant the back channel or to design a body contact.

Only simple (although deeply illuminating) cases can be modeled analytically and these will be detailed in the forthcoming sections. Computers are asked to deal with more complex situations. Listed below are practical consequences of interface coupling that can be used for tuning the device performance:

- decrease of the threshold voltage as the opposite gate moves from accumulation to inversion,
- sharpest subthreshold slope achieved for depletion at the opposite interface,
- distortion of the transconductance curve when the opposite interface is inverted,
- excess signals induced by the opposite interface in various characterization techniques (charge pumping, noise, etc.).

A more fundamental consequence of interface coupling is illustrated in Figure 8.1. In bulk-Si MOSFETs, the potential bending at the interface results in carrier confinement to a "triangular" potential well (Fig. 8.1a). The quantum energy levels are separated enough for 2–D quantum magnetotransport effects to be observable at very low temperatures. The situation definitely differs in ultrathin SOI films, where the potential is almost *flat* if both interfaces are biased in inversion, just above threshold (Fig. 8.1b).

In fact, the film potential is a parabolic function of distance. The potential reduction in the middle of the film, as compared to the surface potential, depends linearly on doping and quadratically on film thickness.[1] For example, in SOI films thinner than 100 nm and undoped ($< 10^{15}\,\mathrm{cm}^{-3}$), the potential variation is smaller than 5 meV (Fig. 8.1d). As a result, the minority carriers are no longer confined to a potential well but rather to a spatial (geometric) box that is actually the film itself, delimited by the two interfaces. In other words, inversion at the two interfaces automatically initiates *volume inversion* in the whole film.[2]

The consistency of this new concept of volume inversion has been confirmed by quantum mechanical calculations.[1] Coupling the Poisson and Schrödinger equations proves that the energy separation of the lowest subbands is extremely narrow (0.1 meV) in low-doped SOI films 100 nm thick. According to Figure 8.2, in normal SOI films (thicker than 50 nm),

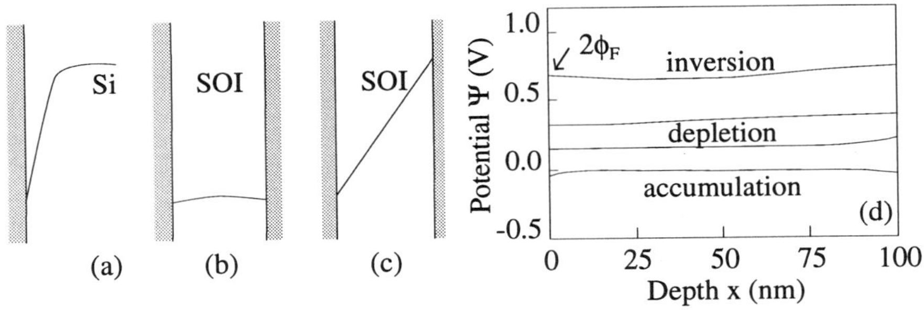

Figure 8.1 Schematic representation of the conduction band bending: (a) potential well in bulk Si or partially depleted SOI MOSFETs, (b) "geometrical box" in a totally inverted thin SOI film, and (c) geometry-limited potential well in a thin SOI film with the front and back interfaces biased in inversion and accumulation, respectively. (d) Potential profiles in an SOI film driven from volume accumulation to volume inversion by the simultaneous biasing of the two gates.

tens of subbands are occupied above 10–20 K and several subbands remain occupied even below 1 K. With the exception of 10–20-nm-thick films, which only recently became manufacturable by SIMOX technology, the density of states and the transport properties correspond to a 3–D system rather than to a 2–D gas. The arrows in Figure 8.2 indicate the extent of the low-temperature domain where quantum phenomena may be expected in very thin films.

The flexibility of thin dual-gate SOI MOSFETs is very attractive for studying quantum effects. If the back-interface bias is varied from inversion into accumulation, the electric field, essentially defined by the film thickness ($E \simeq 2\phi_F/t_{si}$), becomes significant. A triangular potential well is formed (Fig. 8.1c), in which quantization may occur, leading to 2–D quantum effects similar to those observed in bulk-Si transistors. An exciting experiment would be to monitor the transition from 3–D to 2–D systems, in a single device, just by playing with the back-gate bias.

Measurements performed at high magnetic field (B = 0–17 T) and 4 K did show large quantum conductance oscillations, $G(B)$ and $G(V_{G_1})$, at the front channel of moderately thin and doped SIMOX MOSFETs (Fig. 3.18b).[3] Unfortunately, the inferior crystal quality of the buried interface prevented such large oscillations from emerging at the back

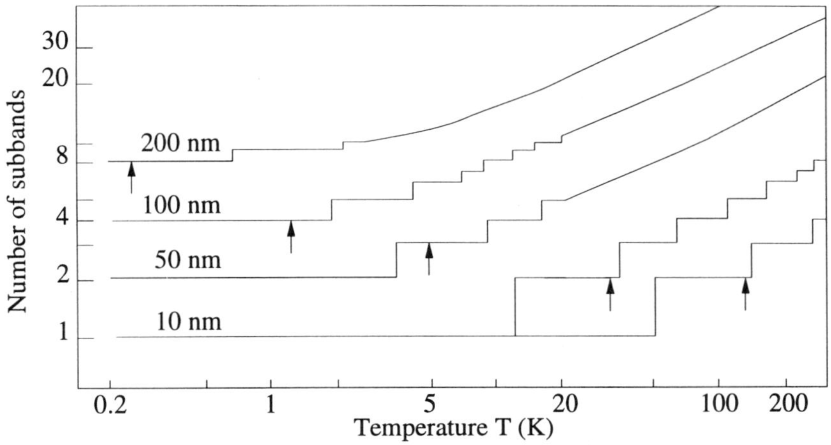

Figure 8.2 The number of occupied subbands in SOI films totally inverted ($N_{inv} = 10^{17}\,\text{cm}^{-3}$) for various temperatures and thicknesses. A subband is considered to be occupied if its population exceeds 10 % of the first subband population. The arrows indicate the equivalent temperature corresponding to the energy separation between the first two subbands (after Cristoloveanu and Ioannou[1]).

channel. No evidence has been obtained so far on the transition from 2–D to 3–D systems, probably because the interface coupling has been too weak.

There is good reason to believe that the concept of volume inversion—currently a property belonging solely to the SOI device family—is a basis for improved performance. The increased number of minority carriers allow higher currents to flow, whereas the influence of interface defects and related scattering mechanisms is attenuated, thereby offering higher mobilities for "volume" minority carriers.[2] Precursors of such devices and encouraging preliminary experimental results have been presented in Section 3.8.1.

8.2 Floating-Body Effects

The major parasitic effects in SOI–MOSFETs are related to the build-up of a film charge originated from impact ionization. This charge cannot be removed rapidly enough, primarily because no contact with the Si film

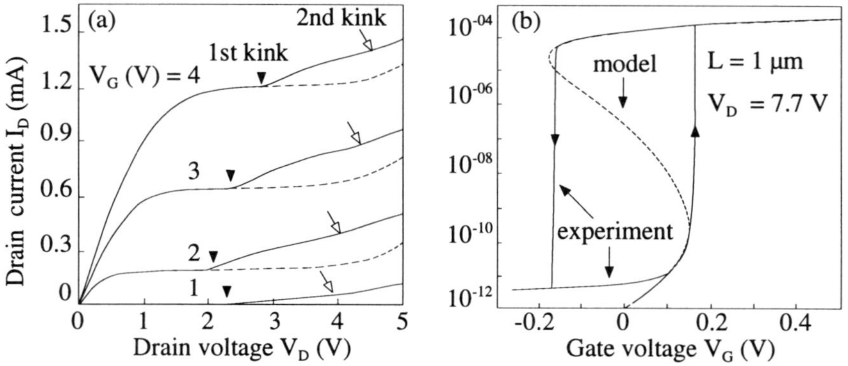

Figure 8.3 Experimental $I_D(V_D)$ and $I_D(V_G)$ characteristics in SOI MOSFETs showing (a) the kink effect (——), which may be canceled by grounding the body (- - -) and (b) the hysteresis induced by the negative transconductance region.

(body) is available. There are various consequences of this charge, which are generically referred to as *floating-body effects*: kink effect (Fig. 8.3a), negative conductance and transconductance, hysteresis (Fig. 8.3b) and instabilities, single transistor latch (Fig. 8.4a), bipolar transistor action, premature breakdown, etc.

These typical SOI effects can actually be observed, even in bulk-Si MOSFETs, if the substrate contact is left floating or at low temperature where the carriers freeze out and the substrate becomes semi-insulating. The significance of floating-body effects is accentuated in SOI by the small area of the source junction that restricts the body current flow and the charge removal.

8.2.1 Kink Effect

The kink effect denotes an abrupt increase in the saturation current of partially depleted enhancement-mode SOI MOSFETs operated in strong inversion (Fig. 8.3a). In the saturation region, channel carriers generate electron-hole pairs via impact ionization near the drain terminal. While the minority carriers are collected by the drain, the majority carriers travel towards the source and give rise to a gradual charge storage in the body. The concomitant increase in the body potential induces the lowering of both the threshold voltage and the source-potential barrier. More minority carriers will be able to flow from the source to the channel, thereby causing an excess drain current and producing many more pairs through the avalanche process. This positive feedback results in a sudden increase in I_D or "kink" in $I_D(V_D)$ characteristics (Fig. 8.3a).

The extent of the kink is limited by the enhanced recombination of accumulated carriers, which acts as negative feedback on the current. Beyond the kink, the current increases quadratically with V_D.[4] Eventually, the forward biasing of the body-to-source junction becomes sufficient to turn on the lateral bipolar transistor, giving rise in short-channel MOSFETs to a *second kink*. However, the first kink effect does not rely on the bipolar transistor action.

Measurements performed on five-terminal devices confirm that impact ionization is the common origin for the kink effect and the body current I_B, which is nothing but the substrate current in bulk-Si MOSFETs. If the body contact is left floating, the kink shows up for the same value of V_D as that initiating an abrupt increase in I_B (monitored by reconnecting the body).[5] Augmenting the gate voltage slows down the impact ionization process and delays the onset of the kink (Fig. 8.3a). In SOI films with a high carrier lifetime, the kink is sharper and appears at a lower drain voltage.[4]

The kink effect leads to improved current and transconductance values. It is not really harmful for digital applications. For analog circuit design, the kink must be carefully modeled and included in Spice–like simulators.[6] A kink-related overshoot of the generation–recombination noise ($1/f^2$ spectrum) has been observed for cryogenic operation of SOI transistors.[7] The kink effect may be suppressed by the use of a body contact (dotted lines in Fig. 8.3a) or by heavily damaging the carrier lifetime.

Fully depleted MOSFETs are naturally free from kink effect (Fig. 8.7). Since the potential barrier between source and body is smaller in a depleted film, the majority carriers apparently can penetrate more easily into the source, and excess carrier accumulation is prevented. However, according to Choi and Fossum,[8] the rise of the body potential is in fact accentuated in thinner depleted films because the lateral field is stronger and the recombination rate is constrained to smaller volumes. The disappearance of the kink is simply explained by the insensitivity of the depletion charge (whose expansion is already restricted by the film thickness) to any body-potential increase. Unlike the case of partially

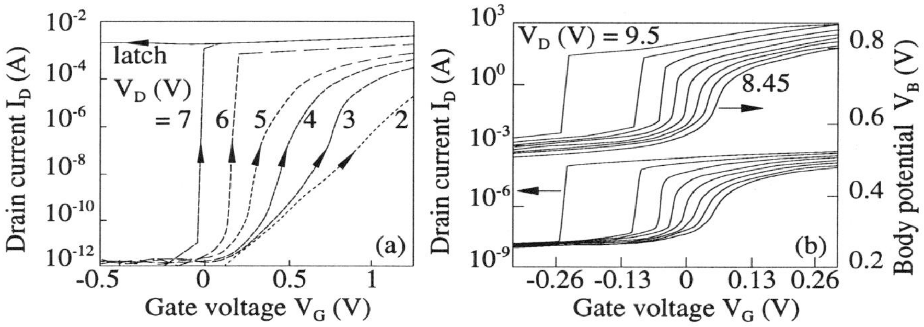

Figure 8.4 (a) Experimental $I_D(V_G)$ curves illustrating the subthreshold slope steepening and the single transistor latch. (b) Simultaneous variations of the drain current and floating-body potential versus gate voltage measured in a five-terminal SOI MOSFET (after Ouisse[12]).

depleted MOSFETs, the threshold voltage remains independent of V_B as long as the amount of stored charge is reasonable.

8.2.2 Hysteresis and Latch

In weak inversion, not only is the lateral field stronger for a given V_D but also the drain current depends exponentially on the body potential. The regenerative physical process, initiated when the body floats, now results in more detrimental consequences. For high drain voltages, the subthreshold characteristics $I_D(V_G)$ become S–shaped. This *snapback* effect is reflected experimentally by a *hysteresis* with anomalously sharp subthreshold slopes (Figs. 8.3b and 8.4a). Distinct current jumps are recorded for the two directions of the voltage scan, and their separation is governed by the drain-voltage magnitude. As shown in Figure 8.4a, for a higher V_D and a decreasing V_G sweep, the transistor may fail to turn off even for $V_G \leq 0$. Such a complete loss of gate control is called *single transistor latch.*

These effects can be modeled by considering simple analytical relations for the physical mechanisms involved in the positive feedback loop. The body current I_B is expressed by

$$I_B = (M - 1)I_{ch} \tag{8.1}$$

where I_{ch} is the channel current without impact ionization ($I_D = MI_{ch}$), and the multiplication factor in the saturation region ($V_D > V_{D_{sat}}$) is usually given by[9]

$$M - 1 = A_i(V_D - V_{D_{sat}}) \exp\left(-\frac{l_p B_i}{V_D - V_{D_{sat}}}\right) \tag{8.2}$$

with the ionization coefficients $A_i \simeq 1.2$ and $B_i \simeq 0.4 \div 2.8\,\text{MV/cm}$. The effective length of the pinch-off region, $l_p = 0.22(1.6t_{si})^{1/2}(t_{ox})^{1/3}$, is smaller in thin SOI films than in bulk Si, leading to higher electric-field peaks $E_{max} = (V_D - V_{D_{sat}})/l_p$. It is noted that in deep-submicron transistors the multiplication factor is computed far more accurately from the kinetic energy or effective temperature of carriers (via nonlocal transport models) rather than from the electric-field peak.[10]

The film potential, V_B, is governed by the body current and the properties of the source–film junction

$$V_B = \frac{nkT}{q} \log\left(\frac{I_B}{I_0} + 1\right) \tag{8.3}$$

where I_0 is the reverse current and n the ideality factor of the diode.

In partially depleted MOSFETs, the rise in V_B induces, via the lowering of the threshold voltage ($V_T \sim \sqrt{2\phi_F - V_B}$), a sharp increase in the drain current. We first neglect the bipolar transistor activation by focusing on the case of long channels (partially depleted enhancement-mode n-MOSFETs).

The key to the success of the model is to associate an MOS channel current relation valid in all regions of operation, since the rise in body potential can drive the transistor from weak inversion above threshold.[11] Once the set of equations is completed, it is possible to calculate, for any couple (I_D, V_D), the corresponding values of I_B, V_B, and V_G, and to reproduce the experimental characteristics $I_D(V_G)$ or $V_B(V_G)$ of Figure 8.4. There are two types of instabilities:

- For high drain voltages, the snapback effect allows the transconductance to take negative values (Fig. 8.3b).
- Reciprocally, if the gate voltage is fixed, the $I_D(V_D)$ curves are S–shaped too, with negative conductances (see Fig. 8.6).

A unique feature, pointed out by Ouisse et al.,[11] is that the negative values of the transconductance and conductance are proportional

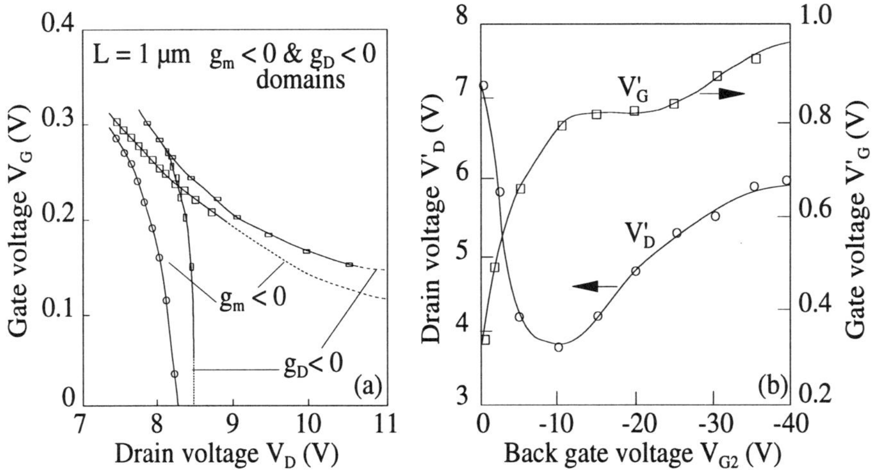

Figure 8.5 (a) Domains of negative conductance and transconductance measured on a 1 μm n-channel SIMOX MOSFET and (b) influence of the back-gate voltage on the limiting biases ($V_D > V_D'$, $V_G < V_G'$) that enable hysteresis to occur (after Ouisse et al.[11]).

and emerge simultaneously. The demonstration of this "floating-body theorem" proceeds from a quite rigorous small-signal analysis, based only on the generic expressions $I_D(V_G - V_T)$, $I_B(V_D, V_B)$, and $V_B(I_B)$, which are free from the assumptions involved in the explicit Equations (8.1–8.3).

A relevant experimental support is shown in Figure 8.5a: the two domains of negative values for conductance and transconductance are almost superposed,[2] in spite of various fluctuations that affect the accurate recording of a hysteresis. It is clearly seen in Figure 8.5b that the limiting biases $V_{G,D}'$, which trigger the onset of the hysteresis, are very sensitive to the back-gate bias.

In normal measurements, when the drain or gate voltage is scanned, S–shaped characteristics are not directly observable, although the onset of a hysteresis is fully reminiscent of their presence. Figure 8.3b shows such an experimental hysteresis, together with the fitting S–curve computed using the above model. Complementary experiments, performed on five-terminal SIMOX MOSFETs, provided the parameters needed for the simulation: impact ionization coefficients, reverse current I_0, and ideality factor n of the source–body diode. Moreover, strikingly parallel variations were measured for the current and body potential as V_G was increased (Fig. 8.4b). The $V_B(V_G)$ characteristics exhibit a negative slope region (i.e., hysteresis and abrupt jumps), and then a "plateau" strongly dependent on V_D.

These curves are fully confirmed by simulations, demonstrating that it is the sharp increase in V_B that triggers the current transition from a low to a high level. For a given surface potential, the negative transconductance occurs when the body potential exceeds a minimum value V'_B:[12]

$$V'_B \simeq \psi_s - \frac{\epsilon_s q N_A}{2} \left(\frac{n-1}{C_{ox}} \right)^2 \tag{8.4}$$

The question now is whether S–shaped $I_D(V_D)$ characteristics can be experimentally accessed. The solution is to impose the current and V_G and then measure the corresponding drain voltage. This works because, in the S–region, a precise V_D bias corresponds to each current value, whereas for a given V_D there are three possible currents. Caution is needed, however, to avoid spontaneous current oscillations brought about by subsisting parasitic capacitors connected to the negative resistor; the use of a load resistor on the drain might sometimes be helpful.

Compared in Figure 8.6 are $I_D(V_D)$ characteristics measured by imposing either the voltage or the current. They respectively exhibit typical hysteresis or S–type features. It is more difficult to visualize the snapback of the transconductance. In fact, the $I_D(V_G)$ curves need to be reconstructed from the experimental set of diagrams $V_D(V_G)$, measured for each value of I_D.

A fair concordance between model and experiments was also found for the variation of the threshold voltage with V_D. When the body charging becomes prevailing, V_T decreases far more rapidly with V_D than would be expected from standard relations (derived for transistors that are free from floating-body effects but are subject to drain-induced barrier lowering (DIBL)), so an ad hoc value of V_T must be defined.

Various groups have proposed more or less similar approaches to predict the snapback effect. Huang et al. demonstrate that if the body

---

[2]The region $g_m < 0$ was determined by drawing, for each bias V_D, the two V_G values corresponding to the jump points of the hysteresis in $I_D(V_G)$ characteristics. Symmetrically, the region $g_D < 0$ is defined by measuring, for each V_G, the two V_D values leading to current jumps in $I_D(V_D)$ curves.

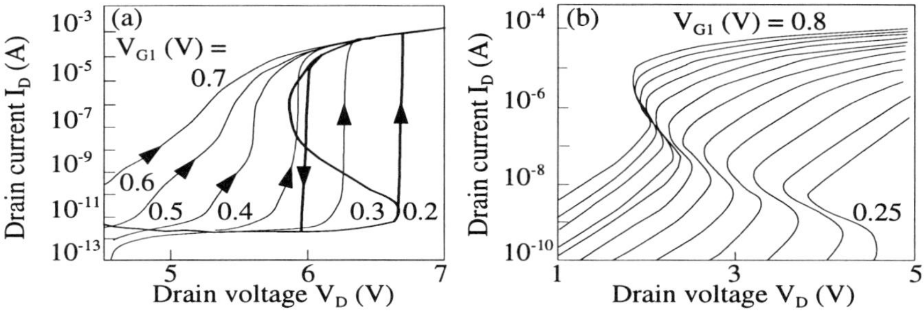

Figure 8.6 Experimental $I_D(V_D)$ characteristics in 1 μm MOSFETs measured by either (a) increasing the drain voltage (except for the three bold curves) or (b) by imposing the current (after Ouisse et al.[11]).

or contact resistances are large, an attenuated snapback subsists even if the body is tied to the source.[13] They alleviate the difficulty of formulating a continuous current equation from weak to strong inversion by selecting a different expression of the threshold voltage according to the amount of stored charge. The transistor latch, observed for reverse V_G scan in Figure 8.4a, is explained by the fast increase of the lateral field in weak inversion, which overwhelms the gate-induced drop of the carrier concentration.

Starting from universal charge-sheet model, Matloubian et al. have established an equivalent electric circuit.[14] They successfully simulate the steepening of the subthreshold slope, illustrated in Figure 8.4a, as well as the threshold voltage lowering for high drain bias and positive gate voltage sweep.

In short-channel MOSFETs, the floating-body effects are exacerbated by the activation of the bipolar transistor, which will be discussed next.

8.2.3 Bipolar Transistor Action

When the source-to-body junction is on, the lateral bipolar transistor comes into play by injecting additional minority carriers from the source (emitter) into the body (base). Conventionally, the gain of a long bipolar transistor, $\beta \simeq 2(L_n/L)^2 - 1$, depends on channel length and diffusion length L_n, but is rather independent of V_G.[15]

The collector current (βI_B) contributes to the drain current and augments the impact ionization

$$I_B = (M-1)(I_{ch} + \beta I_B) = \frac{M-1}{1-\beta(M-1)} I_{ch} \qquad (8.5)$$

$$I_D = M(I_{ch} + \beta I_B) = \frac{M}{1-\beta(M-1)} I_{ch} \qquad (8.6)$$

The above equation depicts the amplification of the channel current by impact ionization and bipolar action. Breakdown occurs when the drain voltage becomes high enough to satisfy the condition

$$\beta \times (M-1) = 1 \qquad (8.7)$$

This relation demonstrates that bipolar-transistor action is fully responsible for the premature breakdown in SOI as compared to bulk-Si MOSFETs. The feedback mechanism is self-limiting because an increased body potential leads to a higher saturation voltage, which in turn lowers the peak of the lateral field and $(M-1)$.

Refinements to the above first-order model are necessary for computing the actual 2–D profiles of the surface and body potentials. They account for (1) distinct multiplication factors for MOS channel and bipolar body currents, (2) depth-dependent ionization coefficients, (3) bipolar-gain dependence on gate or body bias and channel length, (4) bandgap narrowing, which reduces the predicted bipolar gain, (5) direct band-to-band Auger recombination, etc. Plain device simulations confirm that the SOI breakdown voltage, BV, is well below that of the isolated drain junction. Neither does it originate from punch-through, since for $M = 1$ a normal breakdown voltage is recovered.[8]

In *partially depleted* transistors, BV is improved by a factor proportional to $\beta^{1/n}$ (with $3 < n < 6$) simply by grounding the body contact.[15] This is consistent with the ratio between the common emitter breakdown voltages, BV_{CBO} and BV_{CEO} (measured with the base grounded or open), given by the bipolar transistor theory.

Even though no kink is discernible, a substantial bipolar action occurs in *fully depleted* SOI. This is because the source-to-body barrier potential (0.2 eV) is lower than in thicker films; hence, it is easier to turn on the source junction. Moreover, the rapid increase of the multiplication factor with V_D prevails over the exponential decrease of β.[3]

---

[3]Under high-injection conditions, the recombination rate is enhanced and offsets the base-current increase; hence, $\beta \propto \exp(-qV_{BE}/2kT)$.[8]

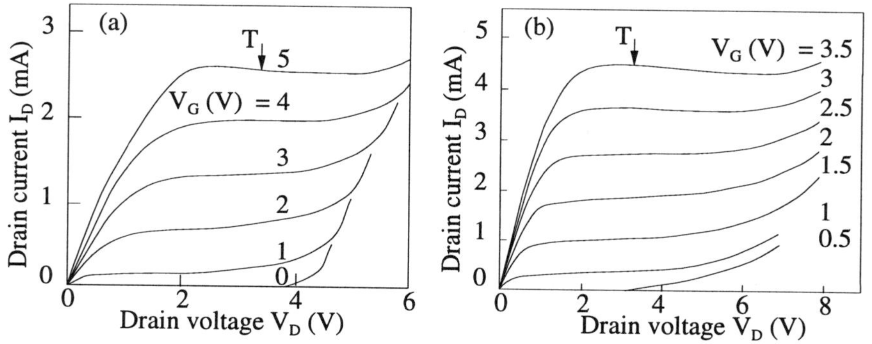

Figure 8.7 Current–voltage characteristics in (a) 150-nm-thick enhancement-mode and (b) 80-nm-thick accumulation-mode fully depleted SIMOX transistors ($L = 1\,\mu$m). The arrows indicate the negative output conductance induced by the rise in body temperature (after Armstrong et al.[16] and Faynot et al.[17]).

Figure 8.7 shows that BV degrades with decreasing V_G, which is opposite to the behavior of bulk Si or thick SOI MOSFETs.[16, 17] To understand this difference, we have to examine separately the two conditions defining the breakdown: (1) source junction turn-on and (2) avalanche multiplication.

For thin SOI MOSFETs operated in a strong inversion and saturation region, the source junction is on before Equation (8.7) is fulfilled (Fig. 8.8a). Avalanche multiplication acts as a limiting mechanism for breakdown, which is experimentally confirmed in Figure 8.7 by rather parallel variations of the saturation and breakdown voltages. This is why, at low V_G when $(M - 1)$ is large, premature breakdown occurs for relatively low drain voltage. In contrast, the limiting process in bulk Si is the source junction turn-on. Therefore, a much stronger inversion is required for a given V_D bias, or a higher drain voltage is needed at moderate V_G.[18]

Another special feature of thin SOI MOSFETs is that, beyond the breakdown voltage, the current increases more gently with V_D than in bulk Si, being controlled by the space-charge region, which covers the whole film.

Illustrated in Figure 8.9a is BV degradation in ultrathin films and for

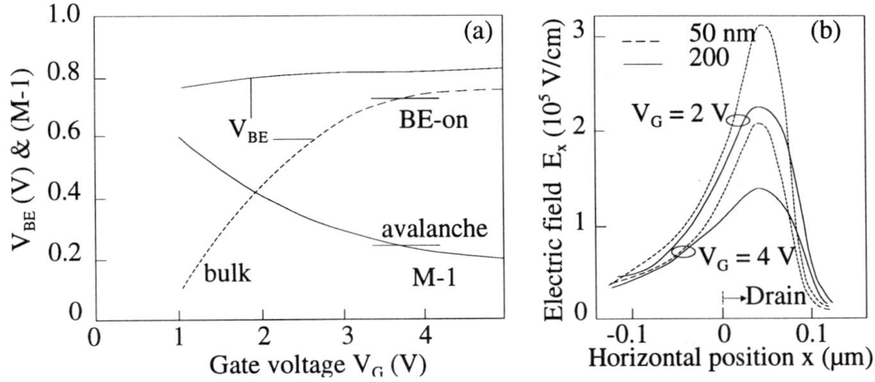

Figure 8.8 (a) Body-to-source potential V_{BE} and current multiplication factor $(M-1)$ versus gate voltage in enhancement-mode transistors ($L = 2\,\mu$m, after Young and Burns[18]). (b) Lateral profile of electric field for various film thicknesses and gate voltages (after Yoshimi et al.[19]).

decreasing V_G.[4] These properties are explained in Figure 8.8b by the rise in the multiplication factor,[9] brought about by a stronger lateral field.[19] Fortunately, β decreases in thinner films primarily as a result of an increased forward bias of the source junction, which implies a higher injection level. Also, lifetime is likely to be poorer in thinner films, which further attenuates the premature breakdown.

In submicron devices, both $(M-1)$ and β are higher, producing an even more drastic BV lowering (Fig. 8.9b). It is clear that $(M-1)$ increases because carrier velocity saturation occurs and diminishes $V_{D_{sat}}$, whereas β is enhanced by the base narrowing.

A typical situation for thin SOI films takes place in the subthreshold region where, for a given bias V_D, the lateral field is very large, since $V_{D_{sat}} \leq 0.1$ V. Here, the source turn-on condition is not satisfied naturally. But, when $\beta(M-1)$ reaches 1, the current increases and the junction is turned on. Since β rises, a smaller multiplication factor, and hence a lower drain voltage, is needed to sustain the avalanche condition. A snapback happens in the breakdown region. No such regenerative breakdown is detectable when the source junction is on before avalanche multiplication occurs (for example, in very short channels) and β decreases with V_D.[9, 20]

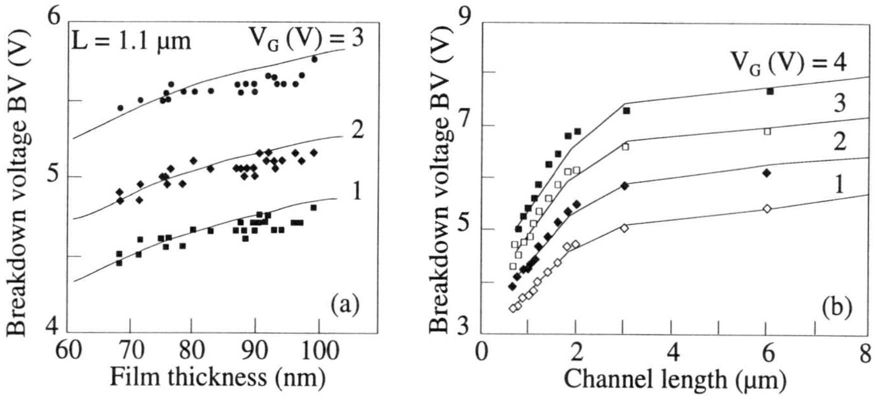

Figure 8.9 Measured and simulated values of the breakdown voltage versus (a) film thickness and (b) channel length in enhancement-mode SIMOX transistors (after Chen et al.[9]).

As discussed in the previous section, a hysteresis-related latch state is experienced when the transistor cannot be turned off by reducing V_G. This drawback is reinforced by the carrier injection from the bipolar transistor, which opposes the gate effort to eliminate channel carriers.

Even more bothersome is the latch occurring for the forward V_G scan. Since there is no channel current for $V_G < 0$, the bipolar transistor here is triggered by subsisting parasitic currents, which play the same role as I_{ch} in Equations (8.5) and (8.6). Bipolar transistor carriers cause regenerative action when the impact ionization current exceeds the leakage current. Two sources of leakage have been incriminated: back-channel current due to drain-induced barrier lowering (DIBL) at the back interface, and thermal generation of electron-hole pairs in the drain junction space charge. As V_G becomes more negative, the gate-induced drain leakage (GIDL) increases and lowers the breakdown voltage.[9]

The maximum applicable drain voltage is therefore dictated by the subthreshold region where premature breakdown and latch take place, resulting in prohibitive standby current and power consumption. The *holding* voltage, defined as the minimum V_D value for which the device is still turned on at $V_G = 0$, is indicated by a dot in Figure 8.10a.

Bias conditions corresponding to the single transistor latch ($V_G \simeq 0$, $V_D > BV$) cause a rapid device aging. Many defects are generated at the front and back interfaces near the drain junction. This results in a locally damaged carrier lifetime, which is reflected by an *improvement* of the breakdown voltage for the reverse mode of operation.[21] When the back interface is selectively degraded, by means of deliberate hot electron injection from the back channel, no modification of the snapback behavior is observed.[11] Therefore, it is the quality of the film above the interface, and not the back interface itself, which seems to be the important factor.

Summarized below are the roles played by the main device parameters. Only by considering their separate contributions as well as their interactions can the floating-body effects be fully understood and alleviated.

- *Doping*—The lower the doping is, the more attenuated the kink and hysteresis effects will be. The peak electric field and impact ionization rate are reduced, the reverse current of the source junction is increased, and interface coupling is enhanced. Diminishing the doping tenfold increases by several orders of magnitude the body current needed to sustain the body potential that triggers the snapback.[11] Note, however, that the bipolar action is reinforced since β increases for lower doping.

- *Gate oxide*—The thinner the oxide is, the better the gate control and smaller the floating-body influence will be (see Eq. (8.4)).

- *Source-to-body junction*—Optimized diodes have low ideality factors that delay the onset of the hysteresis (Eq. (8.4)). The bipolar action (i.e., β coefficient) can be reduced by maximizing the recombination rate in the source. Effective solutions are the source silicidation[22] or the tailoring of a "lightly doped source" (LDS): a highly doped surface region, enabling small series resistances, over a lower-doped ($10^{17}\,\mathrm{cm}^{-3}$) volume.[8]

- *Drain engineering*—The doping and configuration of the spacer are the most flexible parameters that the design engineer can play with in order to reduce the longitudinal electric field. Breakdown voltage improves with reduced LDD dose at the expense of higher series resistances and lower current drivability.[16] An asymmetrical

[4] Experimentally, BV is taken to be the value of V_D for which the current slope is $dI_D/dV_D = 0.05(W/L)\,\mathrm{mS}$.

transistor with a thin body and a thickened drain region has been demonstrated to yield a higher BV via the reduction of both E_{max} and $(M-1)$.[19]

- *Film thickness*—In thinner fully depleted films, contrasting trends of $(M-1)$ and β lead to overall weaker holding and breakdown voltages (Fig. 8.9a). This inconvenience is balanced by a better control of short-channel effects, so a design trade-off is suitable.

- *Channel length*—According to Figure 8.9b, more detrimental floating-body effects are expected in deep-submicron transistors.

- *Carrier lifetime*—The poorer the lifetime is, the less annoying the floating-body effects will be. Lifetime killing does not represent a realistic solution, not only because it causes excessive leakage but also because it runs counter to the effort produced to improve SOI quality.

- *Back-gate bias*—The limiting value V_D' of the drain voltage, above which a hysteresis appears (Fig. 8.5b), can vary by several volts. It is a minimum when V_{G_2} is biased in weak accumulation and saturates in strong inversion. The limiting front-gate voltage V_G', below which instabilities show up, increases by about 0.6 V from depletion to weak inversion at the back interface and saturates in strong inversion.[11] It is speculated that the growth of a back accumulation layer modifies the effective doping, reverse current, and ideality factor of the source–body junction. Also, for high V_D, interface coupling occurs, even in partially depleted MOSFETs, in the drain controlled depletion region. In fully depleted MOSFETs, accumulation at the back interface can lower the breakdown voltage and reinstate the kink effect.[18]

- *Body ties*—Body contact design is the straightforward way to cancel floating body effects in partially depleted SOI, since they provide a path for majority carriers to be evacuated. The body can be tied to the source or independently biased (five-terminal MOSFETs). Body ties may be T-, H-, or Japanese-letter-shaped (see also Fig. 8.14).[23, 24] A "smart" contact has a low resistance in the subthreshold region to keep the transistor off and a very high resistance in strong inversion, thus preserving the advantage of an increased drive current induced by the floating body. A more exotic solution

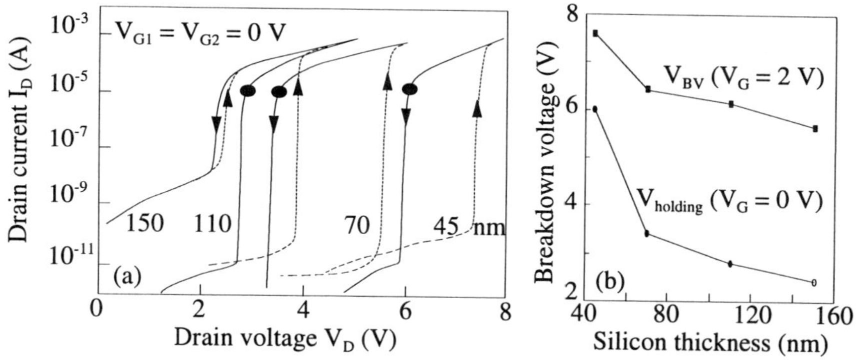

Figure 8.10 (a) Hysteresis in $I_D(V_D)$ characteristics. (b) Variations of holding voltage (defined by dots in Figure 8.10a) and breakdown voltage versus film thickness in accumulation-mode SIMOX transistors ($L = 1 \mu m$, after Faynot et al.[17]).

is to use a discontinuous buried-oxide, available by masked oxygen implantation or ZMR. A small aperture in the oxide, located beside the active channel region, provides a vertical body contact to the underlying substrate. The transistors are kink and snapback free and exhibit improved breakdown.[25] The common-gate twin-MOSFET structure, described in Section 3.8 (Fig. 3.20a), is also effective in deleting floating-body effects.[26] The penalty incurred by these devices is either to hamper integration density and design simplicity, or to require a customized synthesis of SOI wafers.

A totally different approach consists of fabricating ultrathin accumulation-mode MOSFETs. These transistors are attractive because they are kink free (Fig. 8.7b) and offer better threshold control and higher current. Although n^+–n–n^+ structures a priori do not accommodate any bipolar action, there is still snapback in the subthreshold region (Fig. 8.10a). Indeed, a hole inversion layer, whose density is controlled by the body potential rise, is built up and plays the role of the base.[27] The regenerative feedback results from combined effects of this quasi-bipolar transistor, back-channel leakage, and impact ionization. Figure 8.10b illustrates an outstanding merit of depletion-mode MOSFETs: the breakdown and holding voltages are both dramatically improved as the films become thinner.[17]

8.2.4 SOI and Phase Transitions

Further analysis of fundamental physical interest has demonstrated that the floating-body instabilities in SOI transistors belong to the family of cusp catastrophes[28] and can be ascribed to the phase transition theory. When crossing the boundary of the (V_G, V_D) domain of Figure 8.5a, a *critical curve* is defined as the transition from the characteristics with $g_m > 0$ to those with $g_m < 0$. In a *critical point* on this critical curve, g_m and g_D become infinite. Several external parameters $(V_D, V_{G_1}, V_{G_2}, T)$ are candidates for controlling the phase transition, their influence being summarized by a *critical exponent.*

The fascinating feature is that the experimentally determined critical exponents for floating-body effects are surprisingly close to those predicted by theory in the general context of phase transitions in thermodynamics.[11] To some extent, an analogy between SOI MOSFETs and Van der Waals gases can be made. A similar example in microelectronics was offered by the unijunction transistor.[29] The following expression was found for the critical curve in SOI transistors:[11]

$$K_1 \left(\log \frac{I_D}{I_{Dc}}\right)^3 - K_2 \frac{V_D - V_{Dc}}{V_{Dc}} \log \frac{I_D}{I_{Dc}} = V_G - V_{Gc} \qquad (8.8)$$

where $K_{1,2}$ are numerical constants, and the subscript c indicates the values corresponding to the critical point.

It is speculated that the position of the critical point represents a signature of floating-body effects and can be used as a criterion for technology optimization. Moreover, operating the device close to the critical point makes it highly sensitive to any external parameter (temperature, pressure, magnetic field, etc.) and opens broad microsensor prospects. Yet another application is to take advantage of the negative conductance of the transistor in order to generate oscillations.[11]

The bottom line is that phase transitions, oscillators, and other microtransducers should not distract us from the fact that floating-body effects still represent a severe problem in SOI. The simple models presented above do not pretend to provide more than a guideline on the physical mechanisms to be incorporated into accurate 2–D simulators, and on the cautious experiments that must be conducted to validate computed data for actual technologies. A few trends emerge from this discussion: (1) reduced $(M-1)$ by proper drain design, (2) lower β, for instance by the use of thinner films, (3) increased recombination rate by source engineering. However, most likely the floating-body drawbacks will disappear quite naturally by bias scaling, which is unavoidable in the context of ULSI microelectronics.

8.3 Transients Effects

8.3.1 Bias-Induced Effects

In addition to the above-cited floating-body effects, the total isolation of the transistor body sometimes is responsible for long transients. Once the body is charged by impact ionization or gate voltage pulses, a long relaxation time is needed to recombine the charge and reach equilibrium. Conversely, if a prompt supply of holes is demanded in an n-channel MOSFET, there is no solution other than the generation mechanism to create them.

The amplitude of transient effects is, therefore, governed by the carrier lifetime in the material. This family of effects will be taken advantage of, for characterization purposes, in Section 9.5. Our immediate concern is to identify the incidental transients and to find out how to avoid them under normal conditions of circuit operation.

The terminology *transient* and *static* implies that no interaction could occur between these two types of experiments. However, the so-called "static" characteristics are, in general, measured with automated systems. The data acquisition is much faster than the generation–recombination time constants in good-quality SOI films. Therefore, $I_D(V_G)$ curves can be completely misunderstood if the transients effects are not properly accounted for.

In a partially depleted *depletion-mode* n^+–n–n^+ MOSFET, the *ohmic* $I_D(V_G)$ characteristics may depend on the direction of voltage scanning. If V_G is varied from accumulation to inversion (negative sweep—dotted curve), small deep-depletion regions are formed at each bias step. Instead of obtaining a steady-state depletion region, the current continues to drop (Fig. 8.11a) and inaccurate curves are recorded, unless very long delay times are allowed between the measured points.

For this reason, a positive V_G sweep is normally used, but even so there is no certitude of accuracy. Figure 8.11a shows that the characteristics recorded *immediately* after a starting negative voltage was applied to the gate are typical for a totally depleted MOSFET. This is simply because a large, deep depletion region was initially formed and temporarily

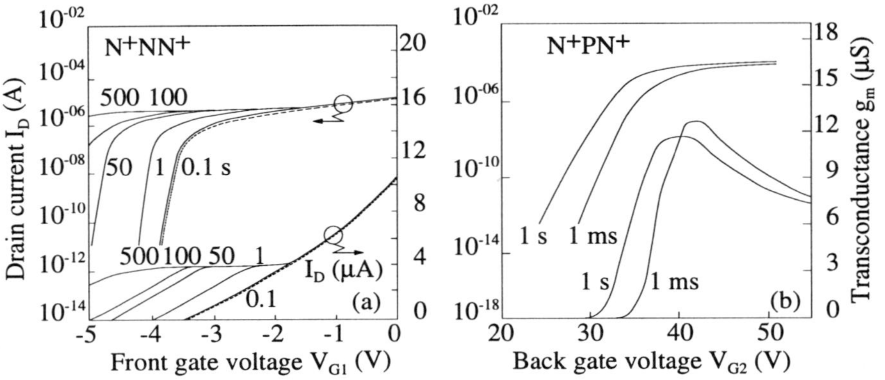

Figure 8.11 $I_D(V_G)$ characteristics showing the influence of (a) the hold time in accumulation-mode MOSFETs (only the dotted curve was scanned from positive to negative bias) and (b) the delay time in enhancement-mode partially depleted MOSFETs (after Cristoloveanu[30]).

covers the whole film. Indeed, there is no reservoir of holes to compensate promptly for it. If, once the bias was applied, a hold time[5] as long as 500 seconds is allowed before the curve is recorded, the minority carrier generation does succeed in restoring equilibrium: at this point, the transistor turns out to be partially depleted.

The confusion between a partially and a fully depleted transistor is terribly misleading. For instance, the doping of the film is straightforwardly underestimated. Subsequently, the use of the right resistivity value yields of course an overestimated mobility. Some enthusiastic reports in the past had the misfortune to claim exaggerated low levels of residual doping (contamination-related) and carrier mobilities underscoring those in bulk Si. Stress effects and other elegant theories were vainly invoked to explain the enhanced mobility in SOI. Nonetheless, these mistakes did not come from a particularly careless experiment. This example only demonstrates how dangerous may be the routine gained on bulk-Si MOSFETs, once applied to SOI.

There are symmetrical situations in *enhancement-mode* transistors, this time due to the absence of majority carrier reservoirs. The "static"

[5] Also called *expresso* time by coffee promoters.

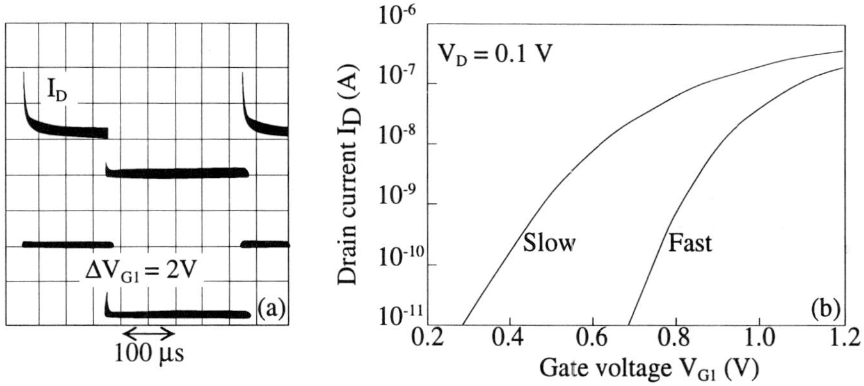

Figure 8.12 (a) Switching characteristics showing the drain current overshoot (after Kato and Taniguchi[31]). (b) Front-channel characteristics measured with slow (1 mV/s) and fast (100 mV/s) positive scan rates, after $V_{G_2} = -20\,\text{V}$ was applied to the back gate of fully depleted SOI transistors (after Assaderaghi et al.[33]).

characteristics of n–MOSFETs may be affected by the direction of voltage scanning, hold time, and delay time, as soon as a supply/removal of holes is necessary for maintaining equilibrium. Figure 8.11b shows that the subthreshold slope, threshold voltage, and transconductance depend on how patient the experimentalist was.

A drain current *overshoot* is observed in partially depleted SOI transistors when the gate is switched from accumulation to inversion. As illustrated in Figure 8.12a, the current at the turn-on stage may exceed twice the steady-state value. Time-domain simulations performed by Kato and Taniguchi[31] show that the holes expelled from the surface, when the depletion region forms, have no way to escape and temporarily keep up the body potential. Since the inversion charge and drain current respond very rapidly to any potential modification, the overshoot transient merely reflects the time needed for excess holes to recombine within the body or in the source region. The overshoot amplitude ΔI_D is proportional to the number of excess holes and therefore increases for a higher doping or for a stronger accumulation before turn-on.

Similarly, at turn-off, the potential recovery is dependent on the thermal carrier generation. If the off-time period is too short for equilibrium to be recovered, the overshoot induced by the following pulse is attenuated. Both ΔI_D and decay time are reduced when the drain is biased in the saturation region, since the forward biasing of the body-to-source junction makes easier the recollection of excess holes.

The overshoot is beneficial for the speed of integrated circuits, since the propagation delay is governed by the magnitude of transient current available to drive the following devices. Unfortunately, the overshoot acts as a parasitic effect in current–DLTS measurements, which therefore require five-terminal transistors. The overshoot, like the kink effect, is totally suppressed in fully depleted transistors, where the source potential barrier is lower and the threshold voltage is not sensitive to a small amount of excess holes.[32]

Transient effects become yet more complex when both gates are involved in the measurement. Figure 8.12b shows the case of a fully depleted transistor where the accuracy of front-channel $I_D(V_{G_1})$ curves, recorded for $V_{G_2} < 0$, depends on the delay allowed for the accumulation layer to form at the back interface.[33] Only if a slow scan is used will the subthreshold swing and threshold voltage take the values expected for an accumulated back interface.

Otherwise, for a fast V_G sweep, the swing is close to the theoretical limit of 60 mV/dec, thus indicating that the back interface is still depleted (see Eqs. (8.31) and (8.35)). Moreover, the lack of holes prevents the rapid formation of a screening layer at the back interface, so that the vertical field is higher and the body potential is more negative than at equilibrium.[6] As a result, the front-channel threshold voltage is temporarily increased.

This effect is more pronounced at liquid nitrogen temperature, where the transients become very long and give rise to metastable states of operation.[34] The body isolation is improved because the reverse current through the source junction vanishes according to a thermally activated process. In response to a negative bias on the back gate, the transistor switches to a quasi-permanent high V_T state. The "normal" low V_T state is not recovered until a large pulse is applied on the drain to trigger the hole generation by impact ionization. When the back gate is swept to zero, holes are rapidly expelled from the body through forward-biased junctions, and the initial situation is reset. The experiment is repeatedly

[6]Since the concept of *deep depletion* is meaningless in fully depleted films, their analysis must be conducted in terms of time-dependent surface and body potentials.

reproducible, which means that SOI devices may be used as multistable charge-controlled memory devices.

Similar transients affect the monitoring of hot-carrier-induced transistor degradation at low temperature.[35] During the device stress, many holes are generated by impact ionization. Even after the stress is interrupted, these holes succeed in keeping up the body potential in a metastable state. The extrapolated threshold voltage is underestimated and gives a totally inaccurate image of the extent of interface damage caused by hot carriers. Only after a long period of time or after some pulse maneuvering will the body be discharged and the high value of V_T recovered.

The bipolar transistor and kink effects are responsible for parasitic transients observed in the saturation region.[36] For example, when the drain and gate biases are up, the source is pulsed in the off-state. Rapid evacuation of the previously accumulated body charge through the source acts as a large base current, which is further amplified at the drain terminal by bipolar transistor action.

Different transients are induced by drain pulses applied for a fixed gate-to-source bias in the on-state. When V_D is increased (e.g., from 1 to 5 V), a transient is needed for impact ionization to generate enough carriers and restore equilibrium. Conversely, if V_D is reduced from 5 to 1 V, the time required for the body charge to be removed through the source depends on V_G and may exceed 10 μs. These saturation-mode transients are detrimental for the operation of integrated circuits (lost of precharge in microprocessors and glitches in SRAMs), but can be eliminated by using body ties.

It is obvious that the enumeration of transients effects would be much longer if the collective behavior of several interconnected devices was also considered. The extent to which these phenomena are bothersome or, on the contrary, offer opportunities for improved performance is not yet fully assessed. But even for single transistor operation, it is difficult to recommend universal conditions for reliable characterization. A prudent strategy would consist in checking independently, for each type of experiment and technology, the influence of delay and hold times. Most transient effects are governed by carrier generation–recombination mechanisms, whose rates may be quite different according to the quality of the junctions, and the front and back interfaces. When feasible, device illumination before and during the test improves the accuracy by rapidly adjusting the populations of electron/hole reservoirs.

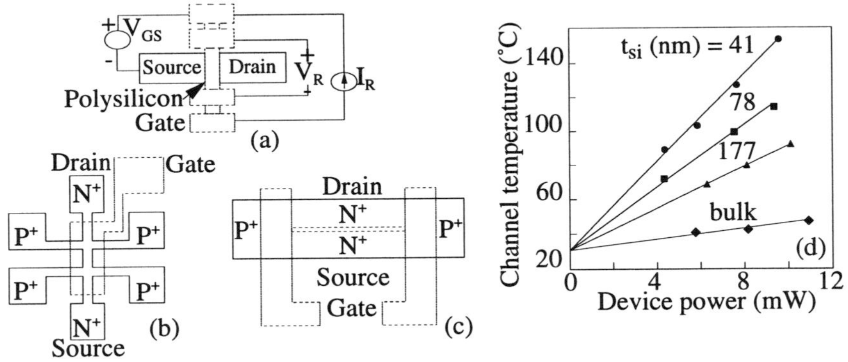

Figure 8.13 (a,b,c) Device configurations for temperature measurements (after Bunyan et al.[39]). (d) Channel temperature versus power in 0.3-μm-long SOI transistors with varying Si-film thickness (after Su et al.[38]).

8.3.2 Thermal Transient Effects

The thermal conductivity of SiO_2 layers that inherently surround any SOI device is about two orders of magnitude poorer than for Si. This inhibits the cooling of SOI transistors and develops self-heating problems. A *negative differential conductance* was observed for high gate voltages in Figure 8.7. As V_D increases in the saturation region, the channel temperature rises and causes the carrier mobility to drop as $\mu \sim T^{-1.5}$.

The rise of the silicon island temperature can be visualized by thermal imaging[37] or by monitoring the subsequent increase in junction leakage current. Figure 8.13 shows device structures especially patterned for in situ temperature measurements. In Figure 8.13a, the resistance of the poly-Si gate is used as a temperature sensor.[38] Preliminary calibration is made with a hot chuck.

Alternately, body contacts can be designed to fit the case of either long-channel (Fig. 8.13b) or short-channel (Fig. 8.13c) transistors. When an accumulation layer is formed at the back interface, its resistance serves as a body temperature monitor during front-channel operation. A more sensitive method is *noise thermometry*, which consists of measuring the thermal noise voltage of the accumulated back-channel resistance ($S_V = 4kTR$).[39] Probing different pairs of contacts shows that the entire island is heated almost uniformly.

The channel temperature is proportional to the power, $I_D V_D$, dissipated in the device. Temperature increases in excess of 100 °C were observed in submicron transistors under static conditions of operation (Fig. 8.13d). The reduction in the saturation current and the negative conductance value depend on the direction of the drain voltage and sweep rate. Systematic measurements show that self-heating is more severe for thinner films, shorter channels, and thicker buried-oxides. If the spacing between channel and metal contacts is shrunk, the device temperature is reduced, whereas the interconnects become warmer. This does not lead to electromigration problems which would adversely affect the metal reliability.

Thermal models have been developed in which the drain, source, gate, and interconnects are treated as cooling fins.[38, 40] Since most of the heat is evacuated through the buried-oxide, thin oxides look very attractive. Self-heating does not appear to limit the *dynamic* operation of ULSI integrated circuits. Although the temperature increase is larger, the loss in drain current drive is attenuated in ultrashort transistors. This is because the saturation velocity is less temperature dependent than the carrier mobility.

Surprisingly, negative conductance is still observed at very low temperatures, where mobility is no longer governed by acoustic phonon scattering and therefore follows a different temperature law.[41] This finding suggests that the consequences of self-heating are not restricted to mobility variations. A more complex mechanism takes place that is based on the interaction of several temperature-related parameters (mobility, threshold voltage, impact ionization, floating-body potential, leakage current, subthreshold swing, etc.).[42]

8.4 Edge Effects

Interest in SOI began with the perfect isolation of silicon islands, which is primarily achieved through local oxidation (LOCOS) or mesa-etching techniques (Figs. 8.14d,e). However, the lateral edges of SOI MOSFETs represent a parasitic conduction path between source and drain. This sidewall transistor operates in parallel with the main transistor, under control of the front and back gates. The sidewall threshold voltage is dictated by strong coupling and charge sharing between the front, back,

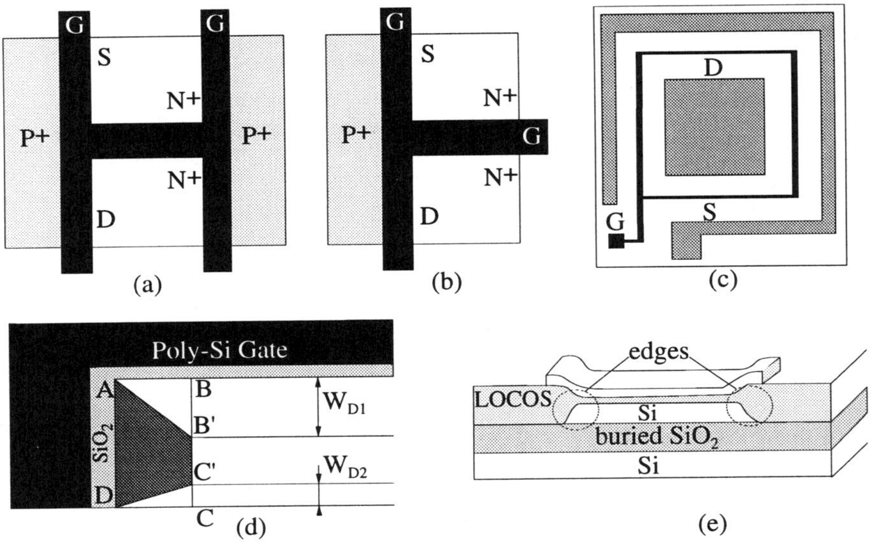

Figure 8.14 Various transistor configurations: (a) H–gate, (b) T–gate, (c) enclosed-edgeless geometry, (d) mesa-isolated, and (e) LOCOS–isolated.

and edge channels. It also accounts for the special topology and different crystalline orientation of the edges, which result in higher defect densities, thicker oxide, and variable film thickness.

The hump observed in the subthreshold curves of Figure 8.15a is representative of the early turn-on of the parasitic transistor. Intolerable consequences include excessive off-state leakage, double subthreshold slope, and lowered pseudothreshold voltage. The sidewall transistor has a very small width, defined by the film thickness, and does not perturb the strong inversion region, except in very narrow devices.

The obvious solution to mask the parasitic transistor consists in shifting its threshold voltage above that of the front channel. This is usually achieved by sidewall overdoping, which is effective as long as the channel doping of deep-submicron MOSFETs is not prohibitive yet. The edge conduction can also be screened by processing thicker-edge oxides, accumulating the back interface, or overdoping the back channel.

These solutions are not fully satisfactory, since they adversely affect the device sensitivity to radiation effects and cancel the benefit of

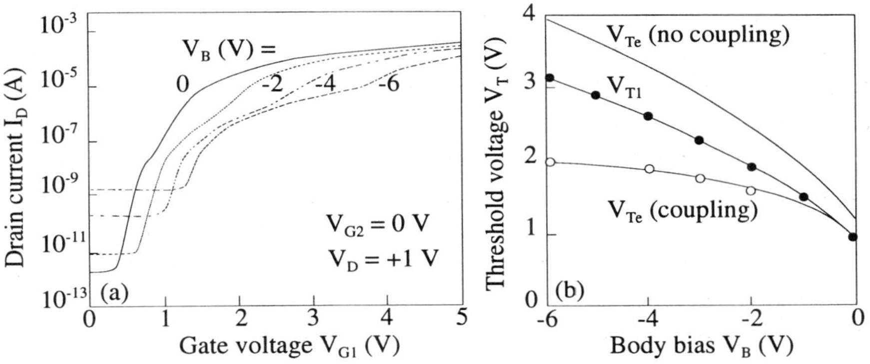

Figure 8.15 (a) Subthreshold characteristics showing the influence of the body bias on the edge component. (b) Sidewall V_{Te} and front-channel V_{T_1} threshold voltages versus body bias: measured points and simulated lines (after Matloubian et al.[44]).

sharp subthreshold slopes. Of course, special edgeless devices may be designed, but this is a space-consuming alternative. The H–gate transistor of Figure 8.14a has two p^+ body contacts that inhibit any conduction path along the sidewalls. It behaves similarly to the edgeless device with enclosed geometry of Figure 8.14c. The unique body contact of the T–gate transistor (Fig. 8.14b) is not efficient in suppressing the parasitic conduction along the opposite edge.

The specifics of the parasitic transistor and the sidewall properties are invaluable inputs for further technology and design optimization of SOI circuits.[43] The question is how to characterize them. The simplest method is to monitor the shift of the sidewall threshold voltage V_{Te}. In five-terminal transistors, the body can be biased to generate a set of $I_D(V_G)$ curves. When a hump is present in the subthreshold characteristics, it becomes possible to discriminate the threshold voltages of the sidewalls and front channel. The threshold voltage is here determined as the gate voltage corresponding to $0.1(W/L)\,\mu$A drain current. On the one hand, this *constant current* definition of V_T is equivalent to the standard value extrapolated from the linear $I_D(V_G)$ characteristics. On the other hand, it is more appropriate for separating the parallel contributions of the two channels, since the channels have very distinct effective widths.

In *mesa-isolated* transistors, the shift of the front-channel threshold voltage with increasing body bias is identical to that measured in bulk-Si or H–gate SOI MOSFETs (Fig. 8.15b). More surprising is the behavior of the sidewall threshold: in spite of a thicker oxide, the shift $V_{Te}(V_B)$ is smaller and tends to saturate.

These features have been explained by the charge sharing with the front and back gates due to overlapping fields at the mesa corners.[44] The schematic cross section of Figure 8.14d was used to determine the fraction F of the total depletion charge (rectangle ABCD) that is actually controlled by the edges (trapezoid AB'C'D): $F = 1 - (w_{d_1} + w_{d_2})/2t_{si}$. Such a first-order correction yields, for the edge threshold voltage,

$$V_{Te} = V_{FBe} + 2\phi_{Fe} + \frac{F}{C_{ox,e}}\sqrt{2\epsilon_{si} q N_{Ae}(2\phi_{Fe} - V_B)} \qquad (8.9)$$

This expression holds as long as the film remains partially depleted and the front and back depletion regions do not merge ($w_{d_1} + w_{d_2} < t_{si}$). For a larger body bias, more depletion charge at the mesa corners escapes the edge control, and the shift in V_{Te} is attenuated. A convincing fit of the experimental data is shown in Figure 8.15b.

Equation (8.9) is useful to evaluate V_{Te} for overdoped sidewalls, where the characteristics are normally hump-free. Only for selected body biases will the hump emerge and will V_{Te} be measurable. Inserting the measured values $V_{Te}(V_B)$ in Equation (8.9) allows the prediction of $V_{Te}(V_B \to 0)$. Moreover, the model remains reliable when the back-gate bias (i.e., w_{d_2}) is varied. This offers the possibility of determining V_{Te} even when no body contact is available and only V_{G_2} can be used to control the presence and position of the hump.

In Figure 8.16a are reproduced subthreshold characteristics of a fully depleted transistor where the mesa hump shows up when the back interface is being accumulated. Although Equation (8.9) does not apply anymore, the concept of charge sharing and residual conduction at the mesa corners can still be invoked to explain the slow shift in V_{Te}. According to another scenario, a very large density of defects may subsist on the edges.[45] This causes the pinning of the Fermi level, which in turn blocks the variation of V_{Te}.

Unfortunately, the above models fail to accommodate the case of *LOCOS–isolated* SOI transistors, which exhibit contrasting characteristics. Figure 8.16b shows that the hump is present for zero back-gate bias, but disappears when the back interface becomes accumulated. While the

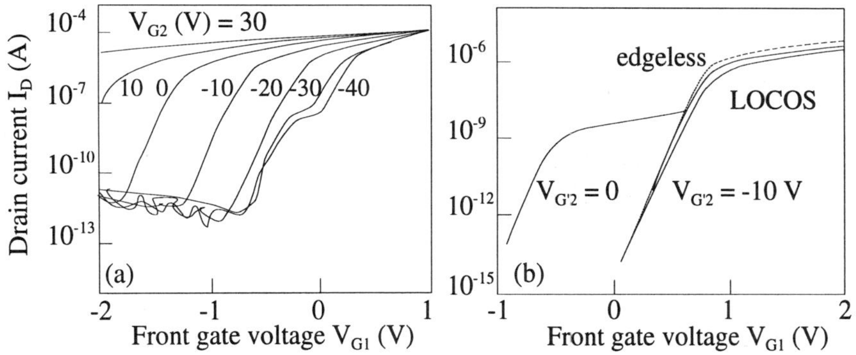

Figure 8.16 Edge-affected front-channel characteristics as a function of back-gate bias in (a) mesa-isolated and (b) LOCOS–isolated SOI MOSFETs (after Mazhari et al.[45] and Elewa et al.[17]).

front-channel threshold voltage is blocked, V_{Te} shifts very rapidly with V_{G_2}. This is because the body is partially depleted, whereas the LOCOS edges are thinner and become totally depleted, thus sensing the back-gate bias. To account for the particular configuration of LOCOS edges (Fig. 8.14e), instead of a single-edge transistor as in Equation (8.9), one has to consider a continuum of parallel transistors that gradually have thinner bodies and thicker gate oxides. According to the back-gate bias, more or less edge transistors are activated.[17]

At this point, a more thorough evaluation of the properties and crystalline quality of the sidewalls becomes unavoidable. This evaluation is feasible by selectively taking aim at the edges with accurate interface characterization techniques. For example, when the transistor is biased in the hump region ($V_{G_1} = -0.7\,\text{V}$ and $V_{G_2} = 0$ in Figure 8.16b), the low-frequency $1/f^2$ noise spectrum (see Section 9.3) is fully evocative of the nature of edge traps. The sidewall noise can either be "extinguished" by accumulating the back interface or mixed with the front-channel noise by varying $V_{G_{1,2}}$.

Other small-signal measurements (dynamic transconductance, for instance) can be carried out in the hump region to segregate the edge properties. In contrast, if large signals are applied on the gate, as for charge pumping, DLTS, or leakage current measurements, the edge and front-channel interfaces are probed simultaneously, and their responses are superposed.[17] The edge contribution may be sorted out by using simple models or by direct comparison with edgeless devices (see also Fig. 9.13).

Complementary data are available by reversing the experiments and looking at the body edges from the back. For characterization purposes, the influence of the sidewalls can voluntarily be reinforced by processing multiedge test devices. They consist of several parallel channels, located in different islands and governed by an unique gate.

Experiments conducted in SIMOX transistors have confirmed that the complex structure of LOCOS sidewalls can be summarized by a continuum of flat-band and threshold voltages. The vertical profile of edge defects is inhomogeneous, increasing by one order of magnitude from the front interface towards the back interface.[17] This profile is reminiscent of the difference in quality between the two oxide interfaces. Also identified were slow edge traps, situated near the back interface close to the source and drain junctions.

An interesting point is that edge defects are still active and can be resolved by charge pumping and noise, even for bias conditions that suppress the subthreshold hump and make the edge current undetectable. It is easy to predict that the processing of SOI circuits will put more emphasis on the design, characterization, and modeling of the sidewalls. ULSI criteria will also define the isolation process:[43] while mesa isolation is very simple and naturally adapted to ultrathin-film MOSFETs, LOCOS is more reliable for thicker films.

8.5 Threshold Voltage

8.5.1 Lim and Fossum Model

The pioneering model developed by Lim and Fossum[47] applies to the case of *fully depleted* enhancement-mode n-channel MOSFETs. It is based on several hypotheses, namely, long channel, low drain bias, constant doping, charge sheet approximation (i.e., negligible thicknesses of inversion and accumulation layers), and no influence from the third interface located below the buried-oxide. Integration of the Poisson equation over the Si film yields

$$E_{s_1} - E_{s_2} = \frac{qN_A t_{si}}{\epsilon_{si}}; \qquad \psi_{s_1} - \psi_{s_2} = t_{si}E_{s_1} - \frac{qN_A t_{si}^2}{2\epsilon_{si}} \tag{8.10}$$

where $E_{s_{1,2}}$ and $\psi_{s_{1,2}}$ stand for the vertical fields and potentials (defined by assuming a virtual neutral point in the film) at the front and back interfaces. The relations between gate voltages and surface potentials are

$$V_{G_{1,2}} = \psi_{s_{1,2}} + \psi_{ox_{1,2}} + \phi_{ms_{1,2}} \tag{8.11}$$

where $\phi_{ms_{1,2}}$ are metal–semiconductor work functions. The potential drops in the oxides $\psi_{ox_{1,2}}$ are obtained by applying Gauss' theorem at both interfaces:

$$\psi_{ox_1} = \frac{1}{C_{ox_1}} \left(\epsilon_{si} E_{s_1} - Q_{f_1} - Q_{c_1} + q D_{it_1} \psi_{s_1} \right) \tag{8.12}$$

$$\psi_{ox_2} = -\frac{1}{C_{ox_2}} \left(\epsilon_{si} E_{s_1} - q N_A t_{si} + Q_{f_2} + Q_{c_2} - q D_{it_2} \psi_{s_2} \right) \tag{8.13}$$

with $Q_{f_{1,2}}$ being fixed charges in the oxides, $Q_{c_{1,2}}$ surface channel charges, and $D_{it_{1,2}}$ average concentrations of interface traps. Eliminating $\psi_{ox_{1,2}}$ and $E_{s_{1,2}}$ in Equations (8.10)–(8.13), we find

$$V_{G_{1,2}} = \phi_{fb_{1,2}} + \psi_{s_{1,2}} \left(1 + \frac{C_{si} + C_{it_{1,2}}}{C_{ox_{1,2}}} \right) - \psi_{s_{2,1}} \frac{C_{si}}{C_{ox_{1,2}}} - \frac{2Q_{c_{1,2}} + Q_{si}}{2C_{ox_{1,2}}} \tag{8.14}$$

where $Q_{si} = -qN_A t_{si}$ is the depletion charge, $C_{si} = \epsilon_{si}/t_{si}$ is the depleted-film capacitance, and $C_{it_{1,2}} = qD_{it_{1,2}}$ are interface-trap capacitances. The flat-band *potentials* $\phi_{fb_{1,2}}$ are defined as in bulk Si:

$$\phi_{fb_{1,2}} = \phi_{ms_{1,2}} - Q_{f_{1,2}}/C_{ox_{1,2}} \tag{8.15}$$

Distinction must be made in SOI between $\phi_{fb_{1,2}}$ and the actual flat-band *voltages* $V_{FB_{1,2}}$ that strictly correspond to zero surface potentials and define the turn-on of accumulation layers.

The standard definition of the front-channel threshold voltage, V_{T_1}, implies that $\psi_{s_1} = 2\phi_F$ and $Q_{inv_1} = 0$ in Equation (8.14).[7] There are three generic cases according to the back-gate bias:

Case 1: Back interface accumulated ($\psi_{s_2} = 0$)

$$V_{T_1}^{acc} = \phi_{fb_1} + \left(1 + \frac{C_{si} + C_{it_1}}{C_{ox_1}} \right) 2\phi_F - \frac{Q_{si}}{2C_{ox_1}} \tag{8.16}$$

[7] A detailed analysis of these conditions, carried out by Mazhari and Ioannou,[48] reveals that the surface potential at threshold may be higher than $2\phi_F$, especially in low-doped films.

Case 2: Back interface inverted ($\psi_{s_2} = 2\phi_F$)

$$V_{T_1}^{inv} = \phi_{fb_1} + \left(1 + \frac{C_{it_1}}{C_{ox_1}}\right) 2\phi_F - \frac{Q_{si}}{2C_{ox_1}} \tag{8.17}$$

Case 3: Back interface depleted ($0 < \psi_{s_2} < 2\phi_F$)

Eliminating ψ_{s_2} in Equations (8.14) yields

$$V_{T_1}^{dep} = V_{T_1}^{acc} - \frac{C_{si}C_{ox_2}}{C_{ox_1}(C_{ox_2} + C_{si} + C_{it_2})}(V_{G_2} - V_{G_2}^{acc}) \tag{8.18}$$

$V_{G_2}^{acc}$ represents the limiting voltage below which (i.e., for $V_{G_2} < V_{G_2}^{acc}$) the back interface is always accumulated, whatever the values of V_{G_1}. It is obtained from Equation (8.14) with $\psi_{s_2} = 0$ and $\psi_{s_1} = 2\phi_F$:

$$V_{G_2}^{acc} = \phi_{fb_2} - \frac{C_{si}}{C_{ox_2}} 2\phi_F - \frac{Q_{si}}{2C_{ox_2}} \tag{8.19}$$

Similarly, the back interface is definitely inverted for $V_{G_2} > V_{G_2}^{inv}$, where the boundary voltage $V_{G_2}^{inv}$ is nothing but the back-channel threshold voltage $V_{T_2}^{inv}$:

$$V_{G_2}^{inv} = V_{T_2}^{inv} = \phi_{fb_2} + \left(1 + \frac{C_{it_2}}{C_{ox_2}}\right) 2\phi_F - \frac{Q_{si}}{2C_{ox_2}} \tag{8.20}$$

The meaning of the various expressions derived above is illustrated in Figure 8.17. The front-channel threshold decreases linearly between two plateaus corresponding to accumulation and inversion at the back interface. The difference between the plateaus, $\Delta V_{T_1} = (C_{si}/C_{ox_1})2\phi_F$, is slightly dependent on doping, whereas the slope of $V_{T_1}(V_{G_2})$ curves is not. None of these parameters depends on the front-interface quality, but all of them can be controlled by adjusting the thicknesses of the SOI structure. In ultrathin films, the slope is merely given by the ratio t_{ox_1}/t_{ox_2}. Conversely, $V_{T_1}(V_{G_2})$ characteristics are useful to determine the film and buried-oxide thicknesses as well as the density of back-interface traps (Section 9.7).

We must stress the polyvalence of Equations (8.16)–(8.18) as compared to the simpler case of partially depleted films, where the threshold

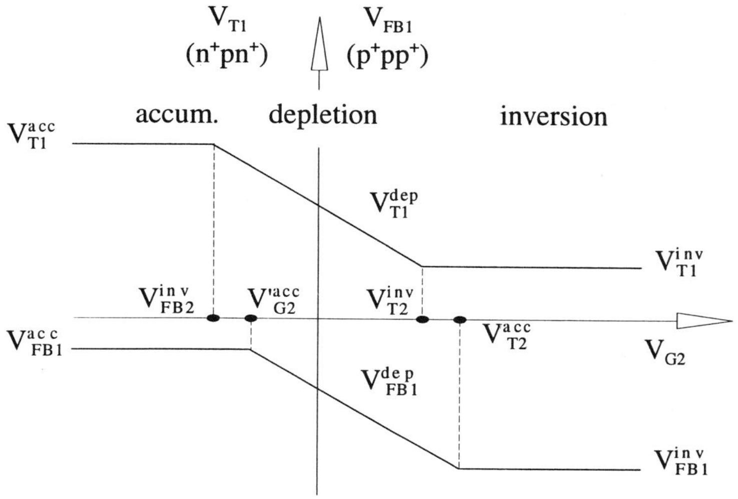

Figure 8.17 Schematic influence of back-gate bias on the front-channel threshold voltage of enhancement-mode n-channel MOSFETs (n^+–p–n^+) and flat-band voltage of accumulation-mode p-channel MOSFETs (p^+–p–p^+) in fully depleted SOI MOSFETs.

voltage is insensitive to V_{G_2} bias and takes the same form as in bulk-Si MOSFETs:

$$V_{T_1} = \phi_{fb_1} + \left(1 + \frac{C_{it_1}}{C_{ox_1}}\right) 2\phi_F + \frac{\sqrt{4q\epsilon_{si}N_A\phi_F}}{C_{ox_1}} \tag{8.21}$$

It is clear that symmetrical expressions, obtained by interchanging the subscripts 1 and 2, hold for the back-channel threshold voltage. The extension to p-channel SOI–MOSFETs is also straightforward. Different relationships have been derived for the special case of *intermediate* thicknesses ($w_{max} < t_{si} < 2w_{max}$), where full depletion cannot be achieved by single-gate action, but becomes effective when front and back depletion regions merge.[47]

8.5.2 Depletion-Mode Transistors

The "threshold" of depletion-mode (or accumulation-mode) MOSFETs is defined as the onset of the accumulation layer and basically corresponds to the flat-band voltage V_{FB_1}. Since in SOI the bands are anything but flat, the meaning of V_{FB_1} is more relaxed and must be understood as the gate voltage leading to zero potential at the front interface ($\psi_{s_1} = 0$). The treatment of the problem follows the same guideline as above. Using similar notations and Equation (8.14), we have for p$^+$–p–p$^+$ transistors

$$V_{FB_1}^{inv} = \phi_{fb_1} - \frac{C_{si}}{C_{ox1}} 2\phi_F - \frac{Q_{si}}{2C_{ox1}} \tag{8.22}$$

$$V_{FB_1}^{dep} = V_{FB_1}^{inv} - \frac{C_{si}C_{ox2}}{C_{ox1}(C_{ox2} + C_{si} + C_{it_2})} \left(V_{G_2} - V_{T_2}^{acc}\right) \tag{8.23}$$

for $V_{G_2} < V_{T_2}^{acc}$. Here, $V_{T_2}^{acc}$ is given by the reciprocal of Equation (8.16) and represents the minimum value V_{G_2} for which the back interface is always inverted (even if the front interface is accumulated). Also, the comparison of Equations (8.19) and (8.22) confirms the intuitive identity $V_{G_2}^{acc} = V_{FB_2}^{inv}$.

Accumulation at the back interface is the unique situation in which the energy bands are really flat in the film. The flat-band voltage now coincides with the standard flat-band potential:

$$V_{FB_1}^{acc} = \phi_{fb_1} \tag{8.24}$$

It is interesting to note that the variation of $V_{FB_1}^{dep}$ with V_{G_2} is parallel with that of the threshold voltage $V_{T_1}^{dep}(V_{G_2})$. However, the regions where these parameters can be modulated by the back-gate bias do not superpose: $V_{G_2}^{\prime acc} < V_{G_2} < V_{T_2}^{acc}$ for $V_{FB_1}^{dep}$, and $V_{FB_2}^{inv} < V_{G_2} < V_{G_2}^{inv}$ for $V_{T_1}^{dep}$, where

$$V_{G_2}^{\prime acc} = \phi_{fb_2} - \frac{C_{ox2} + 2C_{si} + C_{it_2}}{2C_{si}C_{ox2}} Q_{si} \tag{8.25}$$

Equations (8.22) to (8.24) hold directly for the threshold of very thin[8] accumulation-mode p-channel MOSFETs. They can easily be adapted to n$^+$–n–n$^+$ MOSFETs and to back-channel transistors.

8.5.3 Experimental Results

The threshold voltage is by no means the most valuable monitor for any technological development. In fully depleted transistors, the threshold

[8] In thicker films, as V_{G_2} becomes more negative, the depletion region gradually shrinks, starting from the back interface. The apparent transistor threshold coincides with the onset of the body current, which is no longer negligible and leads to a quadratic dependence $V_{FB_1}^{dep}(V_{G_2})$.[49].

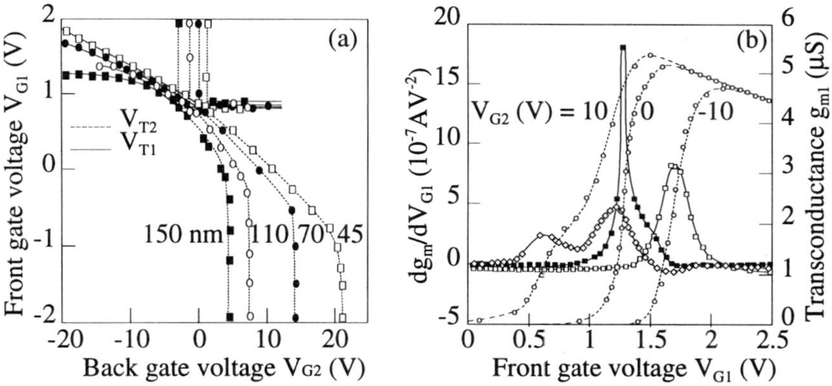

Figure 8.18 (a) Front-channel and back-channel threshold voltages versus opposite-gate bias measured in accumulation-mode SIMOX MOSFETs with various thicknesses (after Faynot et al.[17]). (b) Front-channel threshold voltage evaluation from the transconductance inflection point (second-order current derivative) for various back-gate biases (after Ionescu et al.[52]).

voltage is not restricted to a unique value and must therefore be measured as a function of the back-gate bias.

Figure 8.18a shows typical "dual–V_T" curves, where the front and back threshold voltages are plotted against opposite gate biases. In agreement with the above model, we notice that the thinner the film, the wider the V_T variation range. The linear decrease of $V_{T_1}(V_{G_2})$ from accumulation to inversion at the back interface may be used to extract the thicknesses of the film (from ΔV_{T_1}) and buried oxide (from the slope). Additional information is found from the parameters $V_{G_{1,2}}^{acc}$ and $V_{G_{1,2}}^{inv}$. Apparently similar curves are obtained in fully depleted enhancement-mode and accumulation-mode MOSFETs. In fact, in thicker accumulation-mode transistors where the volume conduction is significant, the decreasing region of $V_{T_{1,2}}(V_{G_{2,1}})$ curves may be composed of several distinct zones.[49]

At this point, it becomes necessary to discuss the efficiency of the various characterization methods that are frequently used to determine the threshold voltage.

$I_D(V_G)$ extrapolation — Once the $I_D(V_G)$ curve has been measured for

the ohmic region of operation, the tangent is drawn at the inflection point, where V_G corresponds to the maximum transconductance. The intercept of the tangent with the horizontal axis gives the "extrapolated" threshold voltage. The accuracy may be improved by repeating the measurement for several V_D values (from 10 to 50 mV) and reextrapolating to $V_D = 0$. The simplicity of this method has rendered it extremely popular for bulk Si and partially depleted SOI. Among the critical aspects, we note that, especially in fully depleted SOI, the extrapolated V_T suffers from series-resistance effects and, in addition, does not correspond to the usual definition based on the surface band bending ($\psi_s = 2\phi_F$).[48] On the practical side, when the back interface is being inverted, extrapolation to $I_D = 0$ leads to inaccurate results. The extrapolation must be actually done to zero front-channel current, but this is not obvious when a current flows at the back channel.

Constant-current V_T — The threshold voltage is here defined as the gate voltage allowing a certain amount of current (typically 0.1 μA per μm width for a 1-μm-long channel and $V_D = 50$ mV). This ad hoc definition has the merit of avoiding any extrapolation error and may be used to monitor very small V_T fluctuations or shifts (such as those induced by short time radiation or hot-carrier degradations). Attention is called to the extensive use of this method in fully depleted SOI, where totally erroneous results may be derived. When the back interface is inverted, a current larger than the reference value (0.1 μA/μm) flows in the device even if the front interface is still depleted. This appears to mean that the front V_T steadily decreases, instead of being constant, when the back interface becomes more and more inverted. Additionally, the current may decrease, even at constant V_T, as a consequence of a mobility degradation.

$I_D/\sqrt{g_m}(V_G)$ *extrapolation* — Drawing this curve in the ohmic region results in a straight line that intercepts the horizontal axis at $V_G = V_T$.[50] This method, discussed in more detail in Section 8.7.4, is less subject to series-resistance effects, and yields quite acceptable values in SOI.

Peak of d^2I_D/dV_G^2 — The peak of the second-order derivative of $I_D(V_G)$

curves, i.e., the inflection point of the transconductance,[9] occurs for a gate voltage that is close to $\psi_s = 2\phi_F$.[51] The method was recently validated by numerically calculating the derivatives of the inversion charge as a function of surface potential for various back-gate voltages and technological options.[52] As compared with the extrapolated V_T, the new method offers an improved accuracy that is fairly insensitive to series resistances and back-gate bias. Its reliability improves further in state-of-the-art devices (thinner gate oxides and higher doping levels). This technique is illustrated in Figure 8.18b for fully depleted MOSFETs. When the back interface is inverted, two peaks are observed; the primary peak corresponds to the front channel V_T and the secondary peak to the back-channel activation. The method also applies for depletion-mode MOSFETs, where several peaks may be observed, reflecting the occurrence of the various conduction regions.[53]

8.5.4 Developments

In short channels, the depletion charge (Fig. 8.19a) is divided into a central trapezoidal region controlled by the gates and two triangular portions governed by source and drain. The *effective* depletion charge associated with front- and back-gate action is

$$Q_{si}^{eff} = (1 - \frac{d}{L}) \times Q_{si} \tag{8.26}$$

where d accounts for the distance over which the back surface potential varies from the source/drain–body potential to the "normal" long-channel value ψ_{s_2}. As L is reduced, charge sharing is accentuated and more depletion charge escapes gate control. The corresponding drop ΔV_{T_1} of the threshold voltage is obtained by inserting Q_{si}^{eff} in Equations (8.16)–(8.19):

$$\Delta V_{T_1} = \frac{d}{L} \times \frac{qN_A t_{si}}{2C_{ox_1}} \tag{8.27}$$

The extension of the critical length d is evaluated by solving the two-dimensional Poisson equation.[54] It is strongly related to the magnitude of the lateral electric field at the back interface, which comprises

[9] A more rapid method of evaluation is to take half the transconductance peak, which roughly corresponds to the transconductance inflection point.

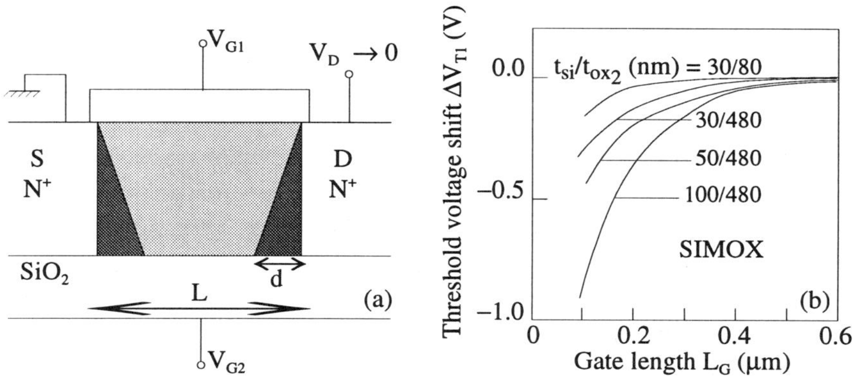

Figure 8.19 (a) Schematics of charge sharing in fully depleted SOI MOSFETs. (b) Experimental data and theoretical threshold voltage reduction versus gate length (after Ohmura et al.[55]).

fringing fields from the back gate through the buried oxide and junction-depletion-region fields. Increasing film doping and/or accumulating the back interface represent obvious solutions for reducing d and minimizing ΔV_{T_1}, but the trade-off is degraded values of the transconductance and saturation current. A far better strategy consists of using thinner films and buried insulators. According to Ohmura et al.,[55] d is proportional to $3t_{ox_2}^2/(t_{ox_2}^2 + L^2)$; hence, t_{ox_2} must be kept smaller than L. It is therefore suitable to include the buried-oxide thickness as an adjustable parameter in any device size-shrinking procedure.

Experiments are unanimous in proving that short-channel effects are much attenuated in fully depleted films as compared to bulk Si or partially depleted SOI, where drain/source junctions are thicker and control a larger portion of the depletion charge. Measurements of threshold voltage reduction with channel length, for various fully depleted SIMOX structures, are reproduced in Figure 8.19b. This clearly demonstrates how efficient ultrathin buried-oxides and Si films are in suppressing short-channel effects, even for subquarter-micron gates.[55]

Three-interfaces SOI models have been constructed by extending the application of Poisson and Gauss equations to the substrate. It is basically found that the substrate influence can be safely ignored for accumulation at the third interface, but may become remarkable for depletion, essentially in the case of very thin buried-oxides. By referring to a two-interface model, the substrate depletion region more or less acts as an artificially increased BOX thickness and a modified body charge.[56] Also the back channel threshold voltage is subject to third-interface-induced corrections.

A major practical difficulty is to provide such refined models with realistic information on the substrate doping (level and profile) and on the nature and amount of defects at the BOX–substrate interface. There are many different combinations of film and substrate dopings or interface defects that obviously lead to distinct results. For example, in n–MOSFETs fabricated on p-type film and substrate, the value of V_{T_1} given by Equation (8.18) is found to be overestimated[57] for $N_{sub} = 10^{16}\,\mathrm{cm}^{-3}$ and underestimated by 50–100 mV for a lower doping ($N_{sub} \simeq 10^{15}\,\mathrm{cm}^{-3}$).[58]

Efforts were also made to model the threshold voltage in thin *dual-gate* MOSFETs with symmetrical ($t_{ox_1} = t_{ox_2}$ and $V_{G_1} = V_{G_2}$, as in gate-all-around or Delta devices) or asymmetrical ($t_{ox_1} \ll t_{ox_2}$, $V_{G_2} \gg V_{G_1}$) configurations. Front- and back-gate voltages are varied together such as threshold occurs simultaneously at both interfaces. Simple relations are obtained for ultrathin or low-doped MOSFETs, where a flat potential may be assumed across the film, leading to a substantial *volume-inversion* effect (see Sections 3.8 and 8.1).[59] Corrections are necessary if the mid-film concavity of the potential distribution (Fig. 8.1d, Section 8.1) exceeds kT/q. Such a model, proposed by Suzuki et al.,[60] suggests an increase in the threshold voltage with decreasing film thickness.

8.6 Subthreshold Slope

8.6.1 Model

In weak inversion, the drain current depends exponentially on gate and drain voltages. The subthreshold slope indicates the sharpness of the current transition from off-state to on-state, and the subthreshold *swing* S is defined as the inverse of the slope. Standard relations for bulk-Si or partially depleted SOI films are[61]

$$
\begin{aligned}
I_D &= I_0 \exp\left(\frac{qV_G}{kT}\frac{C_{ox}}{C_{ox}+C_d+C_{it}}\right) \times \\
&\quad \left[1 - \exp\left(-\frac{qV_D}{kT}\frac{C_{ox}+C_d}{C_{ox}+C_d+C_{it}}\right)\right] \qquad (8.28)
\end{aligned}
$$

and

$$S = \left(\frac{d\log_{10} I_D}{dV_G}\right)^{-1} = 2.3\frac{kT}{q}\frac{dV_G}{d\psi_s} = 2.3\frac{kT}{q}\frac{C_{ox} + C_d + C_{it}}{C_{ox}} \tag{8.29}$$

In fully depleted films, the current depends on both voltages $V_{G_{1,2}}$ and can flow at either interface. It is assumed that the inversion charge in weak inversion, the potential fluctuations and the short-channel effects are negligible. For depletion at the back interface, the differentiation of Equation (8.14) yields

$$C_{ox_{1,2}}(dV_{G_{1,2}} - d\psi_{s_{1,2}}) = C_{it_{1,2}}d\psi_{s_{1,2}} + C_{si}(d\psi_{s_{1,2}} - d\psi_{s_{2,1}}) \tag{8.30}$$

An identical relation is derived by assuming a linear potential profile across the ultrathin film,[45] which is equivalent to neglecting the depletion charge. The front-gate subthreshold swing, S_1^{dep}, is obtained by eliminating $d\psi_{s_2}$ in Equations (8.30)

$$S_1^{dep} = 2.3\frac{kT}{q}\left(1 + \frac{C_{it_1}}{C_{ox_1}} + \alpha_1\frac{C_{si}}{C_{ox_1}}\right) \tag{8.31}$$

where α_1 is an interface coupling coefficient

$$\alpha_1 = \frac{C_{ox_2} + C_{it_2}}{C_{si} + C_{ox_2} + C_{it_2}} < 1 \tag{8.32}$$

which accounts for the influence of back-interface traps and buried-oxide thickness on the front-channel current.[45, 62] It is worth noting that this relation also holds for low doped depletion-mode MOSFETs.[63, 64] Rigorous numerical simulations are presented in Figure 8.20a and show that S_1^{dep} depends slightly on the back surface potential, the minimum swing being reached for $\psi_{s_2} \simeq 0.5\phi_F$.[63, 64]

According to Equation (8.32), thicker buried oxides lead to a reduction in α_1 and an improved swing. In the ideal case, where $D_{it_{1,2}} \simeq 0$ and the buried oxide is much thicker than both the film and the gate oxide, S_1^{dep} becomes close to the theoretical limit

$$S_1^{min} = 2.3\frac{kT}{q}\left(1 + \frac{t_{ox_1}}{t_{ox_2}}\right) \simeq 2.3\frac{kT}{q} \tag{8.33}$$

which corresponds, at room temperature, to 60 mV per decade of current.

The use of fully depleted SOI MOSFETs, where $\alpha_1 < 1$, appears to be a very efficient solution for improving the swing. Two factors can,

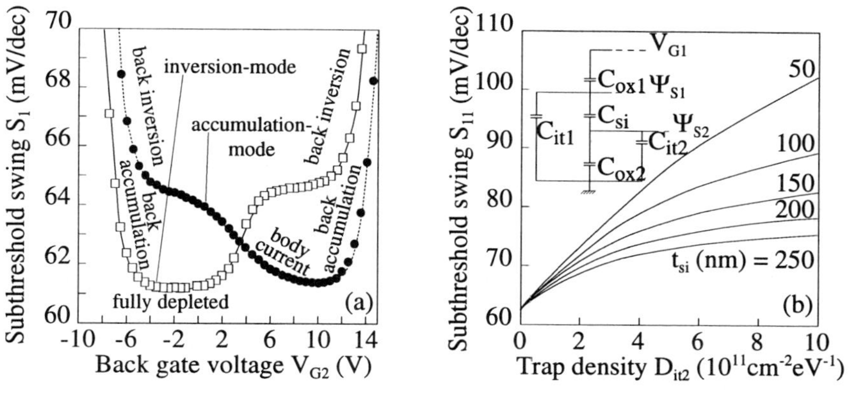

Figure 8.20 (a) Front-channel subthreshold swing versus back-gate bias in inversion-mode and accumulation-mode SIMOX n–MOSFETs (after Colinge et al.[64]). (b) Influence of the back-interface trap density and film thickness on the minimum swing value S_1^{dep} ($t_{ox_1} = 27\,\text{nm}$, $t_{ox_2} = 850\,\text{nm}$, $D_{it_1} = 10^{10}\,\text{cm}^{-2}\text{eV}^{-1}$, after Mazhari et al.[45]).

however, limit this benefit. First, for a constant doping, C_d is smaller in partially depleted than in fully depleted SOI, where the extension of the depletion region is restricted by the film thickness. Second, the front-channel swing may be governed by *back-interface* defects rather than by front-interface traps. In ultrathin films, the condition $\alpha_1 C_{si} > C_{it_1}$ actually corresponds to $D_{it_2} > D_{it_1}$, which is in general fulfilled.

The sensitivity of S_1^{dep} to back-interface traps is illustrated in Figure 8.20b for films with various thicknesses. For small values of D_{it_2}, the swing is nearly independent of film thickness and varies linearly with $D_{it_{1,2}}$. For $D_{it_2} > 2\times10^{11}\,\text{cm}^{-2}\text{eV}^{-1}$, the swing deteriorates more rapidly in thinner films where C_{si} is larger. This implies that thicker, fully depleted films are preferable to ultrathin films if the quality of the back interface is poor. For very large D_{it_2}, the swing saturates to a maximum value corresponding to $\alpha_1 = 1$. This discussion is summarized by the equivalent circuit presented in the insert of Figure 8.20b.

Accumulation at the back interface decouples the front inversion channel from back-interface defects. Indeed, the rapidly increasing capacitance of the accumulation layer is connected in parallel with C_{it_2} and makes α_1 tend to unity:

$$\alpha_1^{acc} = \frac{C_{ox2} + C_{it_2} + C_{acc2}}{C_{si} + C_{ox_2} + C_{it_2} + C_{acc_2}} \longrightarrow 1 \tag{8.34}$$

which gives

$$S_1^{acc} = 2.3\frac{kT}{q}\left(1 + \frac{C_{it_1}}{C_{ox_1}} + \frac{C_{si}}{C_{ox_1}}\right) \tag{8.35}$$

The same expression is obtained by letting $d\psi_{s_2} = 0$ in Equation (8.30). Although the back-interface defects are now masked, the interface decoupling induces a degradation of the swing to its maximum value.

An interesting situation happens when most of the current flows at the weakly inverted back interface ($\psi_{s_2} > \psi_{s_1}$), but the front-gate is still used to control it. By eliminating $d\psi_{s_1}$ in Equations (8.30), it is found that

$$S_1^{w.inv} = 2.3\frac{kT}{q}\frac{dV_{G_1}}{d\psi_{s_2}} = \left(1 + \frac{C_{ox_2} + C_{it_2}}{C_{si}}\right) S_1^{dep} \tag{8.36}$$

This equation holds until either the back channel reaches strong inversion or the front channel regains control ($\psi_{s_1} > \psi_{s_2}$). Such transitions are revealed by a plateau or a hump in $\log I_D(V_{G_1})$ curves ($V_{G_2} = +10$ V and $+20$ V in Figure 8.21a). According to the degree of inversion of the back channel, the corresponding traps are more or less masked. For strong inversion, the back surface potential is almost pinned and the inversion capacitance dominates, which is similar to Equations (8.34) and (8.35).

The present model is fully reversible for back-gate operation. The back-channel swing S_2 is very large, not only because C_{ox_2} is extremely small but also because $\alpha_2 > \alpha_1$. For ideal interfaces ($D_{it_{1,2}} = 0$), the minimum back-channel swing is given by[45]

$$S_2^{dep} = 2.3\frac{kT}{q}\left(1 + \frac{t_{ox2}}{t_{ox_1}}\frac{C_{si}}{C_{si} + C_{ox_1}}\right) \tag{8.37}$$

Several variants of this model have been explored. Considering the spreading of minority or majority carriers into the film does not lead to major modifications. But for relatively thin buried oxide, the third interface (buried oxide–substrate) and the substrate doping may play a nonnegligible role. For example, the front-channel swing is always improved if the substrate is depleted rather than accumulated,[56] since the capacitance of the depleted substrate acts likewise a thicker buried oxide in Equation (8.33). However, the use of sophisticated three-interface models is impractical, because too many parameters are involved and cannot be experimentally checked.

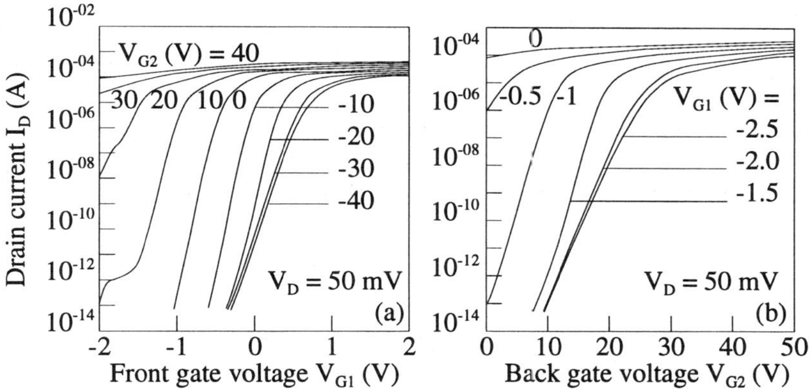

Figure 8.21 Subthreshold drain current measured as a function of (a) the front-gate voltage or (b) the back-gate voltage for different biases on the opposite gate (wafer-bonded SOI, $t_{si} = 70\,\text{nm}$, $V_D = 50\,\text{mV}$, other parameters as in Figure 8.20b, after Mazhari et al.[45]).

8.6.2 Experimental Results

Front- and back-gate subthreshold characteristics are shown in Figure 8.21 for an ultrathin SOI MOSFET. The steepest slopes are always obtained with the opposite interface in depletion. The excellent front-channel swing (69 mV/decade) demonstrates the high quality of the two interfaces, since they both affect S_1^{dep}. In accumulation, the swing degrades to 115 mV/decade, a slightly smaller value than could be expected from Equation (8.35). This means that the coefficient α_1^{acc} does not reach the maximum value $\alpha_1^{acc} = 1$; in other words, the back-interface traps are not totally masked by accumulation.

As noted above, the subthreshold curves become distorted and suffer from a high level of "leakage" current when the back interface is inverted. This makes difficult the accurate appreciation of the swing. Another source of errors is the charge-sharing effect, which causes the swing to be overestimated in short-channel devices.

The subthreshold swing is a straightforward parameter used to monitor the quality of SOI technologies and the damaging effects of irradiations. The measurement of S_2^{acc} is certainly the easiest and most popular method for a rough evaluation of the back-interface trap density. As far as the front interface is concerned, the extraction of small D_{it_1} values is less reliable. As in the case of bulk silicon, there is good reason to suspect the role of potential fluctuations whose influence increases for thin oxides.

An efficient alternative is to make use of the *dynamic conductance* technique.[65] The conductance value, measured in weak inversion ($\psi_s \simeq 1.5\phi_F$) as a function of drain voltage, is obtained from Equation (8.28):

$$g_D = g_D^0 \exp\left(-\frac{q}{kT}\frac{C_{ox}+C_d}{C_{ox}+C_d+C_{it}}V_D\right) \tag{8.38}$$

where g_D^0 stands for the ohmic region conductance ($V_D \rightarrow 0$). Drawing $\log g_D$ versus V_D results in a straight line whose slope yields C_{it}. The advantage of the dynamic conductance method is to eliminate the influence of potential fluctuations by performing the measurements at a fixed gate voltage. Yet more specialized techniques, such as charge pumping, DLTS, and noise spectroscopy (Chapter 9), can be used for accurate interface characterization.

8.7 Transconductance

8.7.1 Mobility Definitions

Straightforward adaptation of bulk-Si relations yields the drain current in a front-channel SOI MOSFET operated in strong inversion and ohmic region

$$I_D = \frac{C_{ox_1}WV_D}{L}\mu_{eff_1}\left(V_{G_1} - V_{T_1}(V_{G_2})\right) \tag{8.39}$$

where μ_{eff} is the *effective* mobility, which is function of the vertical field, surface charge, and interface defects. It is usually expressed in terms of gate voltage

$$\mu_{eff_1} = \frac{\mu_1}{1+\theta_1(V_{G_1} - V_{T_1}(V_{G_2}))} \tag{8.40}$$

which leads to a simple relation for the transconductance g_m:

$$g_{m_1} = \frac{C_{ox_1}WV_D}{L}\mu_{fe_1} = \frac{C_{ox_1}WV_D}{L} \times \frac{\mu_1}{[1+\theta_1(V_{G_1} - V_{T_1}(V_{G_2})]^2} \tag{8.41}$$

In short channels, the *mobility attenuation factor* θ_1 depends on series resistances:

$$\theta_1 = \theta_1^0 + R_{SD}\frac{\mu_1 C_{ox_1} W}{L} \tag{8.42}$$

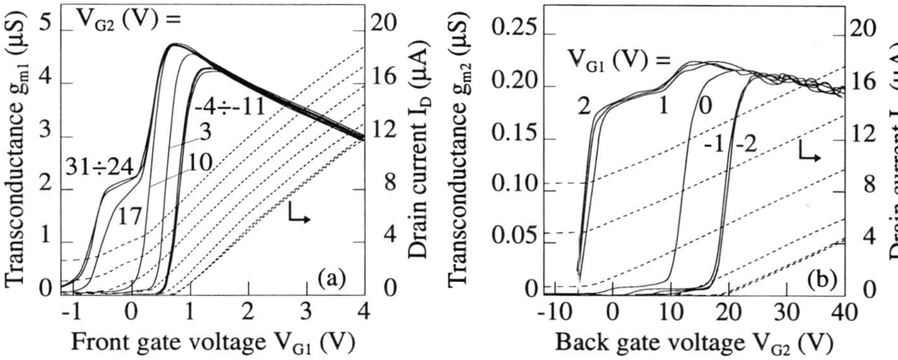

Figure 8.22 Drain current (- - -) and transconductance (——) curves versus (a) front-gate voltage and (b) back-gate voltage in a long SOI MOSFET ($L = 20\,\mu$m, $V_D = 50\,$mV, $t_{ox_1} = 17\,$nm, $t_{ox_2} = 380\,$nm, $t_{si} = 80\,$nm, $N_A \simeq 5 \times 10^{16}\,\text{cm}^{-3}$).

where the intrinsic attenuation factor, $\theta_1^0 \sim 0.1\,\text{V}^{-1}$, is in inverse proportion to the gate oxide thickness.

According to the gate bias, the *field-effect* mobility μ_{fe_1} can be much smaller than the *pure mobility* μ_1, which is empirically related to the amount of interface defects: $\mu_1 \sim (1 + \alpha_1 D_{it_1})^{-1}$, with $\alpha_1 \simeq 0.2 \times 10^{-11}\,\text{cm}^2\text{eV}$.[66] A distinction, therefore, needs to be made between these various mobilities, since they can be accessed by different methods.

8.7.2 Fully Depleted SOI MOSFETs

The complexity of the transconductance curves shown in Figure 8.22a is due to the influence of the back-gate bias. Not only does it govern the front-channel threshold voltage $V_{T_1}(V_{G_2})$ but also it allows a back inversion channel to be activated.

Following the model proposed by Ouisse et al.,[67] four typical situations are distinguished as V_{G_2} is varied from negative to positive values.

Case 1: Back Interface Accumulated

In this case, the back surface potential is nearly constant, and the threshold voltage and vertical field saturate. This explains why a unique transconductance curve is obtained below a certain bias ($V_{G_2} < -4\,$V, in Figure 8.22a).

Case 2: Back Interface Depleted

Since V_{T_1} decreases linearly with increasing V_{G_2} (see Eq. (8.18)), the position of the transconductance peak, $V_{G_1}^{max}$, is laterally shifted.[10] Moreover, the electric field at the front interface decreases, thus allowing the carrier mobility μ_1 and the transconductance peak g_{max_1} to increase.

Case 3: From Depletion to Inversion: $V_{T_2}^{inv} < V_{G_2} < V_{T_2}^{acc}$

For a negative V_{G_1} bias, the front interface is accumulated, the back interface is depleted, and the drain current is zero. Increasing V_{G_1} first opens the back channel. This happens because, for $V_{G_1} > V_{G_1}^{acc}$, the back threshold voltage decreases and eventually crosses the fixed value V_{G_2} (Fig. 8.23a). After the onset of back-channel conduction, the current is given by

$$I_D = I_{D_2} = \frac{C_{ox_2} W V_D}{L} \frac{\mu_2}{1 + \theta_2 (V_{G_2} - V_{T_2})} (V_{G_2} - V_{T_2}(V_{G_1})) \qquad (8.43)$$

where the coefficient $\theta_2 \simeq 5 \times 10^{-3}\,\mathrm{V}^{-1}$ is extremely small owing to the large thickness of the buried oxide. Although V_{G_2} is fixed, the back current linearly varies with V_{G_1} (Fig. 8.23b) via the decrease of $V_{T_2}(V_{G_1})$ as long as the front interface remains depleted. Using the reciprocal of Equation (8.18), I_{D_2} can be explicitly rewritten as a function of V_{G_1}:[67]

$$I_{D_2} = \frac{\gamma_2 C_{ox_2} W V_D}{L} \frac{\mu_2}{1 + \gamma_2 \theta_2 (V_{G_1} - V'_{T_1})} (V_{G_1} - V'_{T_1}) \qquad (8.44)$$

where V'_{T_1} stands for a pseudothreshold voltage

$$V'_{T_1} = V_{G_1}^{acc} + \frac{1}{\gamma_2} (V_{T_2}^{acc} - V_{G_2}) \qquad (8.45)$$

and γ_2 is a threshold-voltage coupling coefficient

$$\gamma_2 = \frac{C_{ox_1} C_{si}}{C_{ox_2} (C_{ox_1} + C_{si} + C_{it_1})} \qquad (8.46)$$

[10]The experiment verifies that $V_{G_1}^{max}$ mimics the variation of V_{T_1}, being slightly larger (by 0.2–0.3 V).

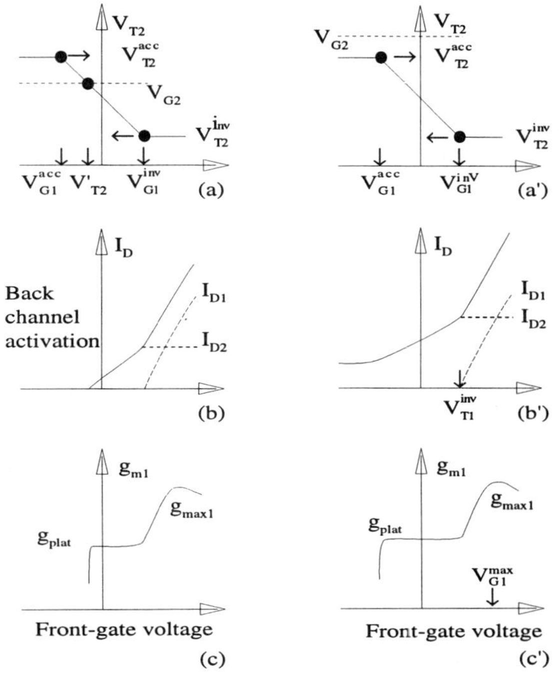

Figure 8.23 Schematic representation of the back-channel threshold voltage (a,a'), the drain current (b,b'), and the transconductance (c,c') as a function of the front-gate bias. Diagrams (a,b,c) correspond to Case 3 (back channel on for $V_{G_1} \geq V'_{T_1}$), whereas (a',b',c') relate to Case 4 (back channel always activated) (after Ouisse et al.[67]).

The transconductance now expresses the variation of the back-channel current under control of V_{G_1}:

$$g_{m_1} = \frac{\gamma_2 C_{ox_2} W V_D}{L} \times \frac{\mu_2}{\left[1 + \gamma_2 \theta_2 (V_{G_1} - V'_{T_1})\right]^2} \tag{8.47}$$

Here, the mobility attenuation factor $\gamma_2 \theta_2$ is very small, so g_{m_1} is represented by a "plateau" (Figs. 8.22a and 8.23c):

$$g_{plat_1} = \frac{\gamma_2 \mu_2 C_{ox_2} W V_D}{L} \tag{8.48}$$

The plateau of the front-channel transconductance is a direct measure of the back-channel mobility.

For $V_{G_1} \geq V_{G_1}^{inv}$, the front channel is activated as well, whereas the values of V_{T_2} and I_{D_2} saturate. The total current now becomes a superposition of I_{D_1} and I_{D_2}, clearly reflected by a slope change in $I_D(V_{G_1})$ characteristics (Figs. 8.22a and 8.23b). The transconductance increases sharply with V_{G_1}, being fully representative for the front-channel mobility. This interesting exchange of prevailing roles between the front and back interfaces is a unique feature of fully depleted SOI transistors. The main consequence is the gradual deformation of transconductance curves (Fig. 8.22a), which become totally different from the conventional curve in bulk Si.

Case 4: Back Interface Inverted: $V_{G_2} > V_{T_2}^{acc}$

The back interface is always inverted, whatever the bias V_{G_1}. When the front-interface channel is accumulated, the current I_{D_2} is a constant (Fig. 8.23b'), and the transconductance is zero. As soon as $V_{G_1} \geq V_{G_1}^{acc}$, the back-channel threshold voltage begins decreasing, and I_{D_2} increases linearly with V_{G_1} according to Equation (8.44). The transconductance plateau of Figure 8.23c' is again given by Equation (8.48), until the front interface reaches strong inversion, for $V_{G_1} > V_{G_1}^{inv}$, and takes the lead.

Cases 3 and 4 look similar, except that in the latter situation the current is never zero and the extension of the transconductance plateau region is maximum: $V_{G_1}^{acc} \leq V_{G_1} \leq V_{G_1}^{inv}$. From the transconductance curve of Figure 8.23c', many useful parameters can directly be deduced: voltages $V_{G_1}^{inv}$, $V_{G_1}^{acc}$, $V_{G_1}^{max} \simeq V_{T_1}$, and front- and back-channel mobilities.

8.7.3 Series-Resistance Effects

The series resistances induce a reduction of the effective voltage drop on the drain and are responsible for the apparent degradation of the transistor parameters. The equivalent circuit of Figure 8.24 shows that the situation is rather complex in SOI MOSFETs, where different series resistances $R_{1,2}$ are associated with the front and back channels. Indeed, the voltage drop at the drain terminal of the front channel depends not only on I_{D_1} but also on the current that may flow in the back channel. The common parasitic resistance R_3 can be adjusted experimentally.

The superscript (^{0}) is used in this section to identify "pure" parameters, unaffected by series resistances. For example, the pure conductances $G_{D_{1,2}}^0$ are given, respectively, by Equations (8.39) and (8.43). The front-channel and back-channel currents are easily calculated from

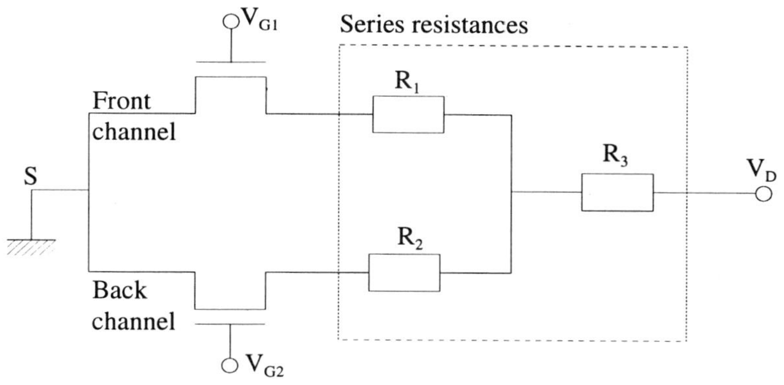

Figure 8.24 Equivalent circuit of an SOI MOSFET with series resistances, operated in the ohmic region (after Ouisse et al.[67]).

Figure 8.24:

$$I_{D_{1,2}} = (1 + R_{2,1} G^0_{D_{2,1}}) G^0_{D_{1,2}} V_D \times [1 + (R_1 + R_3) G^0_{D_1} + (R_2 + R_3) G^0_{D_2} + (R_1 R_2 + R_1 R_3 + R_2 R_3) G^0_{D_1} G^0_{D_2}]^{-1} \tag{8.49}$$

Next, the influence of the series resistances on the transconductance will be examined for each of the above four cases. In Cases 1 and 2, the back interface is either *accumulated* or *depleted*, and $G^0_{D_2} = 0$, giving

$$g_{m_1} = \frac{g^0_{m_1}}{\left[1 + (R_1 + R_3) G^0_{D_1}\right]^2} \tag{8.50}$$

The first-order expansion shows that the degradation of the transconductance peak

$$g_{max_1} = g^0_{max_1} \left[1 - 2(R_1 + R_3) G^0_{D_1}\right] \tag{8.51}$$

is the same as in bulk Si.[68] Equations (8.50) and (8.51) do not account for the case of ultrathin MOSFETs with LDD structure, where the transconductance peak keeps decreasing as the back gate is biased in stronger accumulation. Such an effect is due to an increase in series resistances $R_{1,3}$ brought about by the modification of effective doping in the LDD region.

In Cases 3 and 4, if only the back interface is *inverted* ($G^0_{D_1} = 0$), the degradation of the plateau level is given by a similar approximation

$$g_{plat_1} = g^0_{plat_1}\left[1 - 2(R_2 + R_3)G^0_{D_2}\right] \tag{8.52}$$

When the front interface reaches strong inversion as well, both currents $I_{D_{1,2}}$ contribute to the transconductance. Since $G^0_{D_2}(V_{G_1})$ is now a constant, only $G^0_{D_1}$ is function of V_{G_1}. The first-order degradation of the transconductance peak is given by[67]

$$g_{max_1} = g^0_{max_1}\left[1 - 2R_3G^0_{D_2} - 2(R_1 + R_3)G^0_{D_1}\right] \tag{8.53}$$

Although the reduction rates of the transconductance peak and plateau are quite different, they both depend on the back-channel conductance and bias.

The series-resistance influence on current and transconductance can be expressed in a more condensed form[11] by generalizing the meaning of the mobility attenuation factors defined in Equation (8.42):

$$\theta_{1,2} = \theta^0_{1,2} + (R_{1,2} + R_3)\frac{\mu_{1,2}C_{ox_{1,2}}W}{L} \tag{8.54}$$

The above relations allow the following conclusions about the impact of increasing series resistances on the front-gate transconductance:

- Variable degradation of the transconductance peak, more dramatic when the back interface is inverted

- Reduction of the transconductance plateau, which can even transform, for large values of θ_2, into a secondary peak accounted for by Equation (8.47)

- Sharper decrease of $g_{m_1}(V_{G_1})$ in strong inversion due to the increase in θ_1

- Slight shift of the position, $V^{max}_{G_1}$, of the transconductance peak towards negative values of V_{G_1}, which can be calculated with a first-order development.[68]

[11] Except for the transconductance peak in Cases 3 and 4, where the effective value of θ_1 is more complex.

8.7.4 Experiment and Parameter Extraction

Experimental long-channel transconductance curves, such as those in Figure 8.22a, at a glance deliver several important properties of the SOI structure. The shift of the transconductance peak reflects the relationship $V_{T_1}(V_{G_2})$, whereas the extension of the transconductance plateau yields $V_{G_1}^{acc}$. The front-channel transconductance provides the carrier mobilities, not only at the front interface (from the peak value) but also at the back interface (from the plateau level). In particular, the ratio between the transconductance peak and plateau

$$\frac{g_{max_1}}{g_{plat_1}} \simeq \frac{\mu_1}{\mu_2}\left(1+\frac{C_{ox_1}}{C_{si}}\right) \tag{8.55}$$

offers a rapid comparison of front- and back-channel mobilities. From Figure 8.22a, it is found that $\mu_1 \simeq \mu_2$, which demonstrates the back-interface quality.

This result can be confirmed by performing back-channel transconductance measurements (Fig. 8.22b). Notice that the plateau level is not always easily detectable because it can be comparable or even higher than the transconductance peak: $g_{max_2}/g_{plat_2} \simeq \mu_2/\mu_1$. If the back-channel mobility is poorer, the overlapping plateau and peak may give rise to an apparently higher transconductance peak and, therefore, to an overestimated value μ_2. Earlier claims of μ_2 being higher than μ_1 do not have any technological or physical basis: they presumably come from a confusion between the peak and plateau, and eventually mean that $\mu_2 < \mu_1$!

Among many methods developed for the accurate extraction of the transistor parameters, a very simple alternative is based on Equations (8.39)–(8.41). Drawing, in strong inversion, the function $F(V_G)$[50]

$$F(V_G) = \left(-\frac{d(1/I_D)}{dV_G}\right)^{-0.5} = \frac{I_D}{\sqrt{g_m}} = \sqrt{\frac{\mu C_{ox} W V_D}{L}}\,(V_G - V_T) \tag{8.56}$$

results in a straight line whose intercept with the horizontal axis gives V_T (Fig. 8.25a).

The maximum, "pure" mobility μ is determined from the slope $K = (\mu C_{ox} W V_D/L)^{0.5}$, if the effective channel length is known. Otherwise, the experiment is repeated for various gate lengths and the function K^{-2} vs. L_G is linearized (Fig. 8.25b): the horizontal intercept provides the

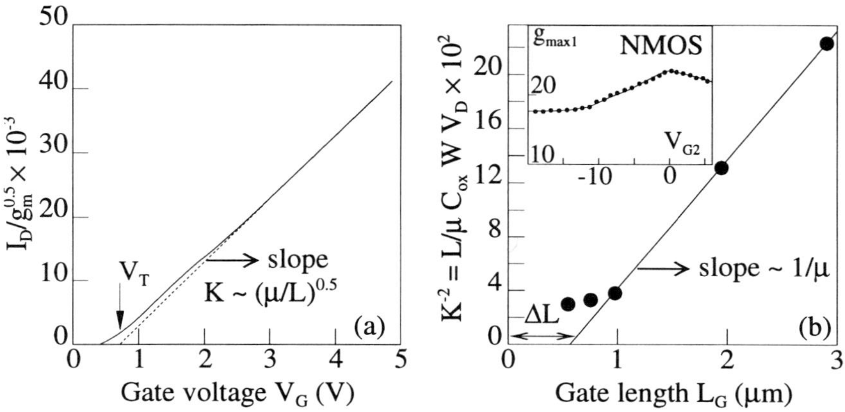

Figure 8.25 Parameter extraction: (a) threshold voltage and parameter K from $I_D/\sqrt{g_m}$ function, (b) channel-length reduction ΔL and low-field mobility. The insert shows the transconductance peak, proportional to the front-channel field-effect mobility, versus back-gate bias.

channel-length correction factor, $\Delta L = L_G - L$, and the slope yields μ. This procedure is, in principle, free from series-resistance effects. It can also be used in the transconductance plateau region to determine the mobility at the opposite interface.

Information on series resistances is contained in the mobility attenuation coefficient θ, whose value comes from Equations (8.39)–(8.42),[50]

$$\theta = \left[\frac{I_D}{g_m(V_G - V_T)} - 1\right] \times \frac{1}{V_G - V_T} \tag{8.57}$$

and does not depend on the gate voltage for $V_G \gg V_T$ (Fig. 8.26a). Furthermore, Equation (8.42) shows that plotting θ vs. $1/L$ results in a new straight line (Fig. 8.26b). The vertical intercept gives θ^0, whereas the slope delivers the series resistance value.

This self-consistent extraction method proves to be efficient for the independent characterization of both channels. Since the technique can easily be implemented in an automatic procedure, it is safe to repeat it for various biases on the opposite gate. Indeed, besides the modulation of the mobility and series resistances, a variation of the effective length of the front channel with V_{G_2} was reported for some technologies.[12]

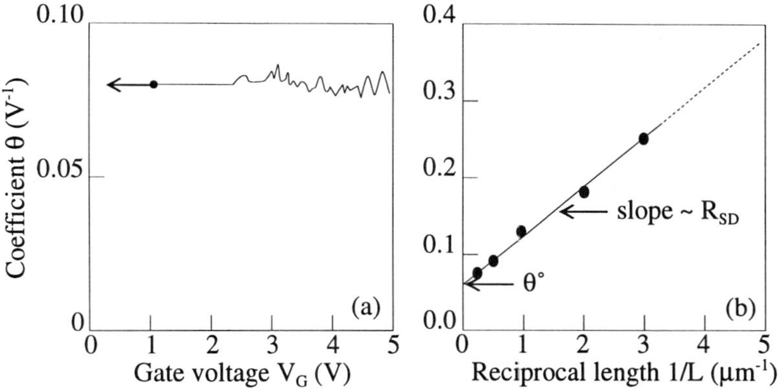

Figure 8.26 (a) Mobility attenuation factor θ (extracted from Equations (8.39) and (8.40)) versus gate voltage and (b) plot of θ versus reciprocal channel length, which yields the values of the intrinsic factor θ^0 and series resistance R_{SD}.

The typical variation of the front-channel mobility as V_{G_2} is increased from accumulation to inversion is shown in the insert of Figure 8.25b. The mobility improvement in depletion reflects the reduction of the vertical field. The slight mobility degradation for strong inversion was explained by invoking (1) the influence of the back-channel series resistance[67] and (2) the spreading of minority carriers in the volume of the film, which may modify the apparent gate capacitance.[45]

At this point, attention must be called to the case of very advanced transistors with extremely thin gate oxides (4–7 nm), where the extraction method must be reconditioned. It was observed that the mobility is mainly dominated by surface roughness scattering, which possibly induces a decrease in drain current with increasing gate bias. The conventional model is unable to account for such negative transconductance effect; therefore, it becomes necessary to include an additional term, quadratically dependent on gate voltage, $\sim \theta'(V_G - V_T)^2$, in the denominator of Equation (8.40). It is an easy exercise to demonstrate that the various parameters (mobility, V_T, θ, θ') can be determined by drawing the first- and second-order derivatives of reciprocal current, $1/I_D$, versus gate voltage.[70]

The *effective* mobility μ_{eff} can directly be extracted from channel

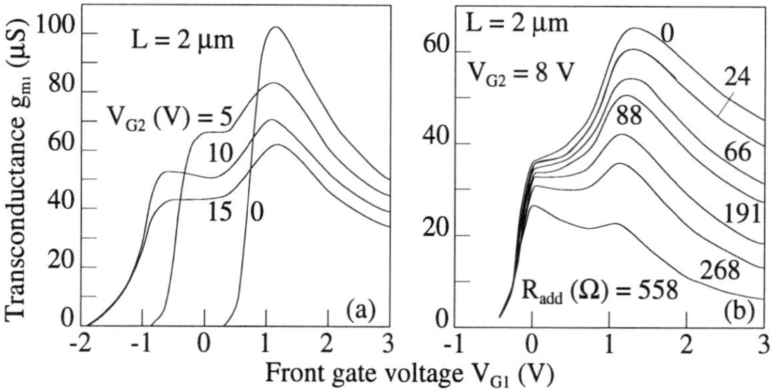

Figure 8.27 Deformation of the front-gate transconductance curves in a short-channel SOI MOSFET for (a) stronger back-channel inversion and (b) increased external series resistances (after Ouisse et al.[67]).

conductance measurements in the ohmic region:

$$\mu_{eff} = \frac{L}{W}\left(\frac{\partial I_D}{\partial V_D}\right)\frac{1}{Q_{inv}} = \frac{L}{W}\left(\frac{\partial I_D}{\partial V_D}\right)\left(\int_{-\infty}^{V_G} C_{inv}\, dV_G\right)^{-1} \tag{8.58}$$

The idea above is to experimentally determine the density of the inversion charge (or the inversion capacitance) with a split C–V technique.[69] The advantage of this approach is to avoid using the approximation $Q_{inv} \simeq C_{ox}(V_G - V_T)$, which is not fully reliable for moderate inversion at low gate bias. The experiment performed on fully depleted SIMOX MOSFETs shows that the effective mobility is still heavily dominated by the vertical field (i.e., front- and back-gate biases), whereas it appears to be rather independent of film thickness and doping (below 5×10^{16} cm^{-3}).

The impact of series resistances on short-channel SOI MOSFETs is illustrated in Figure 8.27a. The transconductance peak and plateau can be degraded by more than 40%, according to the degree of inversion at the back interface.

The theoretical method described in the previous section has been checked by adding external resistors, R_{add}, to the internal series resistance R_3. Figure 8.27b shows, in agreement with Equations (8.52) and (8.53), that the transconductance peak degrades more rapidly than the plateau level. Moreover, for $R_{add} = 558\,\Omega$, the mobility attenuation co-

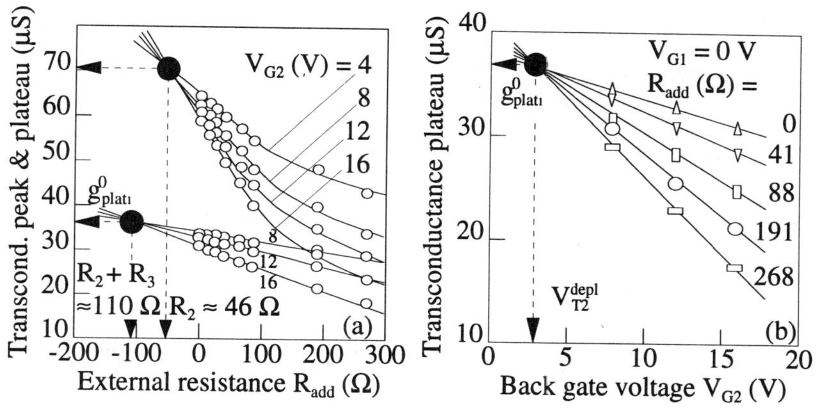

Figure 8.28 Front-gate transconductance peak g_{max_1} and plateau g_{plat_1} versus (a) external series resistances R_{add} and (b) back-gate voltage (SIMOX MOSFET, $L = 2\,\mu$m, after Ouisse et al.[67]).

efficient θ_2 becomes large enough to induce the transformation of the plateau into a secondary peak, defined by μ_2 and V_T' in Equation (8.47). Of course, these curves do not have any equivalent in bulk Si.

The degradation rates of the transconductance peak and plateau, reported in Figure 8.28a, confirm the adequacy of the first-order analysis in Equations (8.51)–(8.53). Only for series resistances exceeding "normal" values (100 Ω) does the peak exhibit a sublinear dependence $g_{max_1}(R_{add})$ resulting from second-order terms. When V_{G_2} is varied, all $g_{max_1}(R_{add})$ curves meet for $R_{add} = -R_3$, according to Equation (8.53).

Similarly, the plateau curves join, for $R_{add} = -R_2 - R_3$, in a unique point whose coordinates give R_2 and the pure plateau value g^0_{plat} (see Eq. (8.52)). The last series resistance, R_1, is extracted using Equation (8.51) from the degradation rate of the peak, measured with the back interface in accumulation.

We have to remember that, in Equation (8.43), $G^0_{D_2}$ is almost proportional with $(V_{G_2} - V_{T_2})$, since the coefficient θ_2 is very weak. As a matter of fact, Figure 8.28b shows that, indeed, the plateau level varies linearly with V_{G_2}. The coordinates of the merging point ($G^0_{D_2} = 0$) are nothing but g^0_{plat} and $V_{T_2}^{depl}(V_{G_1})$.

A striking aspect of this analysis is that many key parameters (μ_2,

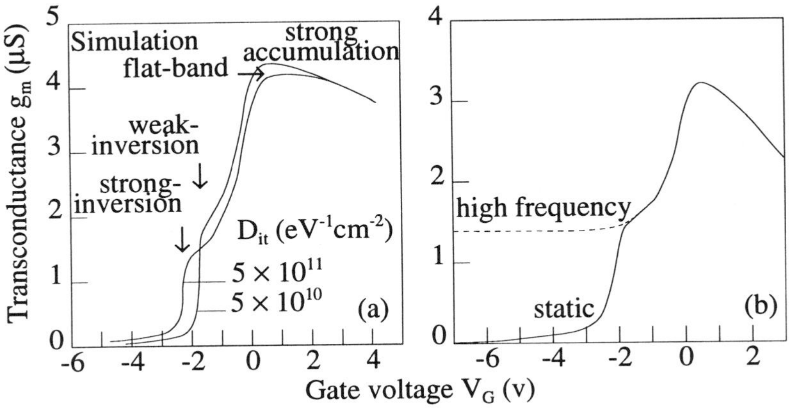

Figure 8.29 Transconductance versus front-gate bias in depletion-mode partially depleted n–channel SOI MOSFETs. (a) Simulations for two different concentrations of back-interface traps ($f = 0$); (b) static and high-frequency measurements ($L = W = 7\,\mu$m, $N_D = 5 \times 10^{16}\,\mathrm{cm}^{-3}$, $t_{si} = 250\,\mathrm{nm}$, $t_{ox_1} = 27\,\mathrm{nm}$, $\theta_1 = 0.03\,\mathrm{V}^{-1}$) (after Haddara et al.[71]).

V_{T_2}, R_2) of the back interface are inferred from front-channel measurements. Once the total set of parameters is extracted, drain current curves such as those of Figure 8.22a can be perfectly reproduced using Equation (8.49), without the need of any adjustable parameter.

8.7.5 Transconductance in Depletion–Mode MOSFETs

Since fully depleted enhancement-mode and depletion-mode transistors behave very much alike, this section will be dedicated to partially depleted n^+–n–n^+ MOSFETs. When the gate is biased in depletion or in inversion and only bulk conduction occurs (see Eq. (9.41)), the transconductance is given by

$$g_m = -\frac{\mu W V_D}{L} \times \frac{dQ_d}{d\psi_s} \times \frac{d\psi_s}{dV_G} \tag{8.59}$$

Here, the doping and mobility are assumed constant across the film. Derivation of the charge-conservation equation

$$V_G = \psi_s + \phi_{fb} - \frac{Q_d + Q_{it} + Q_{inv}}{C_{ox}} \tag{8.60}$$

gives $d\psi_s/dV_G$, so Equation (8.59) becomes[71]

$$g_m = \frac{\mu W V_D}{L} \times \frac{C_{ox} C_d}{C_{ox} + C_d + C_{inv} + C_{it}} \tag{8.61}$$

Shown in Figure 8.29a are numerically simulated curves that can be simply explained by Equation (8.61). The transconductance peak corresponds to a gate voltage slightly larger than V_{FB}. As V_G is reduced, the growth of the depletion region makes C_d and g_m decrease. At the onset of strong inversion, C_d saturates, whereas C_{inv} increases exponentially, leading to a more or less rapid fall of g_m, according to the density of interface traps.[71, 72] For $V_G > V_{FB}$, the transconductance decreases too, but here it is governed by the *surface* mobility degradation in strong accumulation, as mentioned in Equation (8.40).

When the experiment is conducted from the back gate, the small capacitance of the buried-oxide screens most of the influence of C_d. The back-gate transconductance is almost a constant in accumulation, depletion, and weak inversion, before falling to zero in strong inversion.

An interesting extension of this model applies to *high-frequency* measurements. Above 1 MHz, neither the interface traps nor the inversion charge can follow the gate signal. For $V_G < V_{FB}$, the transconductance

$$g_m^{hf} = \frac{\mu W V_D}{L} \times \frac{C_{ox} C_d}{C_{ox} + C_d} \tag{8.62}$$

is nothing but a magnified reproduction of the high-frequency MOS capacitance! Owing to the amplification prefactor (10^2 to 10^4), g_m^{hf} is easily measurable even in small-geometry transistors (Fig. 8.29b) and provides the same information as a large-area capacitor. The measurement of the drain current offers the additional advantage of eliminating the influence of film series resistances and parasitic capacitances.

Making use of the standard expression $C_d(V_G)$, it is found that[71]

$$\left(\frac{g_{min}^{hf}}{g_m^{hf}}\right)^2 = \left(\frac{C_{min}}{C_{ox}}\right)^2 + \frac{2C_{min}^2}{q\epsilon_{si}N_D}(V_G - V_{FB}) \tag{8.63}$$

where g_{min}^{hf} is the minimum transconductance and C_{min} is the series combination of C_{ox} and minimum value of C_d. The linear dependence of the left-hand side of Equation (8.63) versus V_G allows the determination of the average doping from the slope and V_{FB} from the intercept.

In this long chapter, we have tried to depict the major physical phenomena that are specifically involved in the operation of SOI MOSFETs. Partially depleted SOI transistors suffer from floating-body effects, that lead to instabilities and kink in drain current, bipolar action, premature breakdown, transient effects, and so on. The ohmic region characteristics, however, follow the well-known relations established for bulk-Si MOSFETs, so conventional parameter extraction methods can still be applied.

On the other hand, fully depleted transistors are mainly affected by interface coupling problems. We have attempted to illustrate the degree of understanding reached so far and to provide a unified image of available models. The static characterization techniques needed to be enriched in order to become capable of accounting for the complexity of dual-gate effects.

Bibliography

[1] S. Cristoloveanu and D. Ioannou, "Adjustable confinement of the electron gas in dual-gate silicon-on-insulator MOSFETs," *Superlattices Microstruct.*, vol. 8, p. 131, 1990.

[2] F. Balestra, S. Cristoloveanu, M. Benachir, J. Brini, and T. Elewa, "Double-gate silicon on insulator transistor with volume inversion: a new device with greatly enhanced performance," *IEEE Electron Device Lett.*, vol. 8, p. 410, 1987.

[3] T. Ouisse, S. Cristoloveanu, and D.K. Maude, "Experimental investigation of SOI MOSFETs at high magnetic field and low temperature," *J. Appl. Phys.*, vol. 74, p. 408, 1993.

[4] K. Kato, T. Wada, and K. Taniguchi, "Analysis of kink characteristics in silicon on insulator MOSFETs using two-carrier modeling," *IEEE Trans. Electron Devices*, vol. 32, p. 458, 1985.

[5] J. Tihanyi and H. Schlötterer, "Properties of ESFI MOS transistors due to the floating substrate and finite volume," *IEEE Trans. Electron Devices*, vol. 22, p. 1017, 1975.

[6] R. Howes, W. Redman-White, K.G. Nichols, S.J. Murray, and P.J. Mole, "Modeling and simulation of silicon on sapphire MOSFETs for analog circuit design," *ESSDERC'90 Conf. Proc.*, Adam Hilger, Bristol, p. 539, 1990.

[7] E. Simoen, B. Dierickx, and C. Claeys, "Kink-related noise overshoot in SOI n-MOSFETs operating at 4.2 K," *Electron. Lett.*, vol. 28, p. 577, 1992.

[8] J.Y. Choi and J.G. Fossum, "Analysis and control of floating body bipolar effects in fully depleted submicrometer SOI MOSFETs," *IEEE Trans. Electron Devices*, vol. 38, p. 1384, 1991.

[9] J. Chen, F. Assaderaghi, H.J. Wann, P. Ko, C. Hu, P. Cheng, R. Solomon, and T-Y. Chan, "An accurate model of thin film SOI MOSFET breakdown voltage," *IEDM'91 Conf. Proc.*, p. 671, 1991.

[10] J.G. Fossum, S. Krishnan, and P.C. Yeh, "Performance limitations of deep-submicron fully depleted SOI MOSFETs," *1992 IEEE Int. SOI Conf. Proc.*, p. 132, 1992.

[11] T. Ouisse, G. Ghibaudo, J. Brini, S. Cristoloveanu, and G. Borel, "Investigation of floating body effects in silicon-on-insulator metal-oxide-semiconductor field-effect transistors," *J. Appl. Phys.*, vol. 70, p. 3912, 1991.

[12] T. Ouisse, "Modèles physiques et analyse du fonctionnement des composants MOS intégrés sur SIMOX," *Ph.D. Dissertation*, Inst. Nat. Polytechn., Grenoble, 1991.

[13] J.S.T. Huang, J.S. Kueng, and T. Fabian, "An analytical model for snapback in n-channel SOI MOSFETs," *IEEE Trans. Electron Devices*, vol. 38, p. 2082, 1991.

[14] M. Matloubian, C.E.D. Chen, B.Y. Mao, R. Sundaresan, and G.P. Pollack, "Modeling of the subthreshold characteristics of SOI MOSFETs with floating body," *IEEE Trans. Electron Devices*, vol. 37, p. 1985, 1990.

[15] M. Haond and J.P. Colinge, "Analysis of drain breakdown voltage in SOI n-channel MOSFETs," *Electron. Lett.*, vol. 25, p. 1640, 1989.

[16] G.A. Armstrong, J.R. Davis, and A. Doyle, "Characterization of bipolar snapback and breakdown voltage in thin film SOI transistors by two-dimensional simulation," *IEEE Trans. Electron Devices*, vol. 38, p. 328, 1991.

[17] O. Faynot, A.J. Auberton-Hervé, and S. Cristoloveanu, "Investigation of the influence of the film thickness in accumulation-mode fully depleted SIMOX MOSFETs," *ESSDERC'92 Conf. Proc.*, Elsevier, Amsterdam, p. 807, 1992.

[18] K.K. Young and J.A. Burns, "Avalanche-induced drain-source breakdown in silicon on insulator n-MOSFETs," *IEEE Trans. Electron Devices*, vol. 35, p. 426, 1988.

[19] M. Yoshimi, M. Takahashi, T. Wada, K. Kato, S. Kambayashi, M. Kemmochi, and K. Natori, "Analysis of the drain breakdown mechanism in ultra thin film SOI MOSFETs," *IEEE Trans. Electron Devices*, vol. 37, p. 2015, 1990.

[20] L.J. McDaid, S. Hall, K.I. Nuttall, and W. Eccleston, "Comparison of instabilities observed in electrical characteristics of silicon on insulator MOSFETs, both above and below threshold voltage," *Electron. Lett.*, vol. 28, p. 1441, 1992.

[21] B. Zhang, A. Yoshino, and T.P. Ma, "Single transistor latch-induced degradation of front- and back-channel thin film SOI transistors," *IEEE Electron Device Lett.*, vol. 13, p. 282, 1992.

[22] L.J. McDaid, S. Hall, W. Eccleston, and J.C. Alderman, "Reduction of the latch effect in SOI MOSFETs by the silicidation of the source," *ESSDERC'91 Conf. Proc., Microelectron. Eng.*, vol. 15, p. 203, 1991.

[23] M. Matloubian, "Smart body contact for SOI MOSFETs," *1989 IEEE SOS/SOI Technol. Conf. Proc.*, p. 128, 1989.

[24] Y. Ohmura and K. Izumi, "Simplified analysis of body contact effect for MOSFET/SOI," *IEEE Trans. Electron Devices*, vol. 35, p. 1391, 1988.

[25] M. Patel, P. Ratnam, and C.A.T. Salama, "A novel body contact for SIMOX based SOI MOSFETs," *Solid-State Electron.*, vol. 34, p. 1071, 1991.

[26] M. Gao, J.P. Colinge, L. Lauwers, S. Wu, and C. Claeys, "Twin–MOSFET structure for suppression of kink and parasitic bipolar effects in SOI MOSFETs at room and liquid helium temperatures," *Solid-State Electron.*, vol. 35, p. 505, 1992.

[27] J. Gautier and A.J. Auberton-Hervé, "A latch phenomenon in buried n-body SOI NMOSFETs," *IEEE Electron Device Lett.*, vol. 12, p. 372, 1991.

[28] J.M.T. Thomson and H.B. Stewart, in *Nonlinear Dynamics and Chaos*, Wiley, New York, 1986.

[29] J. Brini and G. Kamarinos, "The unijunction transistor used as a high sensitivity magnetic sensor," *Sensors Actuators*, vol. 2, p. 149, 1981/82.

[30] S. Cristoloveanu, "A review of the electrical properties of SIMOX substrates and their impact on device performance," *J. Electrochem. Soc.*, vol. 138, p. 3131, 1991.

[31] K. Kato and K. Taniguchi, "Numerical analysis of switching characteristics in SOI MOSFETs," *IEEE Trans. Electron Devices*, vol. 33, p. 133, 1986.

[32] H. Hazama, M. Yoshimi, M. Takahashi, S. Kambayashi, and H. Tango, "Suppression of drain current overshoot in SOI–MOSFETs using ultrathin SOI substrates," *Electron. Lett.*, vol. 24, p. 1266, 1988.

[33] F. Assaderaghi, J. Chen, R. Solomon, T.Y. Chan, P.K. Ko, and C. Hu, "Transient behavior of subthreshold characteristics of fully depleted SOI MOSFETs," *IEEE Electron Device Lett.*, vol. 12, p. 518, 1991.

[34] M.R. Tack, M. Gao, C.L. Claeys, and G.J. Declerck, "The multistable charge-controlled memory effect in SOI MOS transistors at low temperatures," *IEEE Trans. Electron Devices*, vol. 37, p. 1373, 1990.

[35] J. Wang, N. Kistler, J. Woo, and C.R. Viswanathan, "Threshold voltage instability at low temperatures in partially depleted thin-film SOI MOSFETs," *IEEE Electron Device Lett.*, vol. 12, p. 300, 1991.

[36] C. Leroux, J. Gautier, A.J. Auberton-Hervé, B. Giffard, and M. Spalanzani, "Parasitic transients induced by floating substrate effect and bipolar transistor on SOI technologies," *ESSDERC'91 Conf. Proc., Microelectron. Eng.*, vol. 15, p. 199, 1991.

[37] L.J. McDaid, S. Hall, P.H. Mellor, W. Eccleston, and J.C. Alderman, "Explanation of negative differential resistance in SOI MOSFETs," *ESSDERC'89 Conf. Proc.*, Springer, Berlin, p. 885, 1989.

[38] L.T. Su, K.E. Goodson, D.A. Antoniadis, M.I. Flik, and J.E. Chung, "Measurement and modeling of self-heating effects in SOI n-MOSFETs," *IEDM'92 Conf. Proc.*, p. 357, 1992.

[39] R.J.T. Bunyan, M.J. Uren, J.C. Alderman, and W. Eccleston, "Use of noise thermometry to study the effects of self-heating in submicrometer SOI MOSFETs," *IEEE Electron Device Lett.*, vol. 13, p. 279, 1992.

[40] M. Berger and Z. Chai, "Estimation of heat transfer in SOI MOSFETs," *IEEE Trans. Electron Devices*, vol. 38, p. 871, 1991.

[41] T. Elewa, F. Balestra, S. Cristoloveanu, I. Hafez, J.P. Colinge, A.J. Auberton-Hervé, and J. Davis, "Performance and physical mechanisms in SIMOX MOS transistors operated at very low temperature," *IEEE Trans. Electron Devices*, vol. 37, p. 1007, 1990.

[42] D. Yachou, J. Gautier, and C. Raynaud, "Self-heating effects on static and dynamic SOI operation," *ESSDERC'93 Conf. Proc.*, Frontières, Gif-sur-Yvette, France, p. 695, 1993.

[43] M. Haond and O. Le Néel, "Lateral isolation in SOI CMOS technology," *Solid-State Technol.*, p. 47, 1991.

[44] M. Matloubian, R. Sundaresan, and H. Lu, "Measurement and modeling of the sidewall threshold voltage of mesa-isolated SOI MOSFETs," *IEEE Trans. Electron Devices*, vol. 36, p. 938, 1989.

[45] B. Mazhari, S. Cristoloveanu, D.E. Ioannou, and A.L. Caviglia, "Properties of ultra-thin wafer-bonded silicon on insulator MOSFETs," *IEEE Trans. Electron Devices*, vol. 38, p. 1289, 1991.

[46] T. Elewa, B. Kleveland, S. Cristoloveanu, B. Boukriss, and A. Chovet, "Detailed analysis of edge effects in SIMOX–MOS transistors," *IEEE Trans. Electron Devices*, vol. 39, p. 874, 1992.

[47] H.K. Lim and J.G. Fossum, "Threshold voltage of thin-film silicon-on-insulator (SOI) MOSFETs," *IEEE Trans. Electron Devices*, vol. 30, p. 1244, 1983.

[48] B. Mazhari and D.E. Ioannou, "Surface potential at threshold in thin-film SOI MOSFETs," *IEEE Trans. Electron Devices*, vol. 40, p. 1129, 1993.

[49] D. Flandre and A. Terao, "Extended theoretical analysis of the steady-state linear behavior of accumulation-mode, long-channel p-MOSFETs on SOI substrates," *Solid-State Electron.*, vol. 35, p. 1085, 1992.

[50] G. Ghibaudo, "New method for the extraction of MOSFET parameters," *Electron. Lett.*, vol. 24, p. 543, 1988.

[51] H.S. Wong, M.H. White, T.J. Krutsick, and R.V. Booth, "Modeling of transconductance degradation and extraction of threshold voltage in thin oxide MOSFETs," *Solid-State Electron.*, vol. 30, p. 953, 1987.

[52] A.M. Ionescu, S. Cristoloveanu, and A. Rusu, "Derivative methods for threshold voltage extraction," *16th Ann. Semiconductor Conf. Proc. (CAS'93)*, Sinaia, Romania, p. 197, 1993.

[53] A. Terao, D. Flandre, E. Lora-Tomayo, and F. Van der Wiele, "Measurement of threshold voltages of thin-film accumulation-mode PMOS/SOI transistors," *IEEE Electron Device Lett.*, vol. 12, p. 682, 1991.

[54] S. Veeraraghavan and J.G. Fossum, "Short-channel effects SOI MOSFETs," *IEEE Trans. Electron Devices*, vol. 36, p. 522, 1989.

[55] Y. Ohmura, S. Nakashima, K. Izumi, and T. Ishii, "0.1-μm-gate, ultrathin film CMOS devices using SIMOX substrate with 80-nm-thick buried oxide layer," *Proc. IEDM'91*, p. 675, 1991.

[56] F. Balestra, M. Benachir, J. Brini, and G. Ghibaudo, "Analytical models of subthreshold swing and threshold voltage for thin- and ultra-thin-film SOI MOSFETs," *IEEE Trans. Electron Devices*, vol. 37, p. 2303, 1990.

[57] H.T. Chen and R.S. Huang, "An analytical model for back-gate effects on ultrathin-film SOI MOSFETs," *IEEE Electron Device Lett.*, vol. 12, p. 433, 1991.

[58] J.H. Sim and J.B. Kuo, "An analytical back-gate bias effect model for ultrathin SOI CMOS devices," *IEEE Trans. Electron Devices*, vol. 40, p. 755, 1993.

[59] J. Brini, M. Benachir, G. Ghibaudo, and F. Balestra, "Threshold voltage and subthreshold slope of the volume-inversion MOS transistor," *IEE Proc.-G*, vol. 138, p. 133, 1991.

[60] K. Suzuki, T. Tanaka, Y. Tosaka, H. Horie, Y. Arimoto, and T. Itoh, "Analytical surface potential expression for thin-film double-gate SOI MOSFETs," *Solid-State Electron.*, vol. 37, p. 327, 1994.

[61] R.J. Van Overstraeten, G.J. Declerck, and P.A. Mulls, "Theory of the MOS transistor in weak inversion," *IEEE Trans. Electron Devices*, vol. 22, p. 282, 1975.

[62] D.J. Wouters, J.P. Colinge, and H.E. Maes, "Subthreshold slope in thin-film SOI MOSFETs," *IEEE Trans. Electron Devices*, vol. 37, p. 2022, 1990.

[63] F. Balestra, "Analysis of the dependence of the subthreshold swing and the threshold voltage on the substrate voltage of thin-film SOI MOSFETs. Extraction of the interface state densities," *Solid-State Electron.*, vol. 35, p. 1783, 1992.

[64] J.P. Colinge, F. Van de Wiele, and D. Flandre, "Subthreshold slope of accumulation-mode P–channel SOI MOSFETs," *1993 IEEE Int. SOI Conf. Proc.*, p. 146, 1993.

[65] G. Ghibaudo and B. Cabon, "Influence of hot electron induced aging on the dynamic conductance of short channel MOSFETs," *Solid-State Electron.*, vol. 30, p. 1049, 1987.

[66] S.C. Sun and J.D. Plummer, "Electron mobility in inversion and accumulation layers on thermally oxidized silicon surfaces," *IEEE Trans. Electron Devices*, vol. 27, p. 1497, 1980.

[67] T. Ouisse, S. Cristoloveanu, and G. Borel, "Influence of series resistances and interface coupling on the transconductance of fully depleted silicon-on-insulator MOSFETs," *Solid-State Electron.*, vol. 35, p. 141, 1992.

[68] B. Cabon–Till, G. Ghibaudo, and S. Cristoloveanu, "Influence of source-drain series resistance on MOSFET field-effect mobility," *Electron. Lett.*, vol. 21, p. 457, 1985.

[69] J. Wang, N. Kistler, J. Woo, and C.R. Viswanathan, "Mobility-field behavior of fully depleted SOI MOSFETs," *IEEE Electron Device Lett.*, vol. 15, p. 117, 1994.

[70] P. McLarty, S. Cristoloveanu, and O. Faynot, "Parameter extraction in MOSFETs with ultra-thin gate oxides," *Solid-State Electron.*, in press, 1995.

[71] H. Haddara, T. Elewa and S. Cristoloveanu, "Static and dynamic transconductance model for depletion-mode transistors: a new characterization method for silicon on insulator materials," *IEEE Electron Device Lett.*, vol. 9, p. 35, 1988.

[72] D.P. Vu and J.C. Pfister, "Characterization of beam-recrystallized Si films and their Si–SiO_2 interface in silicon-on-insulator structures," *Appl. Phys. Lett.*, vol. 48, p. 50, 1986.

Chapter 9

Transistor-Based Characterization Techniques

Sophisticated characterization methods have been developed, essentially during the last decade, with the aim of taking a closer look at the electrical properties of SOI materials. Even after complete processing of integrated circuits, the interface quality and some of the crystalline imperfections of the incipient SOI wafer are still transparent in transistor characteristics.

In general, selected pulses or a.c. signals are applied on the gate to provoke individual responses from a given type of defect, which in turn are reflected by the drain current variation. Interface, volume, or edge defects, slow and fast traps, and shallow and deep centers may thus be identified.

Most of these techniques, initially elaborated for bulk-Si technology, had to be carefully tuned to comply with the requirements of SOI structures. A few methods—the dual-gate transient current, for instance—have been especially designed for SOI. This SOI *customization* is not always straightforward. It includes convertibility from capacitance to current measurements (in DLTS and Zerbst methods) as well as methodology to deal with or to exploit interface coupling. In this chapter, details will be given on the most relevant and successful techniques applied to SOI transistors.

9.1 Profiling Vertical Inhomogeneities

The properties of thin SOI films may vary with the distance from the top surface to the buried insulator. Such *vertical inhomogeneities* are often due to enhanced defect generation that takes place near the film–insulator interface during material synthesis (epitaxial growth, oxygen or nitrogen implantation, annealing, etc.). This typically originates from a mismatch between Si and the insulator, in terms of lattice parameters and thermal expansion coefficients. The usual consequences are increased residual stress, contamination, and crystal imperfections near the back interface.

In nonuniform films, the average values of resistivity, mobility, and autodoping are rather meaningless, since they may appear to be very poor even if the properties of the top portion of the layer reach bulk-Si standards. Moreover, contrasting and misleading average parameters may be obtained in SOI layers with identical quality but different thickness. Hence, the evaluation of the electrical depth profiles is a prerequisite to understanding the growth mechanisms and nature of SOI films and has, in the past, enabled optimization of SOS and SIMOX structures. A very efficient method, described below, combines the MOS field effect with magnetoelectric measurements.

9.1.1 Measurement Setup

Figure 9.1a shows the configuration of an MOS–Hall device, which is nothing but a seven-terminal depletion-mode n-channel MOSFET. Besides the gate and the two end contacts (source and drain), there are four small lateral n^+ contacts. The longitudinal voltage drop, measured between contacts $H_{1,2}$ (or $H_{3,4}$), gives the conductance and magnetoresistance values, free of the influence of source/drain junctions. Contacts $H_{2,4}$ (or $H_{1,3}$) serve for Hall voltage measurements. In the case where the device is too large or the Hall contacts are located too close to the terminals, the Hall voltage must be corrected for short-circuiting geometrical effects, as discussed in Section 5.3 (see Fig. 5.4).

The device is operated with a very low drain bias ($V_D < 50\,\mathrm{mV}$) in order to keep the channel thickness constant along the device. The Hall effect and magnetoconductance measurements are performed as a function of V_G. By gradually depleting the film, the thickness w of the conducting region is reduced, and the measured (average) transport

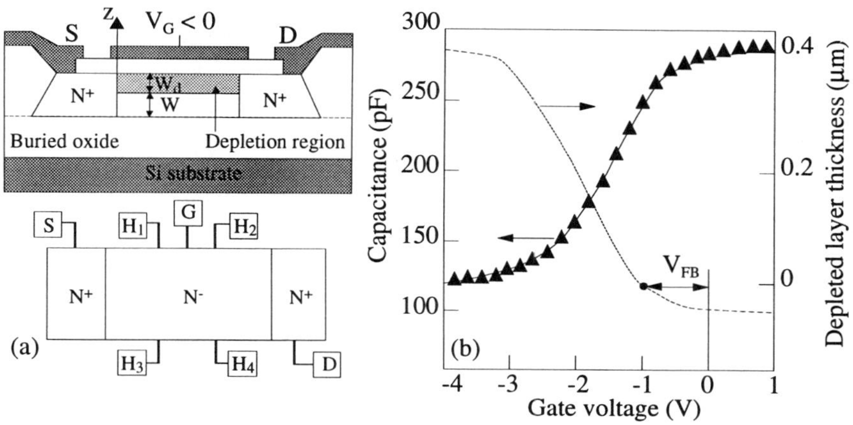

Figure 9.1 (a) Configuration of a seven-terminal MOS–Hall device on SOI and (b) variations of MOS capacitance and depletion depth with gate voltage for n-type SOS film 0.6 μm thick (after Lee et al.[1]).

properties are modified. A procedure involving differentiation provides the contribution of the infinitesimal layer Δw, situated at the limit between depletion and "active" regions.

The gate voltage dependence of the depletion depth w_d is determined from small-signal $C(V_G)$ capacitance measurements (Fig. 9.1b):

$$w_d = \epsilon_{si} l_x l_y \left(\frac{1}{C(V_G)} - \frac{1}{C_{ox}} \right) \tag{9.1}$$

As mentioned in Chapter 6, the accuracy of $C(V_G)$ measurements in thin SOI films is affected, at low frequency, by interface traps and, at high frequency, by series resistances. The experiment is optimized by (1) selecting the frequency range ($\sim$ 100 kHz), (2) compensating series resistances for each bias V_G, and (3) interconnecting all n^+ contacts. Even better is the use of a distinct *interdigitated* capacitor located on the same chip.[1]

The origin of the depth axis corresponds to the flat-band voltage. It is determined from the $C^{-2}(V_G)$ curves, which are quasi-linear and intercept the value $C^{-2} = C_{ox}^{-2}$ for $V_G = V_{FB}$.

9.1.2 Profiling Method

The vertical profile of a parameter A consists of successive local values $A^*(z)$ that are extracted from the experimental average values $\overline{A}(V_G)$. For example, the average resistance (or magnetoresistance) of a conductive sheet of thickness w is

$$\overline{R}(V_G) = \frac{l_x}{l_y} \left(\int_0^w \sigma^*(z)\, dz \right)^{-1} \tag{9.2}$$

Conversely, the local conductivity at $z = w$ is given by

$$\sigma^*(z = w) = \frac{l_x}{l_y} \times \frac{d\overline{R}^{-1}}{dV_G} \times \frac{dV_G}{dw} \tag{9.3}$$

The theory of magnetotransport effects in inhomogeneous films proceeds from magnetoconductivity tensor. In general, galvanomagnetic coefficients are not simple depth-integrals of the homogeneous material coefficients, since they also include weighing factors. We must keep in mind that not only the conductivity tensor but also the Maxwell and conservation equations are depth dependent. The use of boundary conditions will finally determine the expressions to be averaged.[2, 3, 4]

The measured average Hall mobility, given by Equation (5.4), is expressed as

$$\overline{\mu}_H(V_G) = \left(\int_0^w \mu_H^*(z)\sigma_0^*(z)\, dz \right) \left(\int_0^w \sigma_0^*(z)\, dz \right)^{-1} \tag{9.4}$$

where $\sigma_0^*(z)$ is conductivity profile at zero magnetic field.

Differentiating Equation (9.4) and using Equation (9.3) yields for the local Hall mobility

$$\mu_H^*(w) = \frac{l_x}{l_y} \frac{1}{\sigma_0^*} \frac{d\left(\overline{\mu}_H \overline{R}_0^{-1}\right)}{dV_G} \frac{dV_G}{dw} = \overline{\mu}_H - \overline{R}_0(V_G) \frac{d\overline{\mu}_H}{d\overline{R}_0} \tag{9.5}$$

The profiles of $\sigma^*(w)$ and $\mu_H^*(w)$ are free of any theoretical assumption except the validity of Equation (9.1).

It is usual but not always rigorous[1] to infer the *drift* mobility profile $\mu_0^*(w)$ from $\mu_H^*(w)$ by considering $\mu_H = r_H \mu_0$. The scattering coefficient

[1]This relation only holds for isotropic (spheric) energy valleys; this is not the case with n-Si, where the longitudinal and transversal values of the effective mass are different.

r_H is approximated by $r_H = 3\pi/8$ when acoustic phonon scattering dominates (typically at room temperature), and by $r_H = 315\pi/512$ for prevailing ionized impurity scattering.

More accurate is the independent determination of the drift mobility from *longitudinal* magnetoresistance experiments. In n-type bulk Si, the conduction band is composed of six equivalent isoenergetic valleys. The superposed contributions of these six anisotropic valleys lead to a magnetoresistance (MR) even if the applied magnetic field is parallel to the current. The fit of longitudinal MR versus magnetic field curves $\sigma_0^*(w)/\sigma_B^*(w)$ provides, for each depth, the drift mobility and anisotropy coefficient.[1]

The carrier concentration profile $n^*(w)$ may now be resolved by combining the local values of conductivity and drift mobility: $n^*(w) = \sigma_0^*/q\mu_0^*$. By contrast, the profile of ionized impurities is deduced from capacitance measurements

$$N_D^+(w) = -\frac{C(V_G)}{ql_x l_y}\frac{dV_G}{dw} \tag{9.6}$$

It is essential to realize that, in imperfect SOI films, $n^*(w)$ and $N_D^+(w)$ may be quite different. Long-range potential fluctuations, caused by crystalline defects, result in potential wells where carriers may be localized. Only a portion $n^*(w)$ of the total number of electrons available in the conduction band ($n_{tot} = N_D^+$) are really free and contribute to conductance measurements. The average energy Γ of potential fluctuations is evaluated from $\Gamma(w) = kT\ln(N_D^+/n^*)$.[5]

9.1.3 Experimental Data

Most experiments have been conducted on early SOS and SIMOX films, where the vertical homogeneity of the silicon overlay used to be a critical technology issue. Typical profiles, derived using Equation (9.3) and shown in Figure 9.2a, demonstrate the conductivity degradation with depth in SOS layers.

High-magnetic-field longitudinal magnetoresistance profiles are reproduced in Figure 9.2b. The unusual large value of the magnetoresistance coefficient, $\sigma_0^*/\sigma_B^*(w)$, is evidence for the strong stress affecting SOS films. The strain-induced modification of the conduction band (four isoenergetic valleys instead of six valleys) invalidates the use of bulk-Si models. Curve fitting with a proper four-valley model shows that

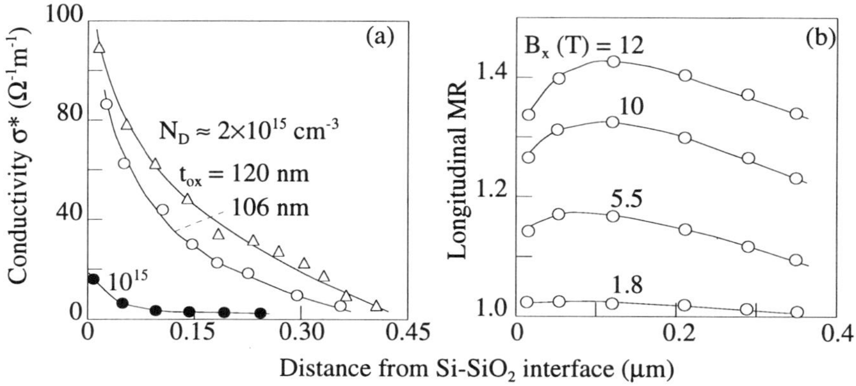

Figure 9.2 In-depth profiles of (a) conductivity and (b) longitudinal magnetoresistance in n-type SOS films (after Lee et al.[1]).

the magnetoresistance decrease is due, near the top interface, to surface scattering and, in the volume, to degraded drift mobility and enhanced anisotropy.

As-measured Hall mobilities versus gate voltage are presented in Figure 9.3a for a p-type SIMOX film.[6] In the accumulation region, the average mobility is larger and increases as the temperature is lowered. In depletion, the saturation of $\overline{\mu}_H$ with gate voltage (i.e., for $V_G > 2\,\mathrm{V}$ at 300 K) does not reflect a limit of mobility degradation, but simply confirms that the thickness of the active region cannot be further reduced because the depletion region has reached a maximum extension.

More physical insight is gained by transposing these curves into mobility profiles (Eq. (9.5)). Several results for early SOS and SIMOX films are collected in Figure 9.3b. The mobility decreases almost linearly with depth in SOS, whereas, in low-temperature annealed SIMOX, it is flat in the upper half of the film and then degrades rapidly close to the buried interface.

The scattering coefficient r_H is determined from the ratio between the drift mobility and Hall mobility profiles of Figure 9.3b, weighted by the anisotropy coefficient.[7] For analysis of the scattering mechanisms, it is very instructive to draw the local mobility at a given depth versus temperature (Fig. 9.4a).

In SIMOX, the mobility is governed, as in bulk Si, by acoustic phonon

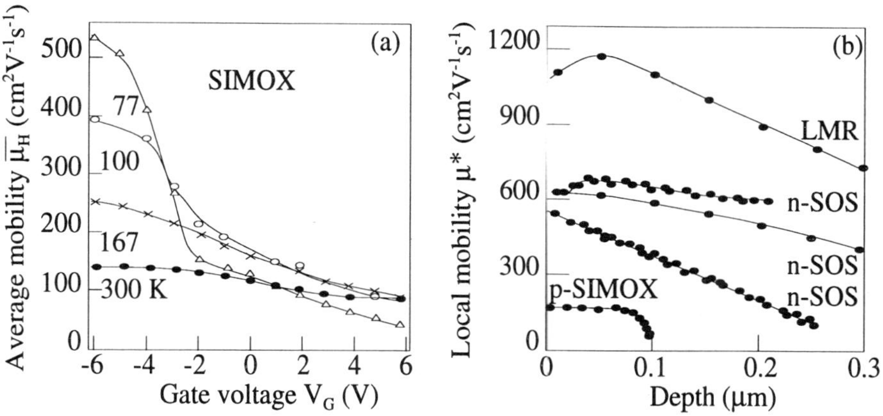

Figure 9.3 (a) Average Hall mobility versus gate voltage for various temperatures (low-temperature annealed p-SIMOX film 0.25 μm thick[6]). (b) In-depth profiles of drift (LMR) and Hall mobilities in various SOS and SIMOX films.[1, 8, 9, 6]

scattering ($\mu_H \sim T^{-1.5}$, above 140 K) and ionized impurities ($\mu_H \sim T^{+1.5}$, below 120 K). The case of SOS wafers is different, since the carrier scattering mechanisms are more complex and have a contrasting temperature behavior according to the depth. For example, ionized impurity scattering events dominate in the middle of the film, even at room temperature.

In Figure 9.4b are compared the vertical profiles of ionized impurities (obtained from $C(V_G)$ data with Equation (9.6)) and mobile carriers (deduced from conductivity and drift mobility values). These profiles coincide at 300 K in the surface region, but show an increasing discrepancy for lower temperatures and deeper regions, due to potential fluctuations and inherent carrier localization. The number of mobile carriers is defined by the position of the Fermi level with respect to the potential well edge. The average energy of potential wells is an electrical image of the crystalline disorder: $\Gamma < 10$ meV in SIMOX and $\Gamma > 20$–40 meV in SOS films.[6, 7]

It is now clear that potential fluctuations render inadequate the use of $C(V_G)$ measurements for reliable evaluations of carrier concentration. The apparently innocent approximation $n^* = N_D^+$ involves an overestimation of the mobile carrier density, which in turn leads to a severe

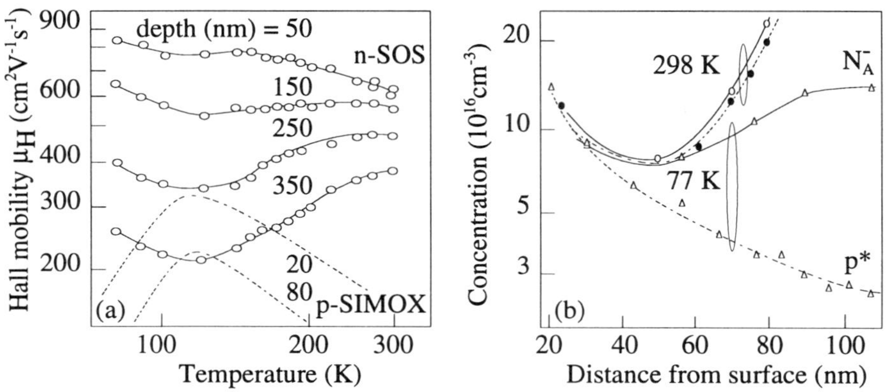

Figure 9.4 (a) Local Hall mobility against temperature for different distances from the top interface in n-SOS (solid curves[10]) and p-SIMOX (dashed curves[6]). (b) In-depth concentration profiles for ionized acceptor impurities (solid curves, deduced from $C(V_G)$ data) and for really free holes (dashed curves) in a SIMOX film (after Cristoloveanu et al.[6]).

underestimation of the drift mobility deduced from experimental conductivity profiles.

In conclusion, MOS–Hall devices offer a rather complete method of investigating in-depth profiles in SOI films. There are three main limitations: (1) the existence of the screening Debye length at the limit between depleted and active regions, (2) the need for full device processing, and (3) the appropriate doping level. The doping must be low enough to enable a deep exploring region and high enough to prevent full depletion.

To answer these problems, thinning techniques have been developed based on the use of the same integral equations (9.2) or (9.4) and differential principles. However, when the gate action is replaced by chemical thinning, sputtered erosion, or anodic oxidation, other drawbacks arise from the control of the surface smoothness, thickness uniformity, damage formation, and contamination. All thinning-type profiles are intrinsically limited by the formation of unforeseeable space-charge regions due to ambient charges. This inconvenience is under control only with MOS–Hall devices.

9.2 Charge-Pumping Technique

9.2.1 Physical Principle and Basic Experimental Setup

Charge pumping (CP) is a very sensitive method for the characterization of low concentrations of interface traps (10^9 cm^{-2}eV^{-1}) in short-channel MOS devices. The principle of conventional CP in enhancement-mode MOSFETs on bulk Si is to repeatedly switch the gate from inversion to accumulation and vice versa, while keeping the source and drain contacts grounded or slightly reverse biased.

In inversion, some of the minority carriers provided by the source and drain reservoirs are trapped on the interface states. During the falling edge of the pulse, the mobile minority carriers are collected rapidly from the inversion layer by the source and drain, and then the trapped carriers recombine with majority carriers provided by the substrate. This recombination gives rise to an *average* charge-pumping current I_{CP} in the substrate terminal, which is a frequency-amplified measure of the density of interface states.

The adaptation of CP to SOI transistors requires a contact with the Si film.[11, 12] Either five-terminal MOSFETs or gate-controlled p–i–n diodes may be used.[13] Figure 9.5a shows that the two terminals of the p^+–p^-–n^+ diode have different functions. The n^+ contact controls the charge of the inversion layer and can be more or less reverse biased. The p^+ terminal, which supplies the majority carriers, is grounded and therefore plays the same role as the body contact in five-terminal transistors.

The gate is pulsed using a trapezoidal or sawtooth waveform generator. An additional amplifier is needed in order to reverse the experiment and pulse the back gate (substrate). The other bias sources and picoammeter are integrated in an HP–4145 or equivalent system. The experiment can be computer controlled, and dedicated software is available.

9.2.2 Model

The CP current is proportional to the average concentration of interface traps D_{it}, frequency f, and surface potential range $\Delta\psi_s$ swept through during the pulse

$$I_{CP} = q^2 f W L D_{it} \Delta\psi_s \tag{9.7}$$

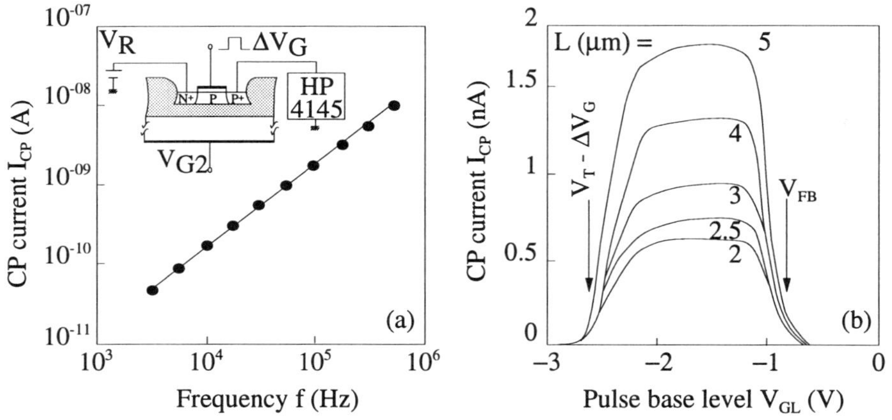

Figure 9.5 (a) Configuration of CP measurements in SOI diodes and variation of CP current versus frequency ($\Delta V_G = 5\,\text{V}$, $V_{GL} = -2\,\text{V}$, $V_R = 0$, $t_r = t_f = 0.1\,\mu\text{s}$); (b) CP current versus base level of the front-gate pulse in SIMOX diodes with different lengths.

By including in the model the direct carrier emission to the conduction and valence bands, Groeseneken et al. found for trapezoidal pulses:[14]

$$\Delta\psi_s = \frac{2kT}{q}\ln\left[v_{th}n_i\sqrt{\sigma_n\sigma_p}\sqrt{t_r t_f}\frac{|\,V_{FB}-V_T\,|}{|\,\Delta V_G\,|}\right] \tag{9.8}$$

where v_{th} is the thermal velocity of carriers, σ_n and σ_p are the capture cross sections for electrons and holes, and t_r and t_f are the rise and fall times, respectively, of the ΔV_G pulse.

If sawtooth rather than trapezoidal waveforms are used to pulse the gate, Equation (9.7) is slightly modified:[14]

$$I_{CP} = 2qfkTWLD_{it}\ln\left[v_{th}n_i\sqrt{\sigma_n\sigma_p}\frac{\sqrt{\alpha(1-\alpha)}}{f}\frac{|\,V_{FB}-V_T\,|}{|\,\Delta V_G\,|}\right] \tag{9.9}$$

where α is the fraction of the period when the V_G signal is rising.

A very meaningful CP curve (Fig. 9.5b) is obtained by varying the bottom level V_{GL} of the trapezoidal pulse while keeping the pulse magnitude ΔV_G and frequency fixed. Three different regions can be distinguished on this *rectangle-like* CP signature: (1) as long as the top level of the pulse does not induce carrier inversion, there is no trapping and

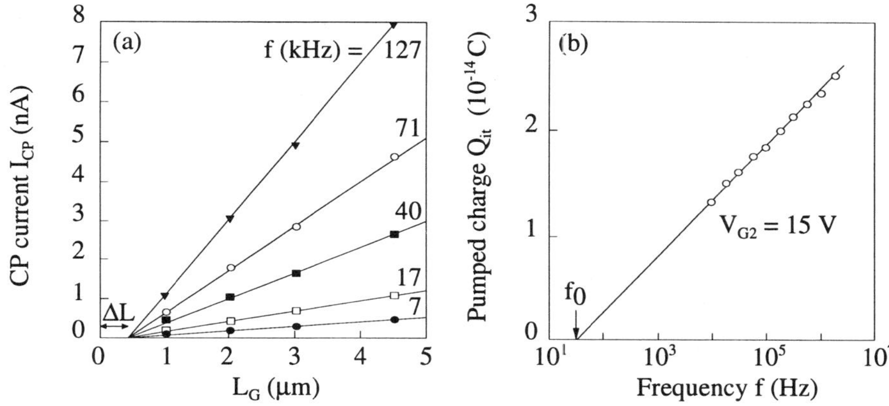

Figure 9.6 (a) Charge-pumping current versus gate length, for various frequencies of rectangular pulses, and (b) pumped charge per period versus frequency for triangular pulses applied on a SIMOX diode.

$I_{CP} \simeq 0$, (2) a maximum CP current, given by Equations (9.7) and (9.8), is obtained when the top level exceeds V_T, and (3) for higher V_{GL} values ($V_{GL} > V_{FB}$), the surface does not return in accumulation; hence, the recombination rate and I_{CP} vanish. It follows that the left- and right-hand edges of the "rectangle" correspond to $(V_T - \Delta V_G)$ and V_{FB}, respectively, whereas the plateau level gives D_{it}.

9.2.3 Parameter Extraction

It is important to determine the *effective channel length L* in a preliminary step of the characterization procedure, in order for this information to be included in the following steps. Diodes with various gate lengths are pulsed with identical trapezoidal waveforms. Drawing I_{CP} as a function of the designed gate length L_G results in a straight line (Fig. 9.6a) in accordance with Equation (9.7). The intercept with the horizontal axis gives the channel-length correction factor $\Delta L = L_G - L$, which primarily originates from lateral diffusion of source/drain terminals beneath the gate. An advantage of CP is that the accuracy can be further improved by repeating the measurement of $I_{CP}(L_G)$ for various frequencies and averaging the extracted values. A similar procedure may be applied to determine the effective width of narrow–channel devices.

The *average concentration of interface traps* D_{it} is extracted with Equation (9.7) from the slope of $I_{CP}(f)$ curves that are linear over several decades (Fig. 9.5a). Alternatively, D_{it} can be deduced from the plateau level of Figure 9.5b or from the slope of the semilogarithmic plot $Q_{it}(\ln f)$, where $Q_{it} = I_{CP}/fWL$ is the normalized charge of interface traps (per cycle and unit area) pulsed with sawtooth waveforms (Eq. (9.9) and Fig. 9.6b).

Information on the *capture cross sections* for electrons and holes $\sigma_{n,p}$ can be obtained from Figure 9.6b. The geometric-mean capture cross section $\sqrt{\sigma_n \sigma_p}$ is found from the intercept f_0 with the horizontal axis, i.e., for $Q_{it} = 0$ in Equation (9.9):

$$\sqrt{\sigma_n \sigma_p} = \frac{1}{v_{th} n_i} \frac{|\Delta V_G|}{|V_{FB} - V_T|} \frac{f_0}{\sqrt{\alpha(1-\alpha)}} \tag{9.10}$$

The values σ_n and σ_p can be separated by performing low-temperature CP measurements[15] or by combining CP with gate-controlled leakage current measurements (Section 7.3).

The *flat-band voltage* V_{FB} and *threshold voltage* V_T are evaluated from the edges of the CP curves of Figure 9.5b. An attractive feature of CP is to provide both V_T and V_{FB} values from a *unique* experimental curve, even if V_T is neither identical to nor as accurate as the "extrapolated" V_T found from the static $I_D(V_G)$ MOSFET characteristics. It has been verified that V_T and V_{FB} are very reliable parameters to monitor the change in substrate bias[13] and the number of defects generated by irradiation.[16]

In the case in which the channel is composed of several regions with distinct trap concentrations and threshold voltages, the shape of the CP curve becomes more complex and reflects the superposition of corresponding "rectangles." The deconvolution of this curve provides the individual parameters of each region and is used to analyze the defects localized on the edges[17] or those induced by hot-carrier injection near the drain.[20]

Increasing the reverse bias V_R causes the CP curves of Figure 9.7a to shrink. Although the right edge, corresponding to V_{FB}, is not modified by V_R, the left edge, related to V_T, is shifted towards a higher V_G bias due to the increase in surface potential ψ_s and maximum thickness $x_{d,max}$ of the depletion region with V_R:

$$x_{d,max} = \sqrt{\frac{2\epsilon_{si}(V_R + 2\phi_F)}{q\overline{N_A}}} \tag{9.11}$$

where ϕ_F is the Fermi potential and ϵ_{si} is the permittivity of silicon.

The decrease of the plateau level is explained by the growth of the junction-controlled depletion region, where the interface traps cannot be pumped any more. The total number of traps contributing to I_{CP} is reduced, which is equivalent to a shorter effective channel length.

The *average doping* $\overline{N_A}$ is determined from the variation of V_T with V_R which, for the case of decoupled interfaces (partial depletion), is expressed by

$$V_T = V_{FB} + V_R + 2\phi_F + \frac{1}{C_{ox}}\sqrt{2\epsilon_{si}q\overline{N_A}(V_R + 2\phi_F)} \tag{9.12}$$

where C_{ox} is the oxide capacitance at the pulsed gate.

Once the values of $V_T(V_R)$ and V_{FB} are deduced from Figure 9.7a, the expression $(V_T - V_R - V_{FB} - 2\phi_F)$ can be plotted versus $\sqrt{V_R + 2\phi_F}$. After the successive iterations needed to calculate ϕ_F, a straight line is obtained, the slope of which is proportional to $(\overline{N_A})^{0.5}$.

The *front-interface* results obtained from Figures 9.5–9.8 are fully consistent with the technological parameters: $\Delta L = 0.7\,\mu\text{m}$, $D_{it} = 1.8 \times 10^{10}\,\text{cm}^{-2}\text{eV}^{-1}$, $\sqrt{\sigma_n\sigma_p} = 4 \times 10^{-16}\,\text{cm}^2$, $V_T = 1\,\text{V}$, $V_{FB} \simeq 0V$, and $\overline{N_A} = 1.2 \times 10^{17}\,\text{cm}^{-3}$.

This set of experiments can equally be conducted by applying very large pulses on the back gate. It was found that the density of traps at the *back* interface is highly sensitive to the processing conditions; for SIMOX, it is one order of magnitude higher ($D_{it_2} = 10^{11} \div 10^{12}\,\text{cm}^{-2}\text{eV}^{-1}$) than at the front interface. When the two gates are pulsed simultaneously (with $\Delta V_{G_2} \simeq (t_{ox_2}/t_{ox_1})\Delta V_{G_1}$), the potential in the volume of very thin films varies as well. If *bulk traps* are present, their pumping yields a low-frequency CP component.[13]

9.2.4 Coupling Effects on Charge Pumping

In the inset of Figure 9.7b, the charge pumping current was measured by applying constant pulses on the front gate and varying V_{G_2}. A sharp peak is observed in the region where the back interface is *depleted*. This

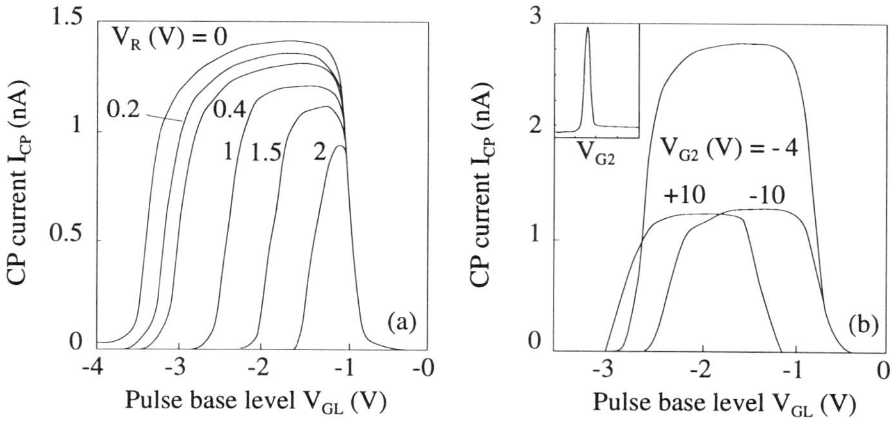

Figure 9.7 Front-gate CP current versus pulse base level for various (a) reverse-junction biases and (b) back-gate voltages.

peak indicates that pulsing the front gate results in a scanning of the back surface potential, which enables the pumping of some of the back-interface traps.

Another consequence of the interface coupling is that the values of the threshold voltage, flat-band voltage, and swept potential range $\Delta\psi_{s_1}$ at the front surface depend on the status (accumulation, depletion, or inversion) of the back interface. This is the reason for the small difference existing between the CP currents measured for inversion and accumulation at the back interface. One might envisage extracting the concentration of back-interface traps D_{it_2} from the peak of I_{CP}, but this would require solving numerically the time-dependent Poisson equation across the film.[18]

Shown in Figure 9.7b is the modification of the "rectangle–like" CP curve as the back interface goes from inversion to depletion and accumulation. The shifts of the left and right edges are used to monitor the variations of V_{T_1} and V_{FB_1} versus V_{G_2}.

Two aspects are of practical relevance for reliable measurements. First, the possibility of extracting V_{T_1} easily, even for inversion at the opposite interface, is extremely useful because the standard extrapolation technique from $I_D(V_{G_1})$ MOSFET curves is no longer accurate when both channels are activated. Second, for the front-interface traps to be

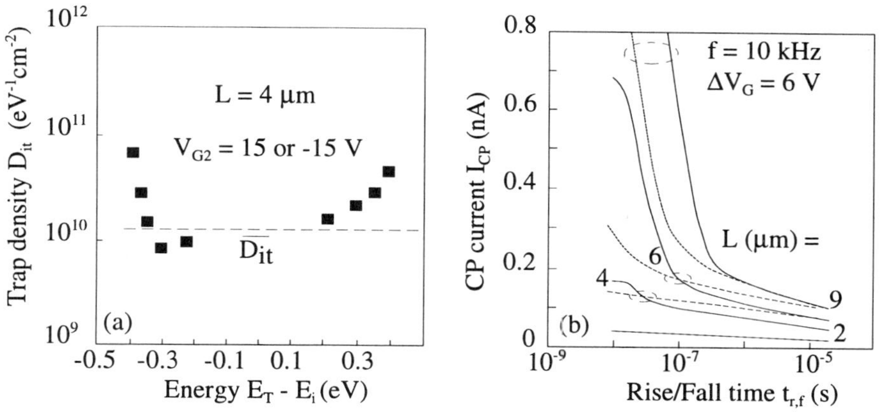

Figure 9.8 (a) Energy profile of front interface traps measured by modifying the rise and fall times of the pulse, and (b) CP current versus rise and fall times ($t_r = t_f$) for various diode lengths, with the back interface in inversion (———) or in accumulation (- - -) (after Ouisse et al.[13]).

accurately characterized, the back gate has to be maintained in strong inversion or accumulation.

9.2.5 Profiling the Interface Traps

The *spatial* distribution of interface traps along the channel is scanned by varying the junction bias. The prevailing idea is that the traps located in the widening depletion region will gradually stop contributing to the recombination process. The CP current is measured, for different V_R, versus the top level of the pulse. A differential technique is used to evaluate the incremental change of the pumped area and the lateral profile.[19]

The interface trap profile $D_{it}(E)$ in the *energy* gap is resolved by varying independently the rise time t_r and fall time t_f of the pulse:[14]

$$D_{it}(E_{h,e}) = -\frac{t_{r,f}}{qfWLkT} \times \frac{dI_{CP}}{dt_{r,f}} \tag{9.13}$$

The corresponding energy levels $E_{e,h}$, situated respectively in the upper or lower half-regions of the gap, are given by

$$E_{e,h} = E_i \mp kT \ln \left\{ v_{th} n_i \sigma_{n,p} \left| \frac{V_{FB} - V_T}{\Delta V_G} \right| t_{f,r} \right\} \tag{9.14}$$

For instance, reducing the rise time shifts E_h towards the valence band but does not affect E_e.

The experimental profile of Figure 9.8a shows that the distribution of $D_{it}(E)$ is almost flat in the midgap region (0.15–0.30 eV below and above the intrinsic energy level) and increases towards the band edges. Very close to the conduction and valence bands, care is needed to avoid dimensional effects (which typically induce distinct front-interface profiles according to the back-gate bias) and to use a more sophisticated CP model and energy-dependent capture cross sections.[13]

9.2.6 Dimensional Effects

Charge pumping depends on several physical time constants (carrier transit times for the formation of the accumulation and inversion channels, carrier lifetime, trapping and emission times) as well as on experimental time constants (period, rise and fall times). The latter parameters have to be longer than the former ones in order to prevent any other source of recombination besides the recombination of trapped minority carriers with majority carriers.

For instance, if all mobile minority carriers are not removed rapidly enough (during the fall of the pulse) before the arrival of majority carriers, an undesirable recombination occurs. This parasitic component of I_{CP}, referred to as a "dimensional effect,"[13] is responsible for an unacceptable overestimation of D_{it}. It obviously happens in long devices (the time for channel formation being proportional to L^2) or if the rise/fall times are too short. Since the shorter the rise/fall times are, the wider the energy range of extracted interface-trap profiles $D_{it}(E)$ will be, it follows that the extension of energy profiles is limited by the occurrence of dimensional effects.

SOI structures are more subjected to dimensional effects because electron or hole reservoirs may exist at the back interface and interact with front-channel carriers. A detailed analysis[13] demonstrates that parasitic recombination takes place essentially during the removal either of majority carriers (if V_{G_2} is in inversion) or of minority carriers (if V_{G_2} is in accumulation). It is clear that, for $V_{G_2} < 0$, the vertical field in the film assists the holes in leaving the surface and reaching the back p^+ reservoir or the p^+ terminal, whereas it is unfavorable for electron evacuation.

The range of confidence for the CP parameters is established by drawing $I_{CP}(\ln t_{r,f})$ curves, which according to Equations (9.7) and (9.8) must be linear. A nonlinearity is evocative for the onset of dimensional effects (Fig. 9.8b). The most critical case, $V_{G_2} > 0$, corresponds to a difficulty in evacuating the holes that are less mobile. Dimensional effects are also revealed by the deformation of the CP peak (inset of Fig. 9.7b), which would become highly asymmetric. However, Figure 9.8b shows that the rise/fall times may actually be small enough: 1 μs for $L = 10\,\mu$m, 100 ns for $L = 4\,\mu$m, 10 ns for $L \leq 2\,\mu$m.

A different dimensional effect occurs in very-short-channel devices. Near the source and drain terminals, the surface potential is higher and its variation, $\Delta\psi_s$, more limited than in the middle of the channel, leading to a less efficient pumping of the corresponding traps. Since Equations (9.7)–(9.9) would lead to an underestimate of D_{it}, it is suitable to couple the CP experiment with 2–D potential simulators.[18] However, the slopes of $I_{CP}(L_G)$ curves are extremely linear (Fig. 9.6a), and the edges of the various curves of Figure 9.5b are length independent, which means that D_{it}, V_T, and V_{FB} are not affected, in diodes longer than 2 μm, by short-channel effects.

In conclusion, CP presents an unrivaled sensitivity and versatility for the interface characterization of short-channel SOI devices. Not only does the frequency act as an internal amplification but also many other experimental parameters can be varied independently to get diversified information. The use of more complex pulse forms (triple-level CP[21]) is under investigation. Dimensional effects do not restrict the technique for normal applications, unless too ambitious profiles are sought in long devices.

9.3 Low-Frequency Noise

9.3.1 Physical and Experimental Principles

The analysis of electrical noise is important in two respects: (1) it provides a practical limit for the operation of integrated circuits, and (2) it represents a powerful tool for the identification and monitoring of oxide, semiconductor, and interface defects. From a mathematical viewpoint, noise is defined as the power spectral density $S_x(f)$ of a time-dependent, but stationary, random function $X(t)$, and it is given by the Fourier transform of the autocorrelation function of $X(t)$.[22]

In a semiconductor device, the noise reflects the current fluctuations brought about by random variations in the carrier mobility and/or concentration. Each physical mechanism involved in current fluctuations is expressed by a specific random distribution (Gaussian, Poisson, Dirac, etc.) and has, in general, a distinct noise spectrum.

In an MOS transistor, the noise level being rather low, caution is needed to eliminate parasitic sources of noise. Some general checkpoints and experimental hints are listed below:

- The measurement system, suggested in Figure 9.9a, should be shielded and the transistor biases $E_{1,2,3}$ supplied by rechargeable batteries.

- The fluctuations originating in the potentiometers $P_{1,2,3}$ are dynamically short-circuited by capacitors.

- The load resistor R_L ($R_L \ll R_{in}$) can be used to determine the drain current.

- A high-performance preamplifier (high gain, low noise) is suitable.

- The computing of the Fourier transform proceeds from the sampling of the random function $V_D(t)$. Both operations are performed by a dedicated spectrum analyzer (HP, Ono Sokki, Advantest, etc.). The maximum frequency of the noise spectrum is less than half of the sampling rate.

The *as-measured* noise $S^\star_{V_D}$, corresponding to drain voltage fluctuations, includes the *pure* drain current noise S_{I_D} and the *background* noise of the system,

$$S^\star_{V_D} = R_E^2 \left[S_{I_D} + S^\star_{1/f} + 4kT \left(\frac{1}{R_L} + \frac{1}{R_{in}} \right) \right] \tag{9.15}$$

where R_E is the parallel combination of R_L, amplifier input resistance R_{in}, and dynamic resistance of the channel, $R_c = \partial V_D / \partial I_D$, known from $I_D(V_D)$ characteristics.

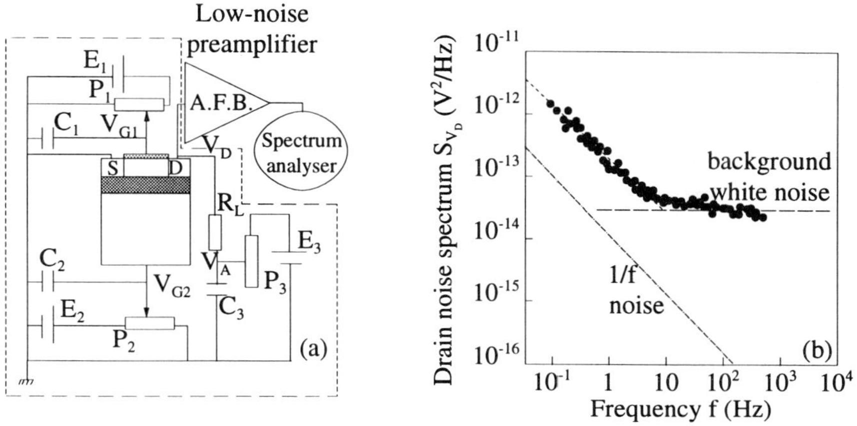

Figure 9.9 (a) Experimental setup for noise measurements and (b) as-measured spectrum of drain voltage fluctuations showing the background $1/f$ and thermal noises of the system.

The background $1/f$ noise $S^{\star}_{1/f}$ and white noise are determined separately by switching off the transistor gate (i.e., $R_c^{-1} \simeq 0$). Figure 9.9b shows the spectra of the total and background noises, from which the pure transistor noise S_{I_D} is extracted using Equation (9.15). For convenience, S_{I_D} can be transformed in equivalent gate voltage noise S_{V_G} or drain voltage noise S_{V_D}:

$$S_{V_G} = S_{I_D}/g_m^2 \qquad\qquad S_{V_D} = R_c^2 \times S_{I_D} \tag{9.16}$$

9.3.2 Model of $1/f$ Noise in MOSFETs

There is strong evidence that the $1/f$ noise of MOS transistors originates from fluctuations in the carrier number and not, as claimed earlier, from mobility fluctuations. The basic mechanism, identified by McWhorter,[23] is the carrier tunneling between the inversion channel and slow oxide traps N_t located at 1–2 nm from the Si–SiO_2 interface. The input-gate voltage spectral density S_{V_G} is given by[23, 24]

$$S_{V_G} = \frac{q^2 \lambda_{ox} k T N_t}{L W C_{ox}^2} \times \frac{1}{f} \tag{9.17}$$

where λ_{ox} is an average tunneling length. The concentrations of slow traps N_t and interface traps D_{it} may be correlated by a first-order approximation $D_{it} \simeq \lambda_{ox} N_t$.[25] S_{V_G} also represents the flat–band voltage spectral density, which results from fluctuations in the oxide charge.[26]

The equivalent drain current noise is obtained by combining Equations (9.16) and (9.17) with the transconductance and current expressions of Chapter 8:[24, 27]

$$\frac{S_{I_D}}{I_D^2} = \frac{q^4 \lambda_{ox}}{\alpha^2 kTLW} \times \frac{N_t}{(C_{ox} + C_d + qD_{it} + C_{inv})^2} \times \frac{1}{f} \tag{9.18}$$

where the inversion layer capacitance is $C_{inv} = (q/\alpha kT)Q_{inv}$, with $\alpha = 1$ in weak inversion and $\alpha = 2$ in strong inversion. In practical cases, a $1/f^{\gamma}$ dependence, with $0.8 \leq \gamma \leq 1.2$, is observed instead of a pure $1/f$ noise.

The advantage of Equation (9.18), also called Reimbold's equation, is to illustrate in detail the influence of the most significant device parameters, from weak inversion to strong inversion and from ohmic region to saturation region. In *weak inversion*, C_{inv} and the variation of C_d are negligible, so the normalized noise factor S_{I_D}/I_D^2 is a constant and does not depend on either drain or gate voltages. In *strong inversion*, however, C_{inv} becomes rapidly prevailing in Equation (9.18). Moreover, a slight increase of the noise factor with V_D is observed due to the decrease in Q_{inv} along the channel, which is summarized by the relation $V_T(V_D) \simeq V_{T_0} + V_D/2$.[28]

A typical noise curve, S_{I_D}/I_D^2 vs. V_G, is composed of a "plateau" in weak inversion, followed by a sharp decrease, proportional to $(V_G - V_T)^{-2}$ in strong inversion. This ideal curve may be altered by inhomogeneous distributions of traps along the channel,[29] in the energy gap,[27] or within the oxide.[30] In case of hot-carrier-induced localized defects near the drain, the noise decreases rapidly as V_D is varied from ohmic to saturation regions and the pinch-off point moves beneath the defective zone.[31]

9.3.3 Other Types of Noise

Several other sources of noise, which may coexist in MOS or bipolar devices, will now be briefly addressed. The *thermal noise* of a resistor R (for instance, the channel resistance) has a *white* spectrum and is expressed in terms of current noise S_I or equivalent voltage noise S_V:

$$S_V = R^2 S_I = 4kTR \tag{9.19}$$

The *diffusion noise* is related to scattering events that modify the direction and magnitude of the carrier velocity:

$$S_I = \frac{4kT}{R} \times \frac{1}{1 + 4\pi^2 \tau_s^2 f^2} \tag{9.20}$$

where the scattering time constant τ_s is very short. There is full equivalence between the diffusion and thermal noises for $f \ll 1/\tau_s$ and for electric fields not leading to carrier heating.[22]

The *generation–recombination (G–R) noise* reflects the fluctuations of the total number of free carriers induced by generation, recombination, or bulk trapping events. The G–R noise is proportional to the square of the current:

$$\frac{S_I}{I^2} = \frac{\overline{\Delta N^2}}{N^2} \times \frac{4\tau_r}{1 + 4\pi^2 \tau_r^2 f^2} \tag{9.21}$$

where τ_r is the carrier lifetime or trapping time constant. The variance $\overline{\Delta N^2}$ of the carrier number N is proportional to the concentration N_{GRT} of generation–recombination–trapping centers and can explicitly be calculated for any type of device. For example, in a thin-film resistor, the thickness of which is delimited by two depletion regions (as in a depletion-mode MOSFET), we have

$$\frac{S_I}{I^2} = \frac{f_t(1 - f_t)N_f}{LWN_D^2(t_{si} - 2w_d)} \times \frac{4\tau_r}{1 + 4\pi^2 \tau_r^2 f^2} \tag{9.22}$$

where $f_t \simeq 0.5$ is the fractional occupancy of traps around the Fermi level and N_f is the volume trap density in the film.

For an accumulation or inversion layer, the G–R noise is given by a similar expression,[32]

$$\frac{S_I}{I_D^2} = \frac{q^2 f_t(1 - f_t)N_s}{LWC_{ox}^2} \left(\frac{g_m}{I_D}\right)^2 \times \frac{4\tau_r}{1 + 4\pi^2 \tau_r^2 f^2} \tag{9.23}$$

with N_s the equivalent surface density of traps located in the Si film. This *Lorentzian spectrum* is therefore composed of a plateau at low frequency followed by a $1/f^2$ decrease. If the G–R noise is not masked by a larger *1/f* noise, the measurement of the cutoff frequency (arrow in Fig. 9.12b) gives the carrier lifetime τ_r (from 1 ns to more than 100 μs in SOI).

The *shot noise* reflects the fact that the current, rather than being a uniform flow of electrons, is formed by a Poisson distribution of random pulses due to carriers leaving the electrode contact. The shot-noise spectrum is "white," except in very short devices where the carrier transit time cannot be neglected and $S_I \sim 1/f^2$. This type of noise emerges in bipolar transistors, p–n diodes, and Schottky junctions, but it is negligible in MOSFETs.

The *Random Telegraph Signal* (RTS) is attributed to individual carrier trapping at the Si–SiO_2 interface. In very-small-area devices, the trapping events are rare and the trapping of one carrier can easily be detected in the time domain as a small pulse superimposed on the "normal" current value (also affected by other sources of noise). The duration of the pulse is nothing but the time constant of the trap. These pulses have rather sharp transitions and constant amplitude (Fig. 9.11). They correspond to the local modification of the flat-band voltage by one trapped carrier.

It is shown that, for small enough fluctuations, the drain current pulse ΔI_D is independent of the trap area[33]

$$\frac{\Delta I_D}{I_D} = \frac{q}{LWC_{ox}} \frac{g_m}{I_D} \left(1 - \frac{d}{t_{ox}}\right) \tag{9.24}$$

where d is the distance of the trap from the interface.

Not only do the trapping events become distinct in very small devices but also the amplitude of the fluctuation increases. In this respect, the back channel of SOI MOSFETs is an ideal tool for studying the properties of RTS fluctuations because C_{ox} is very small and allows larger pulses to be obtained.

The power spectral density of RTS signals, induced by a single type of trap, is similar to that of G–R noise: a plateau at low frequency followed by a $1/f^2$ decrease. When many traps with different time constants coexist, the superposition of their RTS signals results in the standard $1/f$ noise.

9.3.4 Experiment and Parameter Extraction

Most SOI transistors present $1/f$, $1/f^2$, G–R, or RTS noise features. In early SIMOX material, annealed at low temperatures ($< 1250\,^\circ$C), the dominant noises were $1/f$ and Lorentzian (with $1/f^2$ tail).[34] Figure 9.10 shows the typical variations of the normalized $1/f$ noise factor S_I/I_D^2

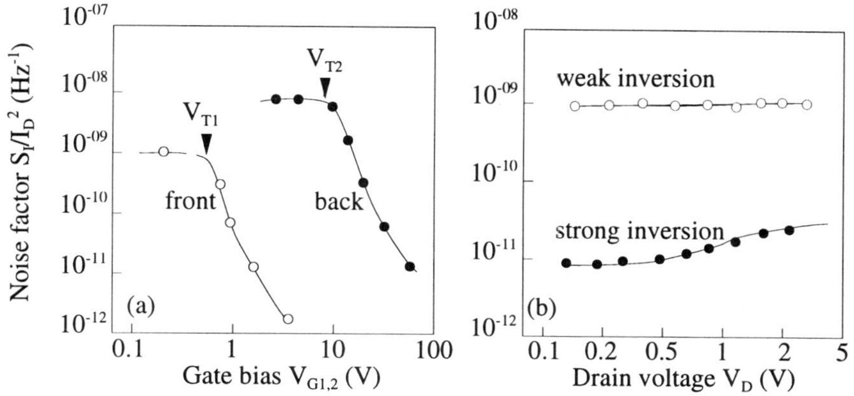

Figure 9.10 Normalized noise factor versus (a) gate voltage (∘∘, front channel; ••, back channel) and (b) drain voltage (∘∘, weak inversion; ••, strong inversion) in a SIMOX n–MOSFET with enclosed geometry.

with the gate and drain voltages. The transition from the weak inversion "plateau" to the strong inversion slope (Fig. 9.10a) corresponds well to the threshold voltage. The density of slow traps is calculated from the plateau level using Equation (9.18).[2] It is found that the buried oxide contains, in general, one order of magnitude more slow traps than the front thermal oxide where $N_{t_1} \simeq 10^{18}\,\mathrm{cm^{-3}eV^{-1}}$.

Evidence for the lateral homogeneity of the interface is obtained if the weak-inversion noise factor is rather independent of the drain voltage (Fig. 9.10b). In strong inversion, as V_D is scanned in the triode region, the noise increases from a plateau corresponding to the ohmic region to another plateau reached when V_D exceeds the saturation voltage.

In SIMOX structures annealed at high temperatures ($\geq 1300\,^\circ$C), the $1/f$ noise is reduced and can be masked by a Lorentzian noise. When probing the back channel, an additional source of G–R noise was sometimes observed for special values of the back-gate voltage. It was attributed to the activation of traps localized near the buried oxide. Their energy level was deduced by computing the position of the Fermi level.[35]

Typical RTS signals have been detected in small-area SIMOX transistors (Fig. 9.11). The pure RTS noise had a Lorentzian spectrum. Once the RTS signal was subtracted from the total current fluctuations, the

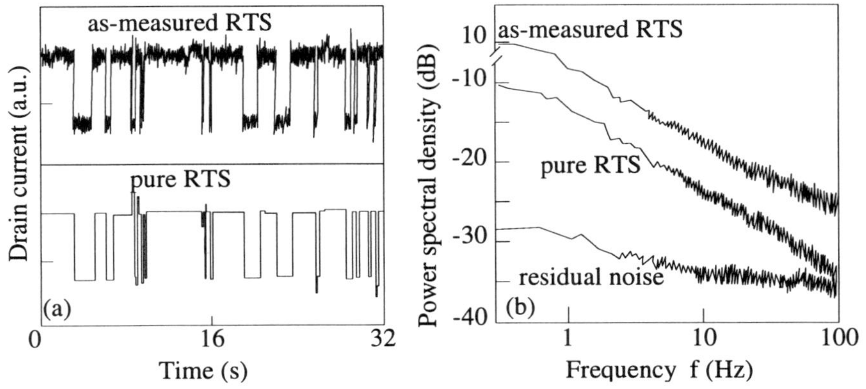

Figure 9.11 (a) As-measured and pure RTS signals in the time domain and (b) corresponding power spectra of a back-channel SIMOX transistor (after Roux et al.[33]).

residual noise had a white or $1/f$ spectrum.[33]

We now consider the case of *depletion-mode* SOI MOSFETs, where the two interface noise sources are complemented by the noise generated in the neutral region of the film. The drain voltage fluctuations, $S^{\star}_{V_D}$, measured in a partially depleted SIMOX n^+–n–n^+ transistor are shown in Figure 9.12.

A very meaningful curve is obtained by varying the back-gate bias while keeping the front interface in strong inversion (Fig. 9.12a). As long as the back interface is inverted too ($V_{G_2} < -20\,\mathrm{V}$), the noise is constant and reflects the unique contribution of the neutral film whose thickness is minimum. When V_{G_2} scans the weak inversion and depletion regions, the noise increases sharply due to the interactions between free carriers and buried-oxide interface traps. The noise peaks at the flat-band voltage. As V_{G_2} goes into accumulation, the interface traps are saturated and the measured noise decreases. This decrease is accentuated by the reduction of the transistor dynamic output resistance and R_E (Eq. (9.15)). A symmetrical behavior is obtained by varying V_{G_1} and maintaining the back interface strongly inverted.

Since the noise exhibits a Lorentzian spectrum, the trap concentra-

[2]The capacitance term in the denominator of Equation (9.18) is the same as in the subthreshold swing (see Equation (8.29)) and can be directly replaced.

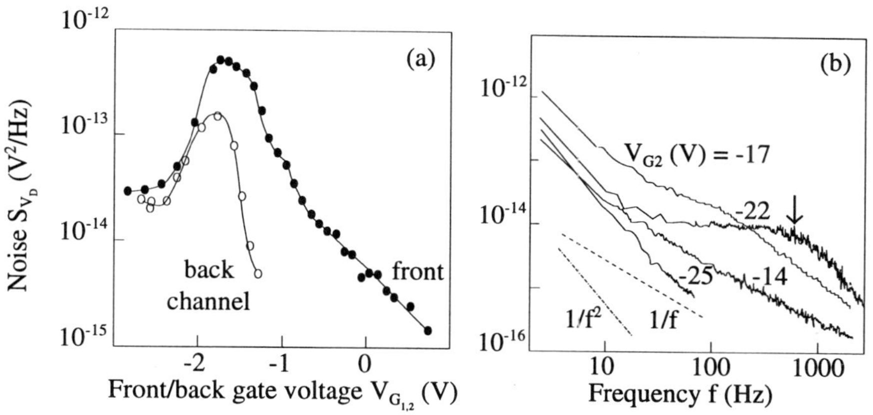

Figure 9.12 Spectral density of drain voltage fluctuations in an n^+–n–n^+ edgeless MOSFET fabricated in a partially depleted SIMOX film. (a) Influence of front- and back-gate voltages with the opposite interface in strong inversion ($f = 10\,\text{Hz}$, $V_{G_2} = 10V_{G_1}$). (b) Spectra showing the activation of a localized trap ($V_{G_1} = -3\,\text{V}$, $V_D = 20\,\text{mV}$) (after Elewa et al.[36]).

tion in the film is extracted using Equation (9.21) from the minimum volume noise. The value, $N_f \simeq 10^{18}\,\text{cm}^{-3}$, suggests the subsistence of dislocations and metal contamination.[36]

It is worth noting that the noise generated in the neutral volume is proportional to the total number of carriers and, therefore, to the volume current I_{vol}. This means that, if the interfaces are not accumulated, the volume noise can be calculated for any biases $V_{G_{1,2}}$:[36]

$$S_{I,vol} = S_{I,min} \times \frac{I_{vol}}{I_{min}} \tag{9.25}$$

It is now possible to subtract the volume noise from the total noise in order to determine the pure contributions of the front and back interfaces and to characterize the related traps.

A qualitative modification of the back-channel noise spectra is observed in Figure 9.12b for $V_{G_2} = -22\,\text{V}$. This again suggests the existence of localized traps, near the back interface, of density $10^{12}\,\text{cm}^{-2}$ and time constant about 1 ms.[36]

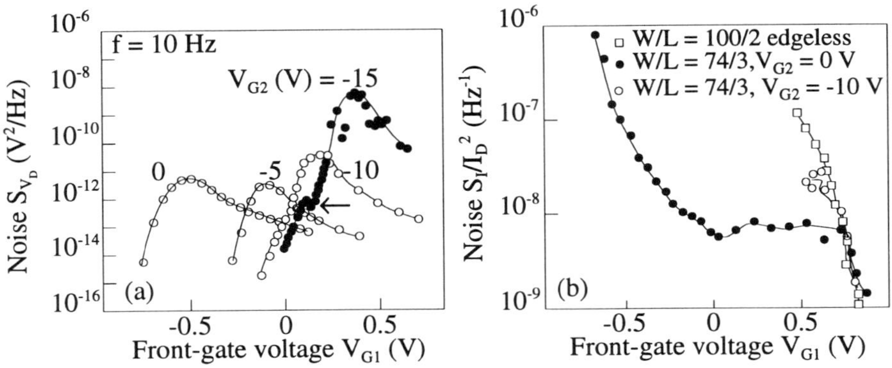

Figure 9.13 Noise factor versus front-gate voltage for various back-gate biases. (a) Fully depleted depletion-mode transistor ($t_{si} = 100\,\text{nm}$, $V_D = 20\,\text{mV}$). (b) Enhancement-mode transistors: □□, edgeless; with LOCOS edges: ••, $V_{G_2} = 0\,\text{V}$; ∘∘, $V_{G_2} = -10\,\text{V}$ (deactivated edges) (after Elewa et al.[36, 17]).

9.3.5 Interface Coupling Effects om Noise

Interface coupling has an impact on noise measurements in SOI transistors in two main ways: *direct* noise generation at the opposite interface and *indirect* modification of the surface potential by the opposite gate bias.

The same noise experiments shown in Figure 9.12a are presented in Figure 9.13 for a thin-film, fully depleted MOSFET. The front-gate-related noise now strongly depends on the back-gate voltage. The shift of the noise peak position perfectly coincides with that of the extrapolated transistor threshold, here $V_{FB_1}(V_{G_2})$. As V_{G_2} becomes more negative, the back interface is gradually driven from accumulation to weak inversion. More noise is therefore generated near the back interface and causes a substantial increase of the peak value.

It is interesting to note the hump appearing for $V_{G_2} = -15\,\text{V}$ on the left side of the peak (pointer in Fig. 9.13a), which illustrates the excess noise generated on the transistor edges. This argument is based on the fact that another typical edge-related hump shows up simultaneously on the subthreshold $\log I_D(V_{G_1})$ characteristics.

Another example showing the influence of the edge sources of noise is given in Figure 9.13b. The untypical black dotted curve is roughly the superposition of noise contributions from two distinct transistors. The edge transistor has a smaller area; hence, it governs the total noise for $V_{G_1} < 0$. The decrease of S_I/I_D^2 with V_{G_1} indicates that the edge transistor, which also has a smaller threshold voltage, has reached the strong inversion region. The plateau observed for $0 \leq V_{G_1} \leq 0.5\,\mathrm{V}$ corresponds to the noise delivered by the front-channel transistor, which still operates in weak inversion.

For higher V_{G_1}, the front channel enters into strong inversion as well, and the noise decreases once more. This latter section of the curve fully coincides with the results obtained if the edge transistor is either absent (edgeless) or intentionally deactivated ($V_{G_2} = -10\,\mathrm{V}$). Unfortunately, noise fitting with such a two-transistor model is not always feasible. In particular, the thickness and quality of a LOCOS oxide is not uniform, so the total noise is given by the parallel association of several uncorrelated sources of noise[17]

$$\frac{S_I}{I_D^2} = \sum_k G_k^2 \frac{S_{I,k}}{I_k^2} \left(\sum_k G_k \right)^{-2} \tag{9.26}$$

where G_k, I_k, and $S_{I,k}$ are the conductances, currents, and noises of distinct transistors.

In conclusion, noise spectroscopy is an elegant method to discriminate and characterize the defects located in different regions of the transistor. The only requirement is to "decouple" a given region from the surroundings, by appropriately biasing the front and back gates, and to measure the noise generated solely by that source.

9.4 Dynamic Transconductance

Dynamic transconductance is a straightforward method for converting MOS capacitance/conductance experiments, which are difficult and less suitable in SOI, into transistor current measurements. The transconductance of MOS transistors is measured in the frequency domain with the aim of extracting the properties of the interface traps. From a fundamental point of view, it is not surprising at all that capacitance and transconductance provide similar information, because they represent two different experimental manifestations of the same physical system, i.e., the MOS structure.

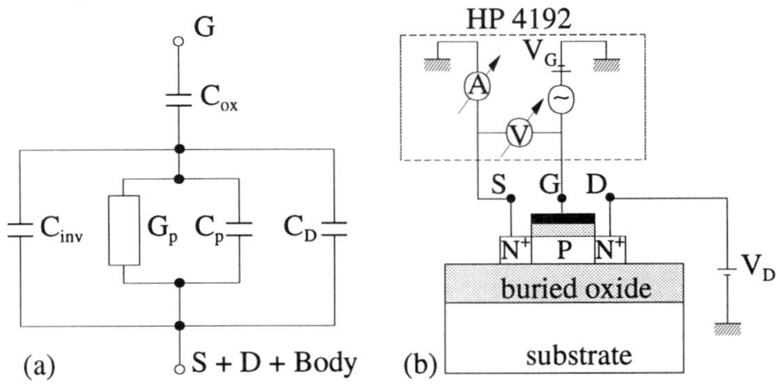

Figure 9.14 (a) Equivalent circuit of a short-channel MOSFET with the source, drain, and body tied together and (b) experimental setup for dynamic transconductance measurements in bulk-silicon or SOI transistors.

The dynamic transconductance is a relatively recent technique that has been developed in response to the difficulty of investigating the interface trap properties in small-area MOSFETs, where capacitance measurements are extremely difficult.[37]

Several models have elaborated the correlation between the frequency dispersion of the transconductance and the interface-trap admittance that governs the equivalent circuit of MOS structures (Fig. 9.14a). These models have made use of either the real part or the imaginary part of the transconductance. The density of interface traps and their energy profile have been determined without knowledge of the doping profile in the channel.

The experimental setup for dynamic transconductance measurements is shown in Figure 9.14b. The transistor is operated in weak inversion, and the drain voltage is maintained at 10–20 mV. The gate voltage is modulated with a small a.c. signal. An HP–4192 impedance analyzer is used to deliver such a modulation and to measure the transfer admittance.[3]

The method may be applied to either channel for the independent characterization of front and back interfaces. There are variants for partially or fully depleted, inversion-mode or accumulation-mode, and n- or p-channel transistors.

9.4.1 Theory and Parameter Extraction

In partially depleted SOI transistors, operated in the ohmic region, the drain current is expressed in terms of inversion charge Q_{inv} and effective mobility μ_{eff}:

$$I_D = \frac{W}{L}\mu_{eff}Q_{inv}V_D \tag{9.27}$$

Under weak inversion, Equation (9.27) is valid for $V_D < kT/q$. By differentiating the drain current with respect to the gate voltage V_G, the transfer admittance is given by[37]

$$g_m = \frac{WV_D}{L}\left(\mu_{eff}\frac{dQ_{inv}}{d\psi_s} + Q_{inv}\frac{d\mu_{eff}}{d\psi_s}\right)\frac{d\psi_s}{dV_G} \tag{9.28}$$

The instantaneous charge balance equation

$$V_G(t) = \phi_{FB} + \psi_s(t) - \frac{Q_{inv}(t) + Q_d(t) + Q_{it}(t)}{C_{ox}} \tag{9.29}$$

is differentiated with respect to time and yields

$$dV_G = d\psi_s + \frac{C_{inv} + C_d + Y_{it}/j\omega}{C_{ox}}d\psi_s \tag{9.30}$$

where ω is the angular frequency of the small a.c. signal modulating V_G, $C_{inv} = -dQ_{inv}/d\psi_s$ is the inversion-channel capacitance, and $C_d = -dQ_d/d\psi_s$ is the depletion-layer capacitance. The parallel admittance Y_{it} of interface traps is deduced from the equivalent circuit of Figure 9.14a, which applies to MOSFETs operated in weak inversion:[37]

$$-\frac{dQ_{it}}{d\psi_s} = \frac{Y_{it}}{j\omega} = \frac{G_p + j\omega C_p}{j\omega} = C_p - j\frac{G_p}{\omega} \tag{9.31}$$

The dynamic transconductance is obtained by combining Equations (9.28)–(9.30):

$$g_m = \frac{WV_D}{L}\frac{\mu_{eff}^2}{\mu_0}\frac{C_{ox}C_{inv}}{C_{ox} + C_d + C_{inv} + Y_{it}/j\omega} \tag{9.32}$$

where μ_0 is the low field mobility and where the usual dependence of μ_{eff} on depletion and inversion charges has been assumed.[38]

---

[3]Several problems have been reported in relation to the inadequacy of the two-port HP–4192 analyzer for three-terminal measurements. For the case of very low densities of traps, other impedance analyzers may be more suitable.

It has been shown that for MOSFETs with channel lengths smaller than 10 μm and operating in the weak-inversion regime, the phase shift due to lateral transport of inversion carriers can be neglected up to frequencies much higher than that at which the interface states become totally inactive.[37] This argument implies that for such a MOSFET, the delay in the inversion layer build-up is negligible. Therefore, C_{inv} is independent of the small a.c. signal frequency and can be written as

$$C_{inv} = \frac{Q_{inv}}{(kT/q)} \tag{9.33}$$

The transfer admittance in weak inversion is, thus, given by

$$g_m = \frac{I_D}{(kT/q)} \frac{C_{ox}}{C_{ox} + C_d + C_{inv} + Y_{it}/j\omega} \tag{9.34}$$

Equation (9.34) was obtained by assuming μ_{eff} equal to μ_0 in the weak-inversion regime and neglecting the influence of surface-potential fluctuations. It is noted that for $\omega = 0$, when $G_p = 0$ and $C_{it} = Y_{it}/j\omega$, Equation (9.34) reduces to the transconductance expression in weak inversion (see also Eq. (8.61)).

The merit of Equation (9.34) is to illustrate that the parallel conductance of interface traps G_p, frequently used for MOS capacitors, is closely linked to the frequency dispersion of the MOSFET transfer admittance g_m. In particular, the imaginary part of $1/g_m$ yields G_p/ω:

$$\frac{G_p}{\omega} = -\frac{C_{ox} I_D}{(kT/q)} \mathrm{Im}\left(\frac{1}{g_m(\omega)}\right) \tag{9.35}$$

Equation (9.35) is the basis of the dynamic transconductance technique. It shows that G_p/ω can be deduced from combined measurements of transconductance and current. More importantly, the multiplying coefficient in Equation (9.35), $C_{ox} I_D/(kT/q)$, is very large due to the built-in transistor gain. This enables the imaginary part of $1/g_m$ to be a highly amplified image of G_p/ω.[37, 39, 40]

The interface trap analysis proceeds from the proportionality between G_p/ω and D_{it}:

$$\frac{G_p}{\omega} = q D_{it} \frac{\omega \tau_{it}}{1 + \omega^2 \tau_{it}^2} \tag{9.36}$$

for a single-level trap of time constant τ_{it}, and

$$\frac{G_p}{\omega} = \frac{q D_{it}}{2\omega \tau_m} \ln(1 + \omega^2 \tau_m^2) \tag{9.37}$$

for a continuum of traps. Here, τ_m is related to the capture cross section and surface carrier concentration: $1/\tau_m = \sigma v_{th} n_s$. If necessary, other expressions of $G_p(D_{it})$ may be used to account for the presence of potential fluctuations.

The curves G_p/ω versus ω are bell shaped (Fig. 9.15). The peak magnitude provides the density of interface traps, whereas their time constant is found from the position of the peak. The measurements are repeatedly performed for various gate biases and then expressed in terms of surface potential and energy using Equation (9.29).

9.4.2 Model Variants

The dynamic transconductance model naturally extends to the case of *fully depleted* MOSFETs.[41] The front-gate voltage now is a function of front and back surface potentials $\psi_{s1,2}$

$$V_{G_1} = \phi_{FB_1} + \left(1 + \frac{C_{si}}{C_{ox_1}}\right)\psi_{s_1} - \frac{C_{si}}{C_{ox_1}}\psi_{s2} - \frac{-qN_A t_{si}/2 + Q_{inv_1} + Q_{it_1}}{C_{ox_1}} \tag{9.38}$$

The simplest way to characterize the front interface is to decouple the two channels. When the back gate is biased into accumulation, the back surface potential is virtually grounded ($\psi_{s_2} \simeq 0$), and Equation (9.38) can be solved for $dV_{G_1}/d\psi_{s_1}$. Making use of Equation (9.28), it is easy to prove that the weak-inversion transfer admittance of the front channel is again given by Equations (9.34) and (9.35); the parameters Y_{it}, D_{it}, and C_{ox} refer to the front interface, and $C_d = C_{si}$ is the capacitance of the fully depleted film.

A wider energy range may be explored by performing the measurement with the back interface in strong inversion and the front interface in accumulation or depletion.[41] The front-gate modulation is sensed by the current, $I_D = \mu_{eff_2}(W/L)V_D Q_{inv_2}$, which now flows through the back channel. By associating Equation (9.38) with its back-gate counterpart (interchanged subscripts 1 and 2), $dV_{G_1}/d\psi_{s_1}$ and $dQ_{inv_2}/d\psi_{s_1}$ are calculated. Since $d\psi_{s_2}/d\psi_{s_1} \simeq 0$, only the front-interface traps are able to respond to the modulation, yielding

$$g_m = \frac{I_D}{C_{ox_2}(V_{G_2} - V_{T_2})} \frac{C_{ox_1} C_{si}}{C_{ox_1} + C_{si} + Y_{it_1}/j\omega} \tag{9.39}$$

and

$$\frac{G_p}{\omega} = -\frac{C_{ox_1} C_{si}}{C_{ox_2}} \frac{I_D}{V_{G_2} - V_{T_2}} \text{Im}\left(\frac{1}{g_m(\omega)}\right) \tag{9.40}$$

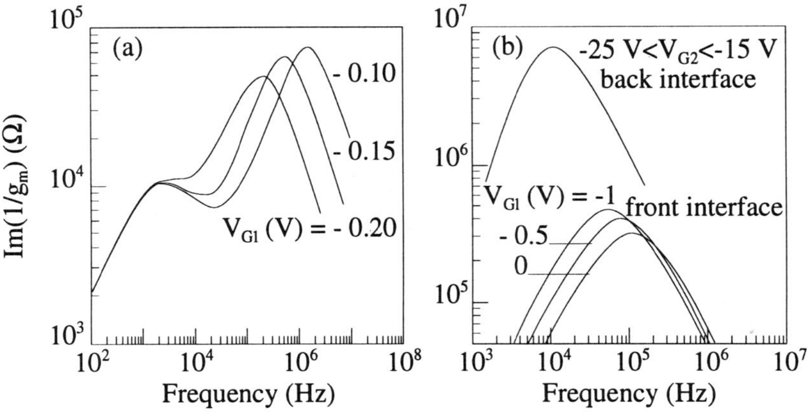

Figure 9.15 Experimental variations of the imaginary part of the reciprocal transconductance $1/g_m$ versus frequency in partially depleted MOSFETs: (a) depletion-mode transistor fabricated on early high-temperature annealed SIMOX and (b) enhancement-mode transistor fabricated on low-temperature annealed SIMOX (after Elewa et al.[39]).

Combining these two types of measurements, most of the energy gap, from accumulation to weak inversion, can be profiled.

When the back interface is depleted, one has to account for coupling effects. Following similar steps to those above, a complex expression of $1/g_m(\omega)$ is found that includes the parameters of both front and back interfaces. There are extreme situations in which the front-gate dynamic transconductance signal is dominated by the transfer admittance of the back interface.

The dynamic transconductance method is equally useful in *accumulation-mode* MOSFETs.[42] We consider a partially depleted n^+–n–n^+ transistor operated in the depletion region, where the equivalent circuit is simple enough and the current is given by

$$I_D = KqN_D(t_{si} - w_d) \tag{9.41}$$

with $K = \mu(W/L)V_D$. Once more, $dV_{G_1}/d\psi_{s_1}$ is obtained from the charge-neutrality law and Equation (9.30), except that now $C_{inv} = 0$.

The final form of the dynamic transconductance equation is[42]

$$g_m = K \frac{C_{ox} C_d}{C_{ox} + C_d + Y_{it}/j\omega} \tag{9.42}$$

and

$$\frac{G_p}{\omega} = -K C_{ox} C_d \mathrm{Im}\left(\frac{1}{g_m(\omega)}\right) \tag{9.43}$$

9.4.3 High–Low-Frequency Transconductance

The *High–Low-Frequency* (HLF) technique belongs to the dynamic transconductance family. The same physical principles and equations are used. The only difference is that the interface-trap profile is deduced from the *real* part, rather than from the imaginary part, of the transconductance.[43] Two measurements, one at low frequency (10 Hz) and other at high frequency (1 MHz), are compared.

For partially depleted MOSFETs, the real part of the inverse low-frequency transconductance is determined from Equation (9.34):

$$\mathrm{Re}\left\{\frac{1}{g_m^{\mathrm{LF}}}\right\} = \frac{(kT/q)}{I_D} \frac{C_{ox} + C_d + C_{inv} + C_{it}}{C_{ox}} \tag{9.44}$$

In the case of high-frequency (≥ 1 MHz) operation, the interface traps cannot follow the a.c. signal, so $C_{it} = 0$ and the real part of the inverse high-frequency transconductance is given by

$$\left\{\frac{1}{g_m^{\mathrm{HF}}}\right\} = \frac{(kT/q)}{I_D} \frac{C_{ox} + C_d + C_{inv}}{C_{ox}} \tag{9.45}$$

The interface trap density is expressed as[44]

$$D_{it} = \frac{I_D C_{ox}}{kT} \left[\mathrm{Re}\left\{\frac{1}{g_m^{\mathrm{LF}}}\right\} - \mathrm{Re}\left\{\frac{1}{g_m^{\mathrm{HF}}}\right\}\right] \tag{9.46}$$

The corresponding energy is determined by noting that, for d.c. operation, Equation (9.30) reduces to

$$\frac{d\psi_s}{dV_G} = \frac{(kT/q)}{I_D} \frac{1}{\mathrm{Re}(1/g_m^{\mathrm{LF}})} = \frac{(kT/q)}{I_D} \frac{dI_D}{dV_G} \tag{9.47}$$

The surface potential can be expressed in terms of drain current by integrating Equation (9.47) from the threshold voltage to the V_G value.

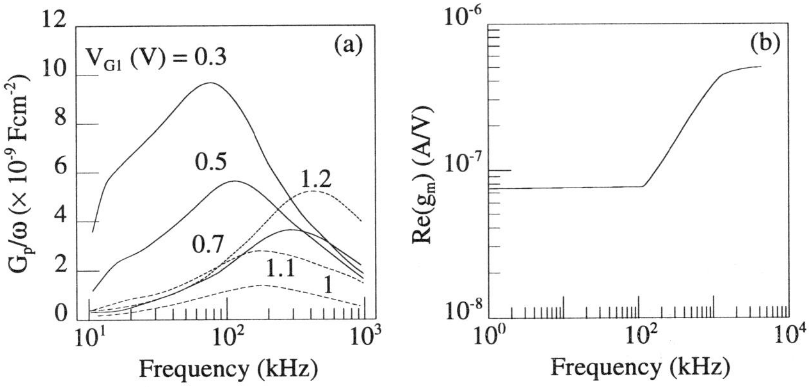

Figure 9.16 (a) Interface state conductance G_p/ω versus frequency in fully depleted SIMOX MOSFETs (- - -, front interface in weak inversion and back interface in accumulation; ——, front interface in depletion and back interface in strong inversion; after Ioannou et al.[41]). (b) The real part of the dynamic transconductance versus frequency.

Finally, the surface potential may be rewritten as a function of the energy E relative to the intrinsic Fermi level E_i:

$$E - E_i \simeq qV_D + 2q\phi_F + kT \ln\left[\frac{I_D(V_G)}{I_D(V_T)}\right] \tag{9.48}$$

It is noted that the relationship between D_{it} and $E - E_i$ can be established without knowing the doping profile. The required device parameter is C_{ox} only, because I_D and V_T are measured from $I_D(V_G)$ curves. Expressions similar to Equation (9.46) are easily obtained from Equations (9.39) or (9.42), and respectively illustrate the case of fully depleted or accumulation-mode MOSFETs.

9.4.4 Experiment

The dynamic transconductance and HLF methods have been applied to a variety of SOI MOSFETs. In Figure 9.15 are reproduced typical frequency-dependent variations of the imaginary part of $1/g_m$. In good-quality SOI, the curves are bell-shaped and well behaved whatever

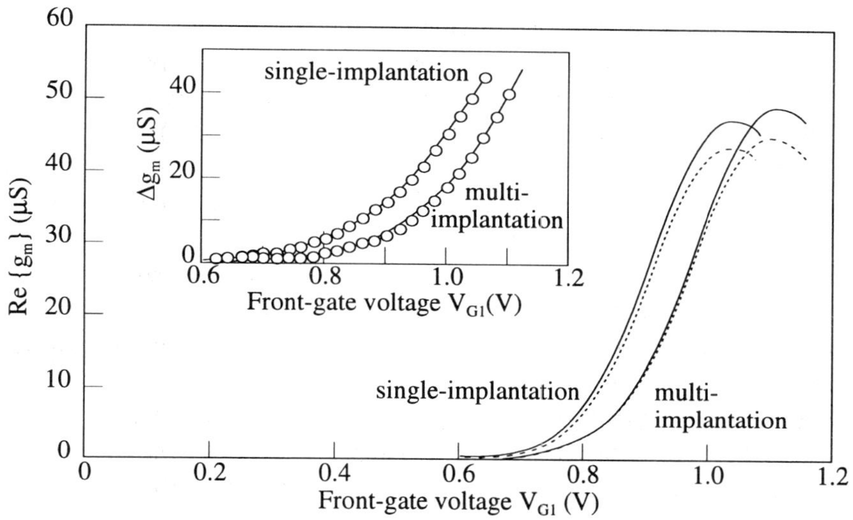

Figure 9.17 The real part of the transconductance (- - -, f = 10 Hz; ——, f = 1 MHz) and $\Delta g_m = \mathrm{Re}\{g_m^{\mathrm{HF}}\} - \mathrm{Re}\{g_m^{\mathrm{LF}}\}$ versus gate bias for single- and multi-implanted SIMOX MOSFETs (after Li et al.[44]).

the transistor type (inversion or accumulation mode, partially or fully depleted).

By interchanging the roles of the two gates, both interfaces can equally be characterized. Back-gate modulation leads to a stronger signal than at the front interface (Fig. 9.15b). This reflects the difference in quality between the two interfaces. In early SIMOX, the back-interface trap density used to be in the mid–$10^{11}\,\mathrm{cm^{-2}eV^{-1}}$ range, exceeding the front density by more than one order of magnitude.[4] Unlike the front-interface peak, the back-interface peak of G_p/ω does not change in position, or in magnitude, with V_{G_2} (Fig. 9.15b). It was presumed that many bulk traps are located close to the buried oxide and dominate the signal.[42]

Bulk traps were also incriminated for the double-peak signal observed in Figure 9.15a.[39] It is well known that annealing of SIMOX at low temperature (< 1200 °C) results in poor quality of the bottom of the film and interface. Only the pure peak of the front interface is sensitive to the gate bias. As V_{G_1} increases, the interface traps situated closer to the conduction band edge are scanned, which is consistent with the peak increase and the shift towards higher frequencies.

Figure 9.16a shows dynamic transconductance data obtained in fully depleted MOSFETs.[41] The two sets of curves respectively correspond to depletion (Eq. (9.41)) or weak inversion (Eq. (9.35)) at the front interface. From the magnitude and position of G_p/ω peaks, the front-interface trap density is found to vary between 2×10^{10} near midgap to $10^{11}\ \mathrm{cm^{-2}eV^{-1}}$ near the band edge.

The variation of the real part of the dynamic transconductance is exemplified in Figure 9.16b. The plateaus observed at low and high frequencies offer experimental support for the HLF transconductance method.

Figure 9.17 shows the transconductance versus gate bias for single- and multi-implanted SIMOX MOS transistors.[44] The transconductance curves of the multi-implanted sample are shifted horizontally for clarity. The inset highlights $\Delta g_m = \mathrm{Re}\{g_m^{\mathrm{HF}}\} - \mathrm{Re}\{g_m^{\mathrm{LF}}\}$ versus gate voltage. It is clear that the difference between the high- and low-frequency transconductance is smaller for the multi-implant sample, which implies a lower interface-state density.

Figure 9.18 shows the distribution of interface traps in the forbidden gap determined from Equations (9.46) and (9.48) for both single- and multi-implant samples. The result is valid only in a limited energy range of the forbidden gap, which extends from the onset of weak inversion to the onset of strong inversion. These data confirm that the interface-state density near midgap can be reduced by using a multi-implant process to the same value as in bulk-Si MOSFETs.[43]

The results of the HLF transconductance method have been compared with charge-pumping (CP) measurements performed on the same devices. The interface state densities determined by the charge-pumping method are superimposed in Figure 9.18. Reasonable agreement is obtained between these two methods near the midgap energy range.

It is concluded that dynamic transconductance is an attractive, although recent and not yet very popular, technique for the characterization of interface traps in short-channel MOSFETs. It has two main advantages as compared to charge pumping: (1) no need for body contacts, and (2) straightforward back-channel measurements.

---

[4] Although SIMOX has been very much improved, a difference still remains between the two interfaces.

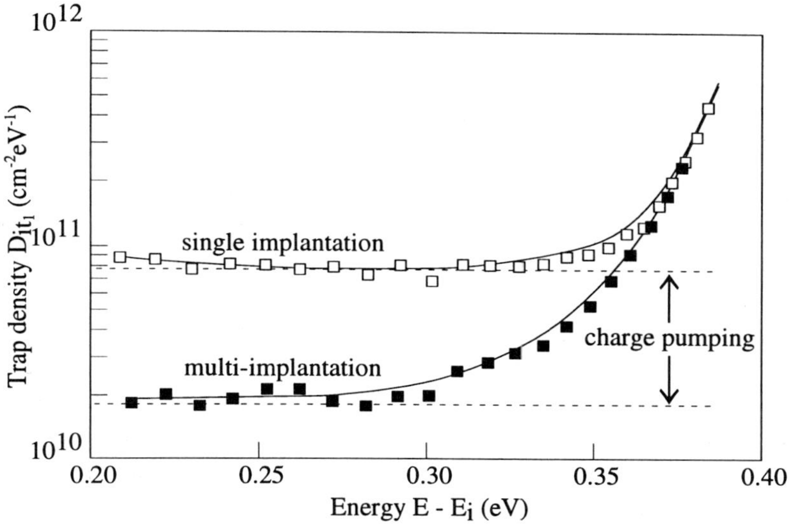

Figure 9.18 The distribution of interface traps in the forbidden gap of Si determined with HLF transconductance method from Equations (9.46) and (9.48) and corresponding charge-pumping data (after Li et al.[44]).

Several models have been presented that establish the correlation between the transfer admittance of MOS transistors and the interface-trap conductance. In all types of transistors, the imaginary part of the reciprocal transconductance is directly proportional to the trap density. Alternatively, the real part of the transconductance can also be exploited according to the HLF model, but the signal sensitivity may be hampered by the values of the depletion and oxide capacitances.

9.5 Drain Current Transient Technique

The carrier generation–recombination properties govern the performance of MOS and bipolar devices (leakage current of p–n junctions, refresh time of dynamic RAMs, floating body effects, bipolar gain, etc.). Moreover, the carrier lifetime depends on residual crystalline defects and stands as a figure of merit for the quality of SOI films. Similarly, the surface generation velocity serves to evaluate of the Si–SiO_2 interface, since it is proportional to the density of interface traps. In this section, the carrier generation properties will be deduced from drain current transients.

The basic idea is to adapt the well-known Zerbst method[45] to SOI films. As discussed in Section 6.4, the Zerbst method consists in applying a pulse on the gate of an MOS capacitor so that a deep-depletion region is formed. Equilibrium is reached through bulk and surface carrier generation processes. The monitoring of the capacitance transient provides the generation lifetime τ (which can substantially exceed the recombination lifetime) as well as the surface velocity S.

Since capacitance measurements are not straightforward in thin SOI films, several modifications will be needed: (1) MOS transistors will serve as test devices, (2) the drain current will be monitored, and (3) the dual-gate configuration will be used to form deep depletion regions and to initiate carrier generation. After describing the variants of the technique for partially depleted accumulation- and inversion-mode MOSFETs, we will discuss its extension to fully depleted SOI and, finally, present the experimental details.

9.5.1 Partially Depleted Accumulation-Mode Transistors

We consider an n^+–n–n^+ transistor biased in the ohmic region (Fig. 9.19a). Applying, at $t = 0$, a large negative voltage on the gate results in a very fast build-up of positive charge (deep depletion region) to balance the gate charge. Then the inversion charge starts forming. This process is much slower, since no reservoir of holes is available. The deep depletion region controls the thickness of the drain current path. The holes supplied by thermal generation allow the deep depletion region to shrink and the current $I_D(t)$ to increase towards an equilibrium value I_∞ (Fig. 9.19a).[46]

The transient variations of the inversion charge Q_{inv}, the depletion charge Q_d, the interface trap charge Q_{it}, and the surface potential ψ_s are governed by the charge balance equation

$$Q_{inv} + Q_{it} = C_{ox}(V_G - \phi_{FB} - \psi_s) - Q_d \tag{9.49}$$

Expressing Q_d, ψ_s, and I_D in terms of depletion depth $w_d(t)$ gives

$$Q_d = q w_d N_D \qquad \psi_s = \frac{q w_d^2 N_D}{2\epsilon_{si}} \tag{9.50}$$

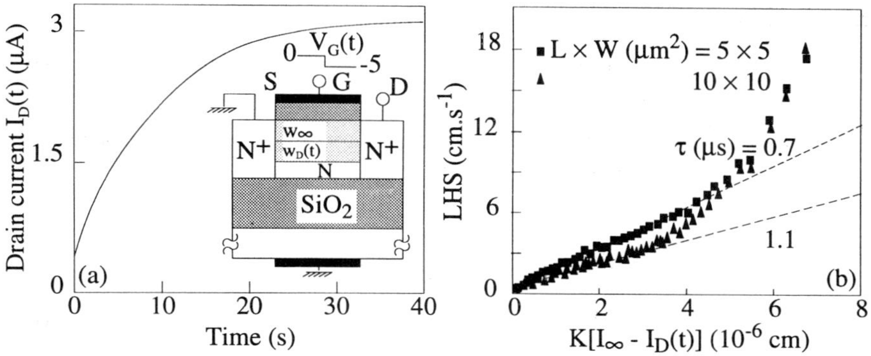

Figure 9.19 Accumulation-mode n^+–n–n^+ transistors pulsed in deep depletion ($V_{G_1}: 0 \rightarrow -5\,\text{V}$). (a) Drain current transient and (b) Zerbst-type analysis for transistors with different sizes (after Elewa et al.[39]).

and

$$I_D = \frac{q\mu N_D W V_D}{L}(t_{si} - w_d) = \frac{1}{K}(t_{si} - w_d) \tag{9.51}$$

Combining Equations (9.49) and (9.50), yields the rate of change of inversion and interface trap charges[39]

$$G = \frac{1}{q}\frac{d(Q_{inv} + Q_{it})}{dt} = -\frac{N_D C_{ox}}{2\epsilon_{si}}\left(1 + \frac{w_d}{\epsilon_{si}}\right)\frac{dw_d}{dt} \tag{9.52}$$

which is perfectly balanced by the total carrier generation rate

$$G = \frac{n_i(w_d(t) - w_\infty)}{\tau} + n_i S + n_i S_e \frac{(w_d(t) - w_\infty)(2L)}{WL} \tag{9.53}$$

The first term stands for volume generation, which is usually assumed to occur in the excess depletion region $(w_d - w_\infty)$, where w_∞ is the depletion depth at equilibrium.[47] The second contribution comes from the Si–SiO_2 interface, and the last term involves the surface generation velocity S_e on the edges. The active-edge area, $2L(w_d - w_\infty)$, has been normalized with respect to the gate area. It is convenient to define an *effective* generation lifetime τ_{eff} by

$$\frac{1}{\tau_{eff}} = \frac{1}{\tau} + \frac{2S_e}{W} \tag{9.54}$$

Equating Equations (9.52) and (9.53) results in a generic Zerbst-type relationship

$$\mathrm{LHS}(t) = \frac{1}{\tau_{eff}} \times \mathrm{RHS}(t) + S \tag{9.55}$$

where the left-hand and right-hand sides can readily be expressed, using Equation (9.51), in terms of drain current

$$\begin{aligned} \mathrm{LHS}(t) &= -\frac{N_D C_{ox}}{2 n_i \epsilon_{si}} \frac{d}{dt} \left[t_{si} + \frac{\epsilon_{si}}{C_{ox}} - K I_D \right]^2 \\ \mathrm{RHS}(t) &= K (I_\infty - I_D) \end{aligned} \tag{9.56}$$

Instead of computing the current transient $I_D(t)$, Equation (9.55) is an elegant way of presenting the experimental data, LHS(t) vs. RHS(t), on a straight line (Fig. 9.19b). The slope and intercept give the values of τ_{eff} and S.

9.5.2 Partially Depleted Inversion-Mode MOSFETs

By contrast to accumulation-mode transistors, in n^+–p–n^+ structures, the drain and source reservoirs do supply minority carriers to form the inversion channel quasi-instantaneously. Since there is no source of majority carriers (no p^+ contact) except the generation process, the experiment is designed to induce a deficit of holes budget.

The front gate is biased in strong inversion, whereas the back gate is pulsed from depletion (or accumulation) to strong accumulation (see Fig. 9.21a). The holes requested in the accumulation layer are released from the neutral region of the film. A deep depletion region forms under the gate, and the film potential drops. To maintain front-gate charge conservation (Eq. (9.49)), the excess depletion charge is compensated for by a temporary drop in the inversion charge and drain current. The current relaxes back to equilibrium through carrier generation.

Since the front surface potential is almost constant in strong inversion ($\psi_{s_1} \simeq 2\phi_F$), it is the film potential $V_B(t)$ that governs the densities of inversion, depletion, and accumulation charges:

$$\begin{aligned} Q_{inv} + Q_{it} &= -C_{ox_1}(V_{G_1} - \phi_{FB_1} - \psi_{s_1}) - Q_d \\ Q_d(t) &= -q w_d N_A = -\sqrt{2 q \epsilon_{si} N_A (\psi_{s_1} - V_B(t))} \\ Q_{acc}(t) &= -C_{ox_2}(V_{G_2} - \phi_{FB_2} - V_B(t)) \end{aligned} \tag{9.57}$$

The charge balance of majority carriers is fulfilled by equating the rate of change of depletion and accumulation charges with Equation (9.53). Eliminating $V_B(t)$ again yields the Zerbst equations (9.55), where the right-hand side RHS(t) is the same as in Equation (9.56) and the left-hand side becomes

$$\text{LHS}(t) = -\frac{N_A C_{ox2}}{2n_i \epsilon_{si}} \frac{d}{dt} \left[K(I_\infty - I_D) + \frac{\epsilon_{si}}{C_{ox2}} + \sqrt{\frac{4\epsilon_{si}\phi_F}{qN_A}} \right]^2 \tag{9.58}$$

In some particular instances,[48] the variation of the accumulation charge may be neglected, and the current will then have a simple exponential transient[49]

$$I_\infty - I_D = I_\infty e^{-t/\tau_R} \tag{9.59}$$

where the recovery time is $\tau_R = \tau N_A / n_i$. This explains the difference of many orders of magnitude between the time constants of the generation process and of the drain current relaxation.

9.5.3 Model Refinements

Several amendments to the above models have been proposed.

- For large pulses, the deep depletion charge may be enough to temporarily suppress the drain current (for instance, when full depletion is achieved in accumulation-mode transistors). During a time t_{off}, freshly generated holes contribute to the relaxation of the front surface potential, which is swept from depletion and weak inversion to strong inversion. The off-state period can be calculated using the above equations.[48] A simpler method is to equate the positive charge generated during t_{off} with the gate charge exceeding that needed to cancel the current.[50] Of course, the Zerbst analysis still holds for $t > t_{off}$.

- The analysis of enhancement-mode transistors can be extended to the case where the front gate is biased in weak inversion. The mathematical derivation proceeds from the same set of Equations (9.57), with the difference that $Q_{inv}(t) \ll Q_d(t)$ and the drain current exponentially depends on the *varying* surface potential $\psi_{s_1}(t)$. In Equation (9.55), we now have[51]

$$\text{LHS}(t) = \frac{C_{ox_1}}{qn_i\beta} \left(1 + \frac{C_{ox2}}{C_{ox_1}} - \frac{C_{ox_1}C_{ox2}}{q\epsilon_{si}\beta N_A} \ln \frac{I_D}{I_K}\right) \frac{1}{I_D}\frac{dI_D}{dt}$$

$$\text{RHS}(t) = \frac{C_{ox1}}{q\beta N_A} \ln \frac{I_\infty}{I_D} \tag{9.60}$$

with $\beta = q/kT$.

- In films that are fully depleted at equilibrium, the concept of deep depletion and the expression $w_d(t)$ are meaningless. However, the dual-gate pulsing technique can still be successful. Numerical computations of the generation rate show that the *effective* generation width $w_{d,eff}$ corresponds to the film region where the electron and hole concentrations are smaller than the intrinsic carrier concentration: $w_{d,eff} = x_2 - x_1$, with $n(x_1) = n_i$ and $p(x_2) = n_i$.[52] Note that as long as the two interfaces are depleted, they play equivalent roles in the generation process. The charge balance equation becomes

$$\frac{dQ_a}{dt} = -\left(1 + \frac{C_{ox2}}{C_{ox1}}\right)\frac{dQ_{inv}}{dt} = qn_i\left(\frac{w_{d,eff}}{\tau_{eff}} + S\right) \tag{9.61}$$

and can be rewritten in terms of $I_D(t)$.

- Pulsing the back gate from accumulation into even stronger accumulation, while maintaining the front interface in strong inversion, has the advantage of keeping constant the two surface potentials and eliminating the corresponding interface generation rates. This greatly simplifies Equations (9.57) and (9.58). However, one has to make sure that the device is in equilibrium, before applying the pulse. This precaution calls for the *double-pulse* procedure illustrated in Figure 9.21a. Only the second transient, induced by the small additional pulse, is used for parameter extraction.[53]

- The surface velocity S may include the residual contribution of carriers generated at the *opposite* interface and diffusing into the space-charge region.[54]

- The surface generation rate depends on the surface potential and, in general, is nonconstant during the transient: S is a maximum for interface depletion and a minimum for strong inversion or accumulation. This makes the Zerbst analysis less accurate at the beginning of the transient.

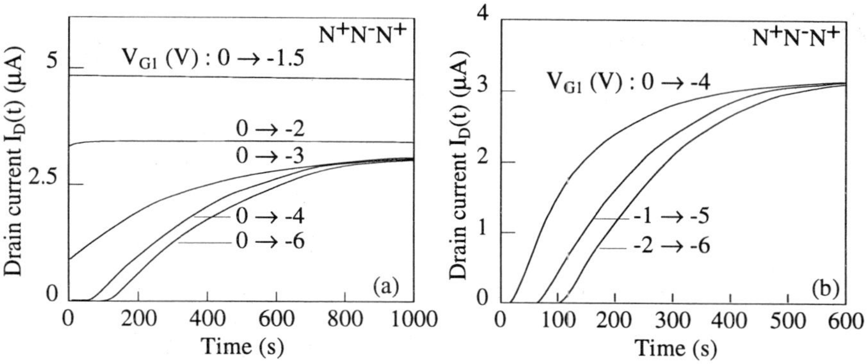

Figure 9.20 Typical transients in accumulation-mode n^+–n–n^+ transistors for different (a) bottom levels and (b) top levels of the pulse.

9.5.4 Experiment and Parameter Extraction

Caution is needed to avoid parasitic sources of carrier generation. The experiment is performed in obscurity, with low voltages (20–100 mV) applied on the drain. Long channels are suitable in order to render the extension of source and drain depletion regions less significant. Junction leakage or BOX leakage currents may jeopardize the accuracy of drain current transients. These parasitic effects may be considered to be negligible if the same set of generation parameters (τ, S) is obtained for several channel lengths.

The measurements are performed with an HP–4145 system and typical transients are shown in Figures 9.19–9.23. These data have been numerically differentiated according to the appropriate expression of LHS inserted in Equation (9.55). As is usual for Zerbst plots, straight lines are obtained except at the beginning of the transient (the right side of the Zerbst curves in Figures 9.19b and 9.21b). This is due to the larger surface generation rate when the interface is depleted.

In Figure 9.20a, the front gate of n^+–n–n^+ transistors is pulsed from zero towards increasing negative values. No transient is observed for small steps in gate bias (-2 V), simply because the depletion depth is narrower than w_∞. The electron population is promptly adjusted by the n^+ contacts, and there is no need at this point for a hole supply. For a step of -3 V, the deep depletion depth lies between w_∞ and t_{si},

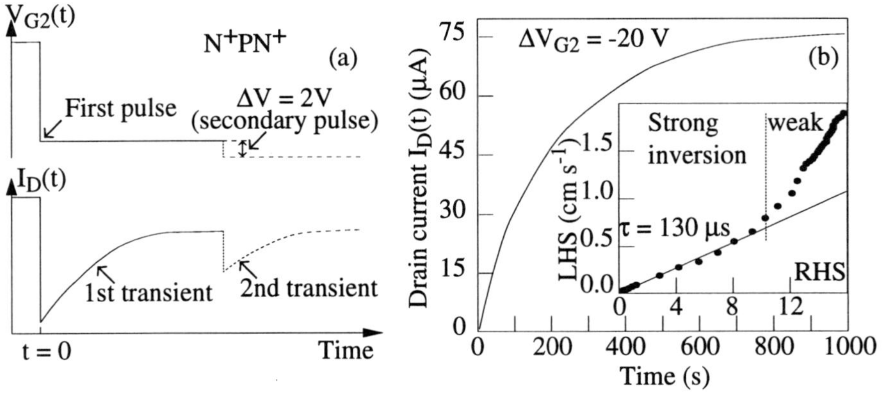

Figure 9.21 Enhancement-mode n^+–p–n^+ transistors pulsed in deep depletion. (a) Single-pulse[48] and double-pulse[53] procedures. (b) Typical current transient and Zerbst-like analysis in strong inversion (after Ioannou et al.[48]).

and the initial current is nonzero. For yet larger steps, the film becomes totally depleted, the current drops to zero, and the off-state region t_{off} increases.

Changing the initial gate bias, while maintaining the same step amplitude, causes the curves $I_D(t)$ to be shifted and t_{off} to vary (Fig. 9.20b). However, the transients still converge to the same saturation current. The dominant parameter is the bottom of the pulse: if it is kept constant and only the top level of the pulse is modified, the transients remain more or less the same.

It is noted that if the experiment is conducted with the high level of the pulse in strong inversion, the device should be allowed a very long time to reach equilibrium before being pulsed in stronger inversion. Finally, the back gate may be biased in accumulation in order to suppress the off-state period and to reach larger saturation currents.

The distinct features of front- and back-gate biases in partially depleted enhancement-mode transistors are illustrated in Figures 9.21–9.23. When the back-gate voltage step is increased, the current variation range is more pronounced and the off-state period is longer, but the steady-state current does not change (Fig. 9.22a). Only in fully depleted MOSFETs may I_∞ be modified by the pulse amplitude via in-

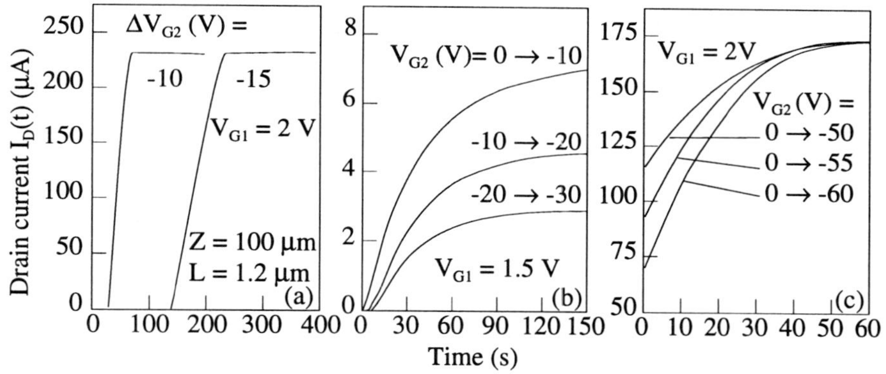

Figure 9.22 Current transients in enhancement-mode n^+–p–n^+ transistors pulsed with various back-gate steps at fixed front-gate bias: (a) partially depleted SIMOX, (b) fully depleted SIMOX pulsed in weak accumulation, (c) fully depleted wafer-bonding MOSFETs pulsed in strong accumulation.

terface coupling (Fig. 9.22b); the bottom of the back-gate pulse acts on the threshold voltage of the front channel. However, once the bottom of the pulse has reached strong accumulation, a further increase of the pulse will no longer modify V_{T_1} and I_∞ (Fig. 9.22c).

It was expected that pulsing the back gate either from depletion to strong accumulation (V_{G_2} : 10 → −30 V) or from weak to strong accumulation (V_{G_2} : −10 → −30 V) would modify the surface generation rate. However, the results are very much the same, revealing that the main mechanism is the carrier generation in the volume of the film.

If the front-gate bias is increased, the steady-state current becomes larger and the off-state region may be suppressed (Fig. 9.23a). This cancels or reduces the transition through the weak-inversion regime and makes the strong-inversion analysis with Equation (9.58) more accurate. For intentionally selected low gate biases, the weak-inversion model of Equation (9.60) may successfully be applied, leading to the same lifetime values.[51]

Measurements performed on various SOI materials show that the generation lifetime is very sensitive to the crystalline quality of the film. In early SIMOX annealed at low temperature, the carrier lifetime was very poor (10–100 ns). High-temperature annealing above 1300 °C was

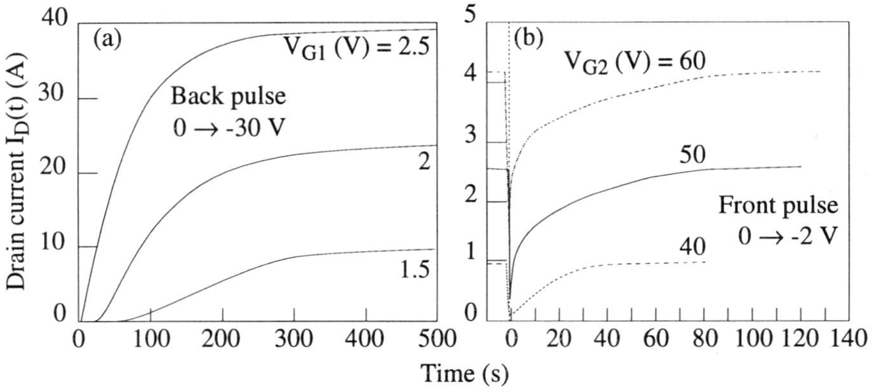

Figure 9.23 (a) Front-channel and (b) back-channel transients showing the influence of the gate bias when a constant pulse is applied on the opposite gate.

very efficient and improved the lifetime by three orders of magnitude (10–100 μs). So far, lifetime in excess of 100 μs has only been achieved in multiple-implanted SIMOX.[48]

The surface generation velocity, currently deduced from Zerbst transients, is extremely small (0.01–0.1 cm/s) and does not reflect the actual generation rate at a *depleted* interface. It was found from weak-inversion analysis using Equation (9.60) that the surface generation velocity decreases by 2–3 orders of magnitude from weak to strong inversion.[51]

The generation velocity on the edges may be substantial. It is evaluated by comparing edged and edgeless structures or by measuring the effective lifetime τ_{eff} in devices with different aspect ratios and making use of Equation (9.54). The edge velocity S_e may exceed 100 cm/s,[39] and thereby dominates the overall generation rate in narrow transistors.

The roles of the two gates can, of course, be interchanged for probing different regions of the film (Fig. 9.23b). Note that the transient duration can be much smaller, owing to the difference in the capacitances of the gate oxide and buried oxide in Equations (9.56), (9.58), and (9.60), without necessarily implying an inhomogeneity of the film quality.

A depth profiling of carrier lifetime is feasible by taking the derivative of the Zerbst curve with respect to w_d. The slopes of the tangents are simply the local values $\tau_{eff}^{-1}(w_d)$ at each point situated beyond w_∞.

Another solution is to monitor the carrier lifetime while gently increasing the bottom level of the pulse. This makes the effective generation region ($w_d - w_\infty$) expand slowly; the local lifetime at the border of the depletion region is found by differentiation.

In summary, the drain current transients are exclusive SOI techniques. They were created to exploit the special dual-gate configuration of SOI transistors and, therefore, there is no chance for corresponding methods to exist in bulk Si. Starting from the same physical principles, it is easy to adapt the method to n- and p-channels, accumulation- and inversion-modes, front and back interfaces, and partially and fully depleted transistors.

9.6 The Current–DLTS Technique

The interface and bulk trap properties in SOI have been widely investigated using various transient capacitance methods such as Zerbst[45] and capacitance DLTS. Recently, the channel-conductance DLTS [55] and drain-current DLTS [56, 57] techniques have been applied to characterize bulk traps in SOI MOSFETs. The use of drain-current or channel-conductance DLTS to extract interface states or bulk traps in SOI MOSFETs is very attractive because the capacitance DLTS measurements are difficult to analyze due to parasitic effects and a small device area. In this section, we describe the drain-current DLTS technique applied to the study of deep-level traps in n-channel enhancement-mode SOI transistors. Defect parameters such as trap energy levels, trap concentrations, and capture cross sections will be determined.

9.6.1 Principle and Experimental Setup

The use of standard capacitance DLTS to study the deep-level defects in thick ($> 1\,\mu$m) SOI materials has been discussed in detail in Section 7.4. However, the very thin nature of the Si overlay and the high series resistance involved in the top Si film preclude the use of the conventional capacitance DLTS method in a normal ($<$ 200-nm-thick) SOI film. In the drain-current DLTS technique,[58] the current transients are not influenced by the large series resistance (which affects only the capacitance transients in thin films). Information thus obtained will give an accurate prediction of the deep-level traps present in the actual SOI circuits.

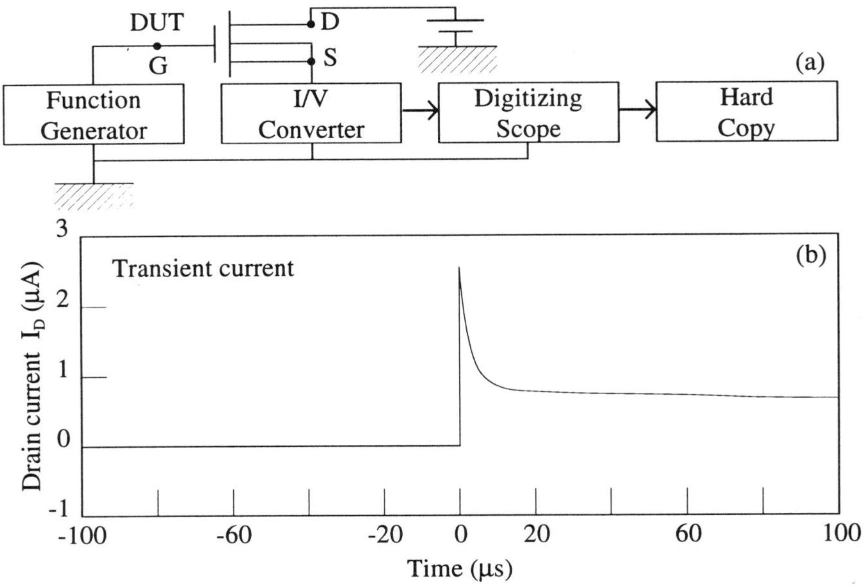

Figure 9.24 (a) Experimental setup for drain-current DLTS, and (b) measured transient drain current for n-channel MOSFET ($V_D = 20\,\mathrm{mV}$, $V_1 = V_T = 0.6\,\mathrm{V}$, $V_2 = -2\,\mathrm{V}$).

Figure 9.24 shows the schematic diagram of the experimental setup and the measured transient drain current for an enhancement-mode n-channel MOSFET. The equipment for the drain-current DLTS measurement consists of an HP function generator that repetitively applies a pulse train to the gate terminal. A current-sensitive low-noise preamplifier (model PAR–181) is used to feed an HP digitizing scope with the voltage signal corresponding to the measured transient drain current. In this setup, both the pulse generator and the scope are controlled via an HP interface bus to vary the pulse parameters and to adjust the sampling points at a fixed fraction of the pulse width.

9.6.2 Theory and Parameter Extraction

We shall describe here the theory of drain-current DLTS for determining the deep donor levels in n-channel enhancement-mode MOSFETs. Haddara et al.[55] have derived a model that relates the changes in the bulk trap charges to the drain-current transient. The model has taken into account the field-dependent mobility and the series-resistance effects in the MOSFETs, and hence it can be used for accurate drain-current or channel-conductance DLTS measurements.

For a bulk or partially depleted SOI n-channel MOSFET operating under strong inversion conditions, the rate-window concept is used to relate the drain current to the emission rate and trap concentration:[55]

$$Q_{D_0}\left(\frac{N_t}{2N_A}\right)[\exp(-t_1/\tau)-\exp(-t_2/\tau)] = \frac{I_c}{\mu_0(W/L)V_D}\times\left[\frac{I_D(t_1)}{I_c-I_D(t_1)}-\frac{I_D(t_2)}{I_c-I_D(t_2)}\right] \tag{9.62}$$

where μ_0 is the low field mobility, I_c is a constant current, and Q_{D_0} is the steady-state depletion charge in strong inversion. The drain-current DLTS signal can be expressed as

$$S = \mu_o(W/L)V_D Q_{D_0}\left(\frac{N_t}{2N_A}\right)[e^{-t_1/\tau}-e^{-t_2/\tau}] \tag{9.63}$$

Quantitative determination of trap parameters such as concentrations and energy levels is made by conducting the current–DLTS measurements. First, the emission time constant τ is found by varying the rate window via pulse-width scanning. The maximum DLTS signal is obtained when the rate window ($t_2 - t_1 = 3T_p/4 - T_p/4$) is of the same order as the time emission rate constant τ. By differentiating S in Equation (9.63) with respect to the period T_p, a maximum S_{max} is obtained at $T_p = T_{p,max}$:[55]

$$S_{max} = \frac{2Q_{D_0}}{3\sqrt{3}}\left(\frac{N_t}{2N_A}\right)\mu_o(W/L)V_D \tag{9.64}$$

$$T_{p,max} = 2\tau\ln 3 \simeq 2.2\tau \tag{9.65}$$

The trap density N_t and the time constant τ can be determined from Equations (9.64) and (9.65) by using pulse-width scanning, whereas a rapid evaluation of the trap-level location may be performed by varying the pulse base level V_2. This measurement procedure does not involve any temperature variation and is therefore simpler and faster than the conventional DLTS measurement scheme.

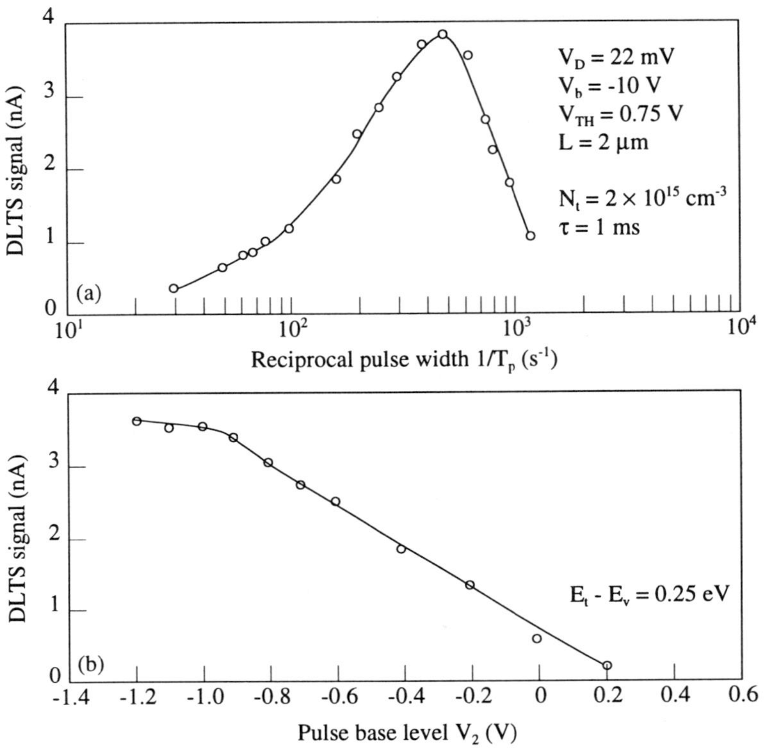

Figure 9.25 (a) Measured DLTS signal versus (a) the inverse of the pulse width T_p and (b) pulse base level V_2 at $T_p = 2\,\text{ms}$ (SIMOX MOSFET with $L = 2\,\mu\text{m}$, $V_T = 0.75\,\text{V}$, $V_D = 22\,\text{mV}$, and $T = 300\,\text{K}$, after Haddara et al.[55]).

Figure 9.25a shows the measured DLTS signal versus the inverse of the pulse width T_p for a partially depleted SIMOX MOSFET. In Figure 9.25b, the DLTS signal is plotted versus V_2 for the same transistor. The DLTS signal reduces to zero for $V_2 = 0.28\,\text{V}$, which corresponds to a trap level situated at $0.25\,\text{eV}$ above the valence band. The values of N_t and τ were found to be $2 \times 10^{15}\,\text{cm}^{-3}$ and $1\,\text{ms}$, respectively.

The energy level E_t for the bulk trap was determined by using a pulse-width scan at room temperature and for a fixed value of the pulse base level V_2. It should be noted that the surface potential has to be known as a function of the gate voltage.

In some cases, the DLTS versus $1/T_p$ exhibited a double-peak behavior, which is attributed to the presence of more than one deep-level trap. This has been observed in SIMOX devices annealed at lower temperatures ($< 1200\,^\circ$C).[55]

The drain-current DLTS technique has also been applied to the study of bulk traps in ultrathin (< 100 nm), fully depleted SOI MOSFETs.[56] In this case, the depletion region extends to the BOX/Si film interface, and hence the drain-current transients will be influenced by the carrier emission from this interface. This effect, however, can be eliminated by keeping the back interface in accumulation during the measurement.

In a fully depleted SOI MOSFET with back surface accumulated, the threshold voltage is given by[56]

$$V_{T_1} \simeq \phi_{FB_1} + \left(1 + \frac{C_{si}}{C_{ox_1}}\right) 2\phi_F - \frac{Q_{si}}{2C_{ox_1}} \tag{9.66}$$

where ϕ_{FB_1} is the front-channel flat-band voltage, C_{si} is the depletion capacitance, C_{ox_1} is the front-gate oxide capacitance, and Q_{si} is the charge per unit surface in the depleted film. The concentration of fast interface traps has been neglected (see also Equation (8.16) in Section 8.5).

Following the analysis of McLarty et al., it can be shown that for an acceptor trap, the threshold voltage V'_{T_1} is modified during the carrier emission after the fill pulse:[58]

$$V'_{T_1} = \phi_{FB_1} + \left(1 + \frac{C_{si}}{C_{ox_1}}\right) 2\phi_F - \frac{Q_{si}}{2C_{ox_1}} \left(1 - \frac{N_t}{N_A} e^{-t/\tau_p}\right) \tag{9.67}$$

where

$$\frac{1}{\tau_p} = e_p = \left(\sigma_p v_{th} \frac{N_V}{g}\right) \exp\left(-\frac{E_t - E_V}{kT}\right) \tag{9.68}$$

The trap concentration N_t is given by

$$N_t = \frac{2V_o N_A}{10^5 \mu_0 V_D (W/L) Q_{si}} [\exp(-t_1/\tau_p) - \exp(-t_2/\tau_p)] \tag{9.69}$$

where V_o represents the DLTS output signal in volts; t_1 and t_2 are the sampling times corresponding to the time constant τ_p. The drain current is expressed as a function of V'_{T_1}:

$$I_D = \mu C_{ox_1} \frac{W}{L} \left(V_G - V'_{T_1} - \frac{V_D}{2}\right) V_D \tag{9.70}$$

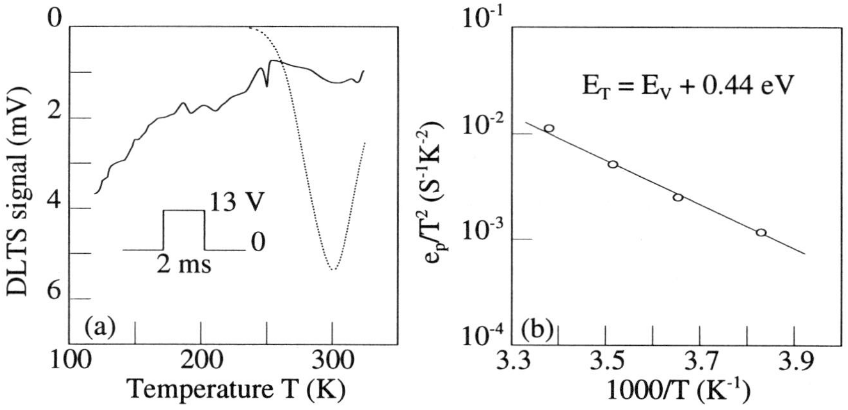

Figure 9.26 (a) DLTS scan of hole traps in an ultrathin fully depleted SOI MOSFET. The solid line is the DLTS scan of the interface states and bulk traps measured at $V_{G_2} = 0\,\mathrm{V}$, while the dashed line is for the bulk hole trap measured at $V_{G_2} = -10\,\mathrm{V}$. (b) Arrhenius plot of the hole trap (after McLarty et al.[56]).

From Equations (9.67) and (9.70), the DLTS signal, $S = I_D(t_2) - I_D(t_1)$, is given by[56]

$$S = KQ_{si}\frac{W}{L}\frac{N_t}{N_A}\left(e^{-t_1/\tau_p} - e^{-t_2/\tau_p}\right) \tag{9.71}$$

In the above equation, K is a slowly varying function of temperature and is nearly constant over the width of the DLTS signal peak. Q_{si} is constant for the fully depleted transistor and is determined by the thickness of the Si film.

A typical current–DLTS scan of a hole trap for an ultrathin n-channel enhancement-mode MOSFET fabricated on SIMOX substrate is shown in Figure 9.26a (solid line) with $V_{G_2} = 0$. Since the layer is fully depleted, both the bulk and back-interface traps will contribute to the DLTS signal. This results in a very broad and complicated spectrum (solid line), which is difficult to analyze.

However, if a constant back-gate bias is applied ($V_{G_2} = -10\,\mathrm{V}$), then the accumulation layer will effectively screen out the back-interface traps, and only the bulk traps will contribute to the DLTS signal; this is illustrated in the dashed line of Figure 9.26a. Figure 9.26b shows the Arrhenius plot, e_p/T^2 vs. $10^3/T$, of this hole trap, which yields an energy level of $E_t = E_V + 0.44\,\text{eV}$; the capture cross section was estimated to be $10^{-16}\,\text{cm}^2$.[56] Using the DLTS spectra and the equivalent of Equation (9.69), the hole trap concentration was found to be $10^{15}\,\text{cm}^{-3}$. Thus, the current–DLTS measurement allows the extraction of relevant defect parameters for the deep-level traps in fully depleted SOI MOSFETs.

9.7 The Threshold-Voltage Method

9.7.1 Principle and Experimental Setup

Determination of the interface-state density profile in a fully depleted SOI MOSFET can also be made by using the threshold-voltage method. Due to the ultrathin, fully depleted film, the electrical properties of the SOI MOSFETs are influenced by the charge coupling between the front and back gate. The capacitance and conductance methods are difficult for characterizing the interface-state density because of the complex multi-interface and the large film series-resistance problems in SOI MOSFETs. Since the threshold voltage of an SOI MOSFET is sensitive to the interface states and gate bias, it can in turn be used for interface-trap analysis.

In this section, we describe an attractive threshold-voltage technique for interface-state profiling in a fully depleted SOI MOSFET operating in the linear region. By measuring the threshold voltage of the front channel under different back-gate bias conditions, the back-interface state profile will be determined. The front interface states can be characterized in a similar way by reversing the polarity of the bias across the gates. The main advantage of this method is the simplicity of the threshold-voltage measurements.

9.7.2 Theory and Parameter Extraction

When the back surface of an SOI MOSFET is depleted, the threshold voltage of the front gate can be written as[59]

$$V_{T_1}^{dep} = V_{T_1}^{acc} - \frac{C_{si}C_{ox_2}}{C_{ox_1}(C_{si} + C_{ox_2} + C_{it_2})}(V_{G_2} - V_{G_2}^{acc}) \qquad (9.72)$$

where $V_{T_1}^{acc}$ denotes the front-channel threshold voltage when the back interface is in accumulation (for $V_{G_2} < V_{G_2}^{acc}$; see Section 8.5), C_{si} is the

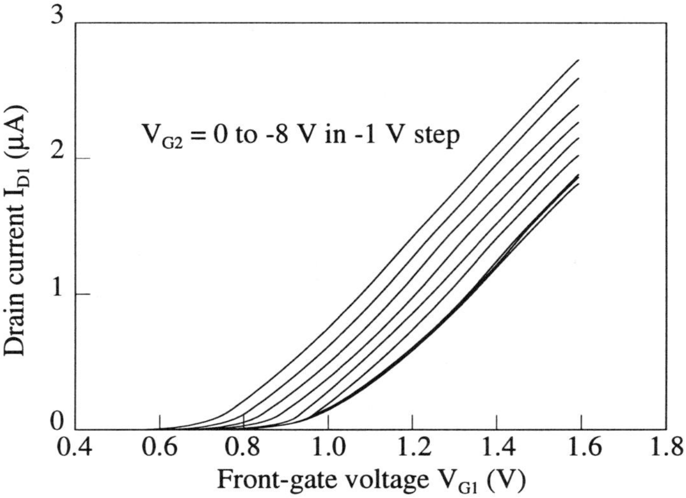

Figure 9.27 The measured front-channel drain current versus front-gate voltage for different back-gate biases. The threshold voltage is determined from linear extrapolation of these curves to the V_{G_1} axis.

depletion capacitance, $C_{ox_{1,2}}$ are the front and back oxide capacitances, and $C_{it_2} = qD_{it_2}$ is the capacitance associated with the back-interface trap charge. By rewriting Equation (9.72), we obtain

$$D_{it_2} = -\frac{1}{q}\left[\left(\frac{C_{si}C_{ox2}}{C_{ox1}}\right)\frac{dV_{G_2}}{dV_{T_1}^{dep}} + C_{si} + C_{ox_2}\right] \tag{9.73}$$

The trap concentration D_{it_2} can be deduced, using Equation (9.73), from the measured $V_{T_1}(V_{G_2})$ curve. It is noted that the device parameters required for calculating D_{it_2} are the thicknesses of the top Si film and the front and back oxides, which can be determined with very good accuracy.[60, 61]

For a strongly inverted front interface, the back surface potential ψ_{s2} versus V_{G_2} is given by:[59]

$$V_{G_2} = V_{G_2}^{acc} - \frac{C_{si}}{mC_{ox_1}}\psi_{s2} \tag{9.74}$$

where $m = dV_{T_1}^{dep}/dV_{G_2} = -C_{si}C_{ox2}/C_{ox_1}(C_{si}+C_{ox_2}+C_{it_2})$ is dependent on V_{G_2}. By expressing ψ_{s_2} in terms of energy relative to the intrinsic

Fermi energy E_i, we have

$$E - E_i = -\frac{qmC_{ox1}}{C_{si}}(V_{G_2} - V_{G_2}^{acc}) - q\phi_F \tag{9.75}$$

where ϕ_F is the Fermi potential in the silicon film.

Thus, the relationship between D_{it_2} and $(E - E_i)$ can be established using Equations (9.73) and (9.75). The above analysis applies equally well for characterizing the interface-trap charge density of the front Si–SiO_2 interface by reversing the polarity of bias voltages applied to the front and back gates.

The threshold-voltage method was tested on n-channel SOI transistors with channel length of 1.8 μm, gate oxide of 15 nm, buried oxide of 350 nm, film thickness of 120 nm, and film-doping concentration of $\sim 4 \times 10^{16}$ cm^{-3}.[61] Static drain-current measurements were performed using an HP–4145B semiconductor parameter analyzer.

The transistor is operated in strong inversion and in the linear region, $V_D \ll V_{G1} - V_{T_1}$, where the relation between the drain current and the threshold voltage of the front channel is well known:

$$I_D \simeq \frac{W}{L}\mu_1 C_{ox_1}(V_{G_1} - V_{T_1})V_D \tag{9.76}$$

with μ_1 the low-field electron mobility. The series resistance and mobility degradation effects are neglected. Equation (9.76) shows that the threshold voltage can be determined from the linear extrapolation[5] of the measured $I_D(V_{G_1})$ curve to the V_{G_1} axis.

Figure 9.27 illustrates the $I_D(V_{G_1})$ plots for different back-gate voltages. Although small deviations exist at higher front-gate voltages, the linear relation, predicted by Equation (9.76), is observed for $V_D + V_{T_1} \ll V_{G_1} \leq 1.5$ V in the present case. The points (solid dots) of maximum slopes on the $I_D(V_{G_1})$ curves correspond to the maximum transconductance values of the $g_m(V_{G_1})$ curves plotted at different values of V_{G_2}. By making tangential lines to the curves at the maximum transconductance points and extrapolating to I_D=0, the threshold voltage V_{T_1} is determined as a function of V_{G_2}. Such an extrapolation is no longer accurate when the back interface reaches strong inversion and the back-channel current becomes significant.

---

[5]The limits of such a threshold-voltage extrapolation have been discussed in Section 8.5.3.

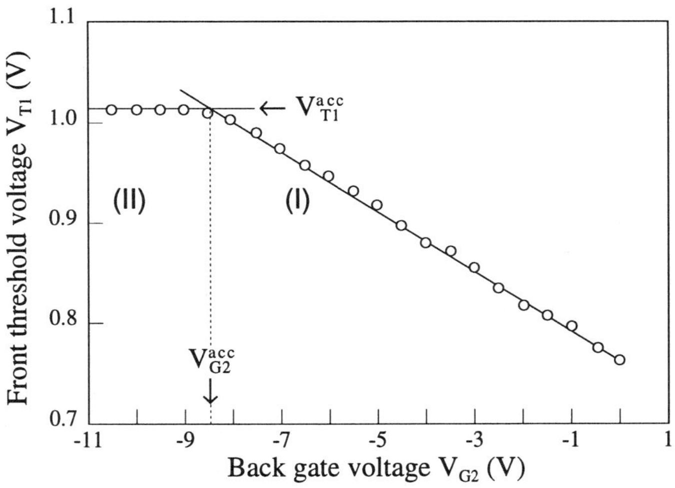

Figure 9.28 Front-channel threshold voltage V_{T_1} versus back-gate voltage V_{G_2} as determined from Figure 9.27.

Figure 9.28 shows that the measured threshold voltage is a quasi-linear function of V_{G_2} when the back surface is depleted (i.e., $V_{G_2}^{acc} < V_{G_2} \leq 0$). For $V_{G_2} < V_{G_2}^{acc}$, V_{T_1} is equal to $V_{T_1}^{acc}$ and is independent of V_{G_2}. Therefore, the back-gate voltage $V_{G_2}^{acc}$ for accumulation at the back interface can be determined from the intersection of the lines fitting the data in regions (I) and (II) of Figure 9.28.

The distribution of the interface-state density in the bandgap for the fully depleted SOI MOSFETs, as determined from Equations (9.73)–(9.75) and Figures 9.27 and 9.28, is similar to the profile shown in Figure 9.18. In the calculation of D_{it_2} from Equation (9.73), $dV_{G_2}/dV_{T_1}^{dep}$ is approximated by $\Delta V_{G_2}/\Delta V_{T_1}^{dep}$, with $\Delta V_{G_2} = 0.1\,\text{V}$. The result is accurate from near the midgap in depletion to near the valence band edge in accumulation. In the present case, the back-interface-state density was found to be $D_{it_2} \simeq 1.5 \times 10^{11}\,\text{cm}^{-2}\text{eV}^{-1}$ near midgap.

By interchanging the roles of the front and back gates, a smaller trap concentration was detected for the front interface ($D_{it_1} \simeq 6 \times 10^{10}\,\text{cm}^{-2}\text{eV}^{-1}$). It should be noted that using the threshold voltage method, D_{it_1} is less accurate than D_{it_2} due to the fact that C_{ox_1} is usually much larger than C_{ox_2}.

9.8 Concluding Remarks

Several methods imported from bulk-silicon research have been revisited in the context of SOI MOS transistors. The adaptation of the experimental setup and basic theory of these methods to thin SOI films is straightforward only if the back interface is maintained in accumulation. Otherwise, unusual and more complex situations are obtained, as compared to the conventional curves obtained in bulk MOSFETs. Full depletion is the condition in which interface coupling effects occur; in other words, the apparent carrier transport and generation properties at one interface can be controlled by the opposite gate. Not only the substrate bias but also the reverse voltage can induce interface coupling.

Charge pumping has the advantage of needing a simple experimental procedure and offering an easy-to-interpret signature of the interface traps. However, it suffers from the necessity for gated diodes or five-terminal MOSFETs. In turn, noise spectroscopy and DLTS are more sophisticated methods that require experimental and theoretical expertise. The nature of the defects they reveal is not always obvious. DLTS is more efficient in the early stage of development of new materials, when the defect concentrations are large. The dynamic transconductance technique is very attractive, in particular in submicron transistors, but needs further validation. Finally, the drain-current transients are exclusive SOI methods. They are mostly devoted to generation lifetime analysis, although there is potential for interface-trap evaluation.

None of the above characterization techniques can override the use of the others. Each of them is focused on a particular feature and presents distinct merits and weaknesses. Only together rather than separately can they provide a complete enough electrical image of SOI structures.

Bibliography

[1] J.H. Lee, S. Cristoloveanu, and A. Chovet, "Non-homogeneous electrical transport through silicon-on-sapphire thin films: evidence of the internal stress influence," *Solid-State Electron.*, vol. 25, p. 947, 1982.

[2] S. Cristoloveanu, J.H. Lee, and A. Chovet, "Effets galvanomagnétiques dans les semiconducteurs anisotropes inhomogènes; application à la caractérisation des films de silicium sur saphir," *Rev. Phys. Appl.*, vol. 15, p. 725, 1980.

[3] A.C. Ipri, "Electrical properties of silicon films on sapphire using the MOS Hall technique," *J. Appl. Phys.*, vol. 43, p. 2770, 1972.

[4] A.B.M. Elliot and J.C. Anderson, "An investigation of carrier transport in thin silicon on sapphire films using MIS deep depletion Hall effect structures," *Solid-State Electron.*, vol. 15, p. 531, 1972.

[5] J.L. Robert, B. Pistoulet, A. Raymond, J.M. Dusseau, and G.M. Martin, "New model of conduction mechanism in semi-insulating GaAs," *J. Appl. Phys.*, vol. 50, p. 349, 1979.

[6] S. Cristoloveanu, J.H. Lee, J. Pumfrey, J.R. Davis, R.P. Arrowsmith, and P.L.F. Hemment, "Profiling of inhomogeneous carrier transport properties with the influence of temperature in silicon-on-insulator films formed by oxygen implantation," *J. Appl. Phys.*, vol. 60, p. 3199, 1986.

[7] S. Cristoloveanu, "Silicon films on sapphire," *Rep. Prog. Phys.*, vol. 50, p. 327, 1987.

[8] S. Onga, K. Hatanaka, S. Kawaji, and Y. Yasuda, "Influence of crystalline defects and residual stress on the electrical characteristics of SOS MOS devices," *Jpn. J. Appl. Phys.*, vol. 17, p. 413, 1978.

[9] A.L. Lin, R. Maddox, and J.E. Mee, "Electrical characterization of ion implanted silicon on sapphire," *Microelectron. J.*, vol. 14, p. 22, 1983.

[10] J.H. Lee, S. Cristoloveanu, and A. Chovet, "Transport profiles, the influence of internal stress and potential fluctuations in thin silicon films on sapphire," *Thin Solid Films*, vol. 89, p. 13, 1982.

[11] T. Elewa, H. Haddara, S. Cristoloveanu, and M. Bruel, "Charge pumping in silicon on insulator structures using P–I–N diodes," in *18th ESSDERC Conf. Proc.* (Montpellier, France), *J. Physique*, vol. 49, suppl. 9, p. C4-137, 1988.

[12] D.J. Wouters, M. Tack, G.V. Groeseneken, H. Maes, and C.L. Claeys, "Characterization of front and back Si–SiO_2 interfaces in thick and thin film silicon on insulator structures by the charge pumping technique," *IEEE Trans. Electron Devices*, vol. 36, p. 1746, 1989.

[13] T. Ouisse, S. Cristoloveanu, T. Elewa, H. Haddara, G. Borel, and D. Ioannou, "Adaptation of the charge pumping technique to gated p–i–n diodes fabricated on silicon on insulator," *IEEE Trans. Electron Devices*, vol. 38, p. 1432, 1991.

[14] G. Groeseneken, H.E. Maes, N. Beltran, and R.F. de Keersmaecker, "A reliable approach to charge pumping measurements in MOS transistors," *IEEE Trans. Electron Devices*, vol. 31, p. 42, 1984.

[15] H. Haddara and S. Cristoloveanu, "Profiling of stress-induced interface states in short-channel MOSFETs using a composite charge pumping technique," *Solid-State Electron.*, vol. 29, p. 767, 1986.

[16] T. Ouisse, S. Cristoloveanu, and G. Borel, "Electron trapping in irradiated SIMOX buried oxides," *IEEE Electron Device Lett.*, vol. 12, p. 312, 1991.

[17] T. Elewa, B. Kleveland, S. Cristoloveanu, B. Boukriss, and A. Chovet, "Detailed analysis of edge effects in SIMOX–MOS transistors," *IEEE Trans. Electron Devices*, vol. 39, p. 874, 1992.

[18] F. Hofmann and W. Hansch, "The charge pumping method: experiment and complete simulation," *J. Appl. Phys.*, vol. 66, p. 3092, 1989.

[19] H.E. Maes and G. Groeseneken, "Determination of the spatial surface state distribution in MOS and SIMOS transistors after channel hot electron injection," *Electron. Lett.*, vol. 18, p. 372, 1982.

[20] P. Heremans, J. Witters, G. Groeseneken, and H. Maes, "Analysis of the charge pumping technique and its application for the evaluation of MOSFET degradation," *IEEE Trans. Electron Devices*, vol. 36, p. 1318, 1989.

[21] W. L. Tseng, "A new charge pumping method of measuring Si-SiO_2 interface states," *J. Appl. Phys.*, vol. 62, p. 591, 1987.

[22] A. Chovet and P. Viktorovitch, "Le bruit électrique," *L'Onde électrique*, vol. 57, p. 699 and 773, 1977, and vol. 58, p. 69, 1978.

[23] A.L. McWhorter, *Semiconductor Surface Physics*, University of Pennsylvania Press, Philadelphia, p. 207, 1957.

[24] G. Ghibaudo, "A simple derivation of Reimbold's drain current spectrum formula for flicker noise in MOSFETs," *Solid-State Electron.*, vol. 30, p. 1037, 1987.

[25] S. Christensson, I. Lundstrom, and C. Svensson, "Low frequency noise in MOS transistors," *Solid-State Electron.*, vol. 11, p. 797, 1968.

[26] G. Ghibaudo, O. Roux, C. Nguyen–Duc, F. Balestra, and J. Brini, "Improved analysis of low frequency noise in field-effect MOS transistors," *Phys. Stat. Sol. (a)*, vol. 124, p. 571, 1991.

[27] G. Reimbold, "Modified $1/f$ trapping noise theory and experiments in MOS transistors biased from weak to strong inversion. Influence of interface states," *IEEE Trans. Electron Devices*, vol. 31, p. 1190, 1984.

[28] Z.H. Fang, A. Chovet, Q.P. Zhu, and J.N. Zhao, "Theory and applications of $1/f$ trapping noise in MOSFETs for the whole biasing ranges," *Solid-State Electron.*, vol. 34, p. 327, 1991.

[29] B. Boukriss, H. Haddara, S. Cristoloveanu, and A. Chovet, "Modeling of the $1/f$ noise overshot in short channel MOSFETs locally degraded by hot carrier injection," *IEEE Electron Device Lett.*, vol. 10, p. 433, 1989.

[30] Z.H. Fang, "Low-frequency pseudo generation-recombination noise of MOSFETs stressed by channel hot electrons in weak inversion," *IEEE Trans. Electron Devices*, vol. 33, p. 516, 1986.

[31] Z.H. Fang, S. Cristoloveanu, and A. Chovet, "Analysis of hot carrier induced aging from $1/f$ noise in short channel MOSFETs," *IEEE Electron Device Lett.*, vol. 7, p. 371, 1986.

[32] P. Gentil and S. Chaussé, "Low frequency noise measurements in silicon–on-sapphire (SOS) MOS transistors," *Solid-State Electron.*, vol. 20, p. 935, 1977.

[33] O. Roux dit Buisson, G. Ghibaudo, J. Brini, and T. Ouisse, "Measurements and analysis of random telegraph signals in small area SOI MOSFETs," in *Silicon–On–Insulator Technology and Devices*, W.E. Bailey (ed.), The Electrochemical Society, Pennington, NJ, vol. 92–13, p. 191, 1992.

[34] A. Chovet, B. Boukriss, T. Elewa, and S. Cristoloveanu, "Low frequency noise of front channel and back channel MOSFETs fabricated on silicon on insulator SIMOX substrates," in *Noise in Physical Systems*, C.M. Van Vliet (ed.), World Scientific, Teaneck, NJ, p. 457, 1987.

[35] T. Elewa, B. Boukriss, A. Chovet, and S. Cristoloveanu, "Low frequency noise spectroscopy in thin SIMOX MOS transistors," *1990 IEEE SOS/SOI Technol. Conf. Proc.*, p. 103, 1990.

[36] T. Elewa, B. Boukriss, H.S. Haddara, A. Chovet, and S. Cristoloveanu, "Low-frequency noise in depletion-mode SIMOX MOS transistors," *IEEE Trans. Electron Devices*, vol. 38, p. 323, 1991.

[37] H. Haddara and G. Ghibaudo, "Analytical modeling of transfer admittance in small MOSFETs and application to interface state characterization," *Solid-State Electron.*, vol. 31, p. 1077, 1988.

[38] G. Ghibaudo, "An analytical model of conductance and transconductance for enhancement mode MOSFETs," *Phys. Stat. Solidi (a)*, vol. 95, p. 323, 1986.

[39] T. Elewa, H. Haddara, and S. Cristoloveanu, "Interface properties and recombination mechanisms in SIMOX structures," in *The Physics and Technology of Amorphous SiO_2*, R.A.B. Devine (ed.), Plenum, NY, p. 553, 1988.

[40] U. Sharma, R.V. H. Booth, and M.H. White, "Static and dynamic transconductance of MOSFETs," *IEEE Trans. Electron Devices*, vol. 36, p. 954, 1989.

[41] D.E. Ioannou, X. Zhong, B. Mazhari, G.J. Campisi, and H.L. Hughes, "Interface characterization of fully depleted SOI MOSFETs by dynamic transconductance technique," *IEEE Electron Device Lett.*, vol. 12, p. 430, 1991.

[42] H. Haddara, T. Elewa, and S. Cristoloveanu, "Static and dynamic transconductance model for depletion-mode transistors: A new characterization method for silicon-on-insulator materials," *IEEE Electron Device Lett.*, vol. 9, p. 35, 1988.

[43] H.S. Chen and S.S. Li, "Determination of interface state density in small-geometry MOSFETs by high-low-frequency transconductance method," *IEEE Electron Device Lett.*, vol. 12, p. 13, 1991.

[44] P.C. Yang, H.S. Chen, and S.S. Li, "Measurements of interface state density in partially- and fully-depleted silicon-on-insulator MOSFETs by a high-low-frequency transconductance method," *Solid-State Electron.*, vol. 35, p. 1031, 1992.

[45] M. Zerbst, "Relaxation effects at semiconductor-insulator interfaces," *Z. Angew. Phys.*, vol. 22, p. 30, 1966.

[46] D.P. Vu and J.C. Pfister, "Determination of minority carrier generation lifetime in beam-recrystallized silicon-on-insulator by using a depletion-mode transistor," *Appl. Phys. Lett.*, vol. 47, p. 950, 1985.

[47] R.F. Pierret, "Rapid interpretation of the MOS–C C–t transient," *IEEE Trans. Electron Devices*, vol. 25, p. 1157, 1978.

[48] D.E. Ioannou, S. Cristoloveanu, M. Mukherjee, and B. Mazhari, "Characterization of carrier generation in enhancement-mode SOI MOSFETs," *IEEE Electron Device Lett.*, vol. 11, p. 409, 1990.

[49] P.W. Barth and J.B. Angell, "A dual-gate deep-depletion technique for generation lifetime measurements," *IEEE Trans. Electron Devices*, vol. 27, p. 2252, 1980.

[50] T.I. Kamins, "Minority-carrier lifetime in dielectrically isolated single-crystal silicon films defined by electrochemical etching," *Solid-State Electon.*, vol. 17, p. 675, 1974.

[51] A. Ionescu and S. Cristoloveanu, "Carrier generation in thin SIMOX films by deep-depletion pulsing of MOS transistors," *Nucl. Instr. Methods Phys. Res.*, vol. B84, p. 265, 1994.

[52] N. Yasuda, K. Taniguchi, C. Hamaguchi, Y. Yamaguchi, and T. Nishimura, "New carrier lifetime measurement method for fully depleted SOI MOSFETs," *IEEE Trans. Electron Devices*, vol. 39, p. 1197, 1992.

[53] D. Kimpton and J. Kerr, "Generation lifetime, interface state density, active defect density and oxide resistivity measurements for SOI–MOSFETs and their radiation dependence," *IEEE Trans. Nucl. Sci.*, vol. 39, p. 2126, 1992.

[54] D.K. Schroder, Y.D. Whitfield, and C.J. Varker, "Recombination lifetime using the pulsed MOS capacitor," *IEEE Trans. Electron Devices*, vol. 31, p. 462, 1984.

[55] H. Haddara, M.T. Elewa, and S. Cristoloveanu, "Measurement and modeling of drain current DLTS in enhancement SOI MOSFETs," *Microelectron. J.*, vol. 24, p. 647, 1993.

[56] P.K. McLarty, D.E. Ioannou, and H.L. Hughes, "Deep states in silicon-on-insulator substrates prepared by oxygen implantation using current deep level transient spectroscopy," *Appl. Phys. Lett.*, vol. 53, p. 871, 1988.

[57] J.W. Chen, R.J. Ko, D.W. Brzezinski, L. Forbes, and C.J. Dell'oca, "Bulk traps in silicon-on-sapphire by conductance DLTS," *IEEE Trans. Electron Devices*, vol. 28, p. 299, 1981.

[58] P.K. McLarty and D.E. Ioannou,"Bulk traps in ultra-thin SIMOX MOSFETs by current DLTS," *IEEE Electron Device Lett.*, vol. 9, p. 545, 1988.

[59] H.K. Lim and J.G. Fossum, "Threshold voltage of thin-film silicon-on-insulator (SOI) MOSFETs," *IEEE Trans. Electron Devices*, vol. 30, p. 1244, 1983.

[60] D.P. Vu, P.M. Zavracky, M.J. Boden, and N.K. Cheong, "A simple electrical method to determine the Si and oxide thicknesses in SOI materials," *IEEE Electron Device Lett.*, vol. 12, p. 427, 1991.

[61] J. Chen, R. Solomon, T.Y. Chan, P.K. Ko, and C. Hu, "A CV technique for measuring thin SOI film thickness," *IEEE Electron Device Lett.*, vol. 12, p. 453, 1991.

Chapter 10

Monitoring Transistor Degradation

In this chapter, two main issues concerning SOI technologies are addressed. On the one hand, SOI circuits are dedicated radiation-hard devices that do not suffer from any competition with bulk Si. On the other hand, the interface degradation, induced by hot-carrier injection into the gate oxide, is an intrinsic problem and a primary challenge for both Si and SOI transistors. The way this problem is alleviated, using simple and inexpensive processing and design solutions, is considered to be a strong argument in favor of a given technology.

An impressive amount of research has been devoted to the analysis of radiation and hot-carrier SOI–MOSFET degradation. We will only briefly review the nature of interface and oxide defects as well as some of the technological options that are envisaged. Emphasis will be placed on discussion of the characterization techniques and procedures used for monitoring transistor degradation. The practical applications of the parameter extraction methods described in the previous chapters will be demonstrated by emphasizing the relations between parameter shifts and defect generation.

10.1 Hot-Carrier-Induced Degradation

The reliability of CMOS circuits depends on the capability of individual transistors to withstand aging effects caused by the injection of hot carriers into the gate oxide. This topic is not new: the tools of investigation have been refined during the past years and successful, yet temporary, technological solutions have been found. As new steps in device scaling are tackled, process engineers are cyclically facing the same set of basic difficulties, which become harder and have more dramatic consequences.

The aging of SOI transistors is an even more challenging problem because not only the gate oxide but also the buried oxide may be degraded. Moreover, special degradation mechanisms arise from interface coupling, floating-body effects, and different electric-field maps. In order to realize the full complexity of SOI devices, we will first give a general introduction on the aging mechanisms.

10.1.1 Basic Degradation Mechanisms

In short-channel transistors, carriers gain enough energy and become *hot*. Some of them are injected into the gate oxide, where another fraction of them generate defects or get trapped. The lateral profiles of carrier energy, injection rate, and defect density mirror the electric-field distribution. The maxima are situated on the drain side, where the lateral field is a maximum as well. The localization of the defective region, close to the drain, is responsible for inhomogeneous properties along the channel. The influence of this localized, damaged region on transistor performances is referred to as "aging effects."

Four groups of phenomena are identified: (1) carrier heating, (2) injection, (3) localized defect formation, and (4) electrical influence of generated defects. The relevance of the physical effects and characterization techniques varies substantially according to the device size and technology.

CARRIER HEATING

Carrier transport in a high electric field is described in terms of mobility reduction, velocity saturation, rise of the electron temperature T_e, and modification of the distribution function. In long-channel MOSFETs, operated at room temperature, the electron mobility starts being field dependent:

$$\mu \simeq \mu_0 / \sqrt{1 + (V_D / E_c L)^2} \tag{10.1}$$

for $V_D/L \simeq 2.5\,\mathrm{kV/cm}$, and the electron velocity saturates ($\simeq 10^7\,\mathrm{cm/s}$), for about $40\,\mathrm{kV/cm}$.

In deep-submicron devices, *nonstationary* transport effects occur as soon as the transit time becomes comparable with the relaxation time constant. Since channel carriers do not experience enough scattering events to reach thermal equilibrium, their mobility and temperature are modified, rendering standard short-channel models obsolete. Not only does the electron velocity overshoot in mesoscopic ($L < 0.1\,\mu$m) devices but also the electron heating requires a stronger lateral field.[1] In other words, for a constant field, carriers are cooler, and hence are expected to cause less damage. The different locations of the peak of the lateral electric field and the peak of the carrier energy must also be taken into account.

CARRIER INJECTION

High-energy channel carriers as well as some of the electrons and holes generated by impact ionization are eligible for injection into the gate oxide. The injection rate and the corresponding formation of defects are nonuniform along the channel, since they strongly depend on the local values of the carrier energy and on the vertical component of the electric field. In order to reach the gate, an electron must be *"lucky"* enough[1] to (i) acquire a higher energy than the oxide barrier, (ii) experience a favorable collision redirecting it towards the interface, (iii) avoid any further scattering event, and (iv) surmount the oxide field (when repulsive).

Although the barrier between the Si conduction band and SiO_2 is 3.2 eV (or 2.8 eV, if barrier lowering by image force is considered), electron injection has been observed even for drain voltages as low as 2–2.5 V (see Fig. 10.7a). Such a low-voltage-injection mechanism involves high-energy carriers associated with the tail of the temperature-based distribution of electrons. The probability for electrons to gain an energy exceeding qV_D is also a function of the lateral profile of the electric field.[2]

The aging of p–MOSFETs is less severe than that of n-channel transistors, simply because the mobility, mean-free path, energy, and ionization rate (i.e., substrate current I_{sub}) are substantially lower for holes than for electrons. Moreover, the oxide barrier for hole injection (4.5 eV)

---

[1] As with any successful theory, the "lucky electron model" has been revisited, improved, and, of course, heavily contested. But never, never were the injected electrons considered to be anything but lucky. With permission from the psychologist of Kluwer, we are now proud to introduce the concept of *unhappy electrons.*

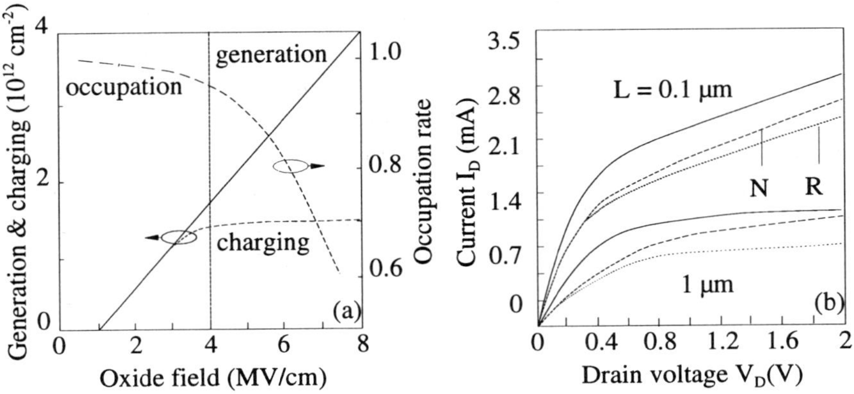

Figure 10.1 (a) Oxide-trap generation and occupation versus vertical field (after Heremans et al.[2]). (b) Asymmetrical $I_D(V_D)$ characteristics in normal (N) and reverse (R) modes induced by defect localization in 0.1 μm and 1 μm stressed n-channel MOSFETs.

is much higher, so most of the degradation is still due to hot-electron injection.

DEFECT FORMATION

Several field-dependent mechanisms are simultaneously involved in the process of defect formation: filling of preexisting oxide traps, generation of new oxide traps, and interface-trap production. Laterally homogeneous Fowler–Nordheim injection as well as *uniform* injection of light-generated carriers are used to determine the distinct features of each mechanism for both electrons and holes. The advantage of the latter experiment is to separate the carrier energy (controlled by the substrate voltage) and the oxide field (V_G–dependent).

Figure 10.1a shows that *electron-trap* generation increases linearly with oxide field E_{ox}, whereas the trap occupancy rate decreases above 4 MV/cm.[2] The balance of these two effects leads to a saturation of the trapped charge for $E_{ox} > 4$ MV/cm. Since the efficiency of electron trapping (10^{-7}–10^{-5} traps/electron) increases with increasing gate current only for $E_{ox} < 4$ MV/cm, it follows that the accelerated stress conducted with higher fields may not be representative for actual device operation.

By comparison, the holes have to surmount a higher potential barrier, but the filling of ***hole traps*** is very efficient (10^{-2}–10^{-1} traps/hole) and does not depend on either E_{ox} or current flow through the oxide. No evidence was found for hole-trap generation.

Several factors are relevant for mesoscopic devices with ultrathin oxides: (i) the threshold field of 4 MV/cm becomes a reality, (ii) tunneling effects will help electrons to reach the gate without damage formation, and (iii) oxide-trap generation does not seem to depend on carrier temperature.

Interface traps D_{it} have been detected subsequent to pure electron injection: the generation rate increases with E_{ox}, the threshold field being 1.5 MV/cm. However, the trapping of holes is a necessary ingredient for D_{it} creation. Their efficiency (one interface state per 40 trapped holes) is 10^3–10^4 times higher than for electrons. Systematical investigations of actual stressing conditions show that D_{it} is a maximum when both holes and electrons are injected. In a.c. stress, the longer the time allowed for hole injection, the larger the amount of D_{it}.

There is no unanimity so far on the physical/chemical mechanisms leading to interface generation (dispersive transport towards the interface of H^+ ions released by hole injection, electron-hole recombination, etc.). Although carrier trapping is subject to poststress relaxation, the build-up of interface traps represents a permanent damage. It is quite a slow process, since D_{it} may increase by one order of magnitude even long after the stress has stopped.

INFLUENCE OF LOCALIZED DEFECTS

The defects are typically responsible for a *local* modification of the carrier mobility, electric field, band bending, inversion charge, threshold voltage, impact ionization rate, etc. The deformation of the surface potential, induced by electron trapping near the drain, is illustrated in Figure 10.2. The potential is locally lowered in n-channels and gives rise to a high resistive region, whereas in p–MOSFETs it is increased, inducing an apparent "channel shortening."

The damage is experimentally visible through the threshold voltage shift, increased leakage, degraded subthreshold slope and modified values of mobility, series resistances, substrate, and gate currents. Evidence for defect localization is found in the asymmetry of $I_D(V_D)$ (Fig. 10.1b), $I_G(V_D)$, $I_{sub}(V_D)$, or $V_T(V_D)$ curves.[3]

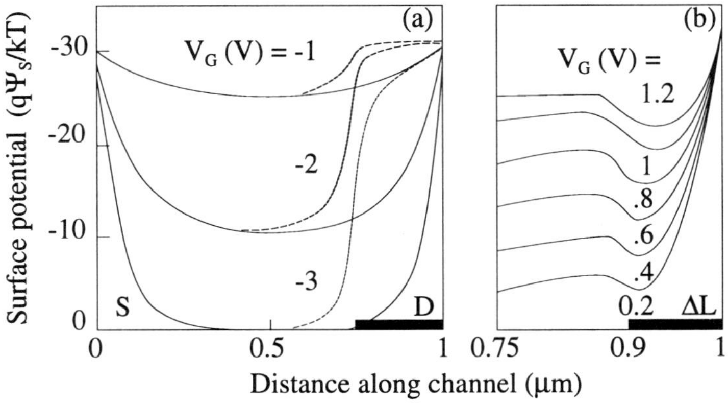

Figure 10.2 Normalized surface-potential profiles showing the influence of localized electron trapping (black region) in 1-μm-long (a) p-channel and (b) n-channel MOSFETs (after Cristoloveanu et al.[3]).

Due to the localized nature of the defects, it is vain to use homogeneous transistor models for damage evaluation and device-lifetime prediction. A simple, yet fully representative image of the physical behavior of the damaged transistor is given by a *two-piece model*.[4] For ohmic region operation, the defective (ΔL, subscript d) and nondefective ($L-\Delta L$, subscript nd) portions of the channel are connected in series, which yields for the transconductance

$$g_{m,a} = \frac{G_d^2}{(G_{nd}+G_d)^2} g_{m,nd} + \frac{G_{nd}^2}{(G_{nd}+G_d)^2} g_{m,d} \tag{10.2}$$

According to Equation (10.2) and Figure 10.3a, the transconductance $g_{m,a}$ of the aged device is governed by the less conductive channel region. At the onset of strong inversion, the damaged region is far more resistive ($G_d \ll G_{nd}$) than the defect-free region (which has already reached strong inversion) and controls the whole transistor. The total transconductance depends on $g_{m,d}$ and ΔL and may *increase* during stress even if the mobility is degraded. The shorter the defective region, the larger the overshoot. In strong inversion, the longer defect-free part regains control of the current, and the total transconductance is lowered.

Two limiting factors tend to balance the transconductance overshoot: the degradation of the carrier mobility in the defective region, and the

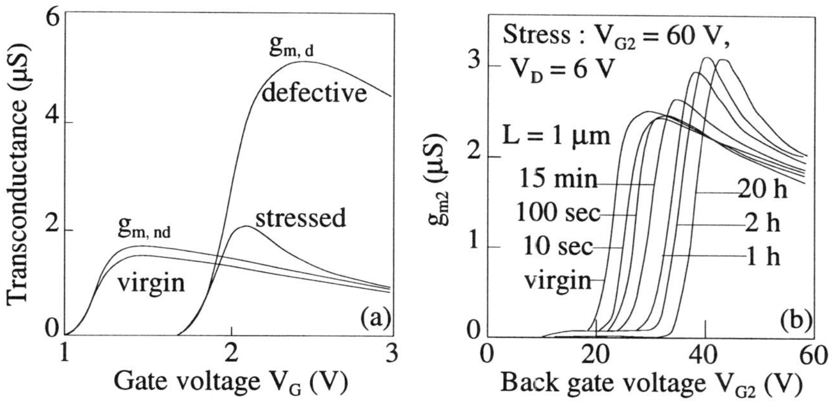

Figure 10.3 Transconductance overshoot induced by defect localization near the drain: (a) model[4] and (b) experiment in back-channel SIMOX n–MOSFET.[5]

reduction of the field-effect mobility with gate bias for $V_G > V_T$. When these factors are not prominent, as in SOI MOSFETs[5] or for low-temperature operation, quite large transconductance overshoots may be measured (Fig. 10.3b). No overshoot was observed in 0.1 μm MOSFETs, since the defective region is rather large ($\Delta L \simeq L$) and always controls the whole channel.[1]

The following are other symptoms of the coupling between the two channel regions:[3]

- The subthreshold slope may change under the influence of localized *fixed* charges and may be insensitive to localized interface traps—a finding that totally contrasts with the viewpoint of the conventional MOS–transistor theory.

- The physical concept of threshold voltage is ambiguous in degraded transistors: it no longer reflects a homogeneous band bending, and can be controlled by either region of the channel or by both of them according to ΔL and defect concentrations.

- The typical rectangle-like charge-pumping curve (Fig. 9.5b, Section 9.2) is distorted by the superposition of two "rectangles" corresponding to the contributions of the two channel regions (Figure 10.4a).

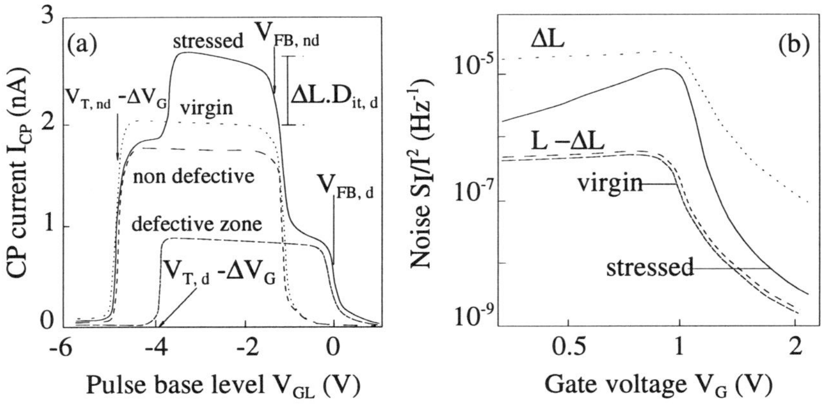

Figure 10.4 (a) Charge pumping and (b) normalized drain-current noise for a virgin device, stressed device, nondefective region, and defective channel region.

- The low-frequency $1/f$ noise (Fig. 9.10, Section 9.3) overshoots due to the excess noise generated in the defective region (Fig. 10.4b).
- A structure is detected in gated-diode leakage $I_R(V_G)$ curves (Section 7.3).

It is worth noting that, in LDD structures, defects are also generated in the spacer and yield increased series resistances.

The above effects occur more frequently in n-channels. In p-channels, the critical source of aging is due to localized electron trapping. The defective region now has a lower threshold voltage and a far larger conductance, so the aged transistor behaves as if it were "shortened" to the length $(L - \Delta L)$ of the nondefective region.

The two-piece model has been refined by replacing the flat distribution of defects with more realistic distributions (gaussian, triangular, etc.). Also, powerful simulators were taught to compute the defective region in accordance with the position-dependent injection rate. Note that in $0.1\,\mu$m MOSFETs, the generic term of *damaged region* is unreliable and must be replaced by the concept of *single-defect-induced aging*. Indeed, 10^{10} defects/cm^2 represent only *one* defect in a region of $0.1 \times 0.1\,\mu m^2$, so the challenge is to sense *individual* defects, as is done in RTS measurements.

10.1.2 Stressing Conditions

Stress tests are normally conducted by choosing the drain voltage in the saturation region. Operation in the breakdown region is not suitable, since peculiar effects come into play. Mostly related to parasitic bipolar action (breakdown, conduction in the volume of the film, etc.), these effects represent supplementary sources of degradation. The extrapolated transistor lifetime is severely underestimated and no longer mirrors the actual reliability of SOI MOSFETs operated under true (normal) circuit conditions.

Although increasing the drain voltage or reducing the channel length leads to more pronounced aging, the extension of the defective region ($\Delta L \simeq 50$–200 nm in 1 μm transistors) depends essentially on gate bias. The following are typical bias conditions and major consequences during the *static* stress of n-MOSFETs:

- $V_G \simeq V_T$—Vertical field reversed near the drain; hole injection and trapping; and generation of interface traps in a narrow region ΔL located near the drain.

- $V_G \simeq V_D/2$ (i.e., peak of I_{sub})—Large impact-ionization rate; injection of both electrons and holes; maximum D_{it} formation; and narrow ΔL.

- $V_G \geq V_D$ (i.e., peak of I_G)—Electron injection (assisted by a favorable vertical field) from a larger section of the channel; trapping; residual D_{it} generation; and a more extended defective region.

During a *dynamic* stress, the above conditions may be combined, leading to more complex situations. *Alternating* stress consists of a superposition of several static stresses—for instance, hole injection followed by electron injection.

Symmetrical situations may be encountered in p-channel MOSFETs; however, most of the degradation is actually created by electrons. For low gate-voltage stress ($|V_G| \ll |V_D|$), the electron gate current in p–MOSFETs exceeds by orders of magnitude the hole gate current in n-channels. The aging of p–MOSFETs consists of a massive electron trapping in the oxide close to the drain. For high gate-voltage stress ($V_G \simeq V_D$), the carrier injection and device aging are attenuated: the electrons experience an opposite vertical field, whereas the holes do not have enough energy.

It is important to secure a rapid retrofit between stressing tests and technology optimization. To this purpose, *accelerated* stress is imposed by increasing V_D (without penetrating the breakdown region), monitoring shorter transistors, or lowering the temperature. For simplicity of analysis, front- and back-channel stresses are conducted independently (with the opposite interface in depletion or accumulation) so as to avoid simultaneous carrier injection from both channels and coupling effects. During the stress, test sessions are scheduled according to a logarithmic time scale.

10.1.3 Extraction of the Defective-Region Parameters

Based on the two-piece model, the difference in channel resistances before and after aging is used to compose a function $F(V_G)$ that yields, for *strong inversion*,[6]

$$F(V_G) = (R_a - R_b)(V_G - V_{T_{nd}}) = \frac{1}{\gamma'(V_G - V_{T_d})} + \frac{\theta'}{\gamma'} \tag{10.3}$$

where $V_{T_{nd,d}}$ and $\mu_{nd,d}$ are the threshold voltages and carrier mobilities of the nondefective and defective channel regions. The constants γ' and θ' are expressed by $\gamma' = \mu_d C_{ox} W / \Delta V_T \Delta L$ and $\theta' = (\mu_{nd} - \mu_d)/\mu_{nd}\Delta V_T$, with $\Delta V_T = V_{T_d} - V_{T_{nd}}$.

Drawing $1/\sqrt{dF/dV_G}$ versus V_G results in a straight line whose intercept and slope give V_{T_d} and γ' (Fig. 10.5a). The coefficient θ' is found from the linear plot $F(V_G)$ vs. $(V_G - V_{T_d})^{-1}$. Since the parameters $V_{T_{nd}}$ and μ_{nd} are known by prior characterization (before stress), the properties of the defective region (μ_d, ΔL, total density of charged defects) are easily determined from θ', γ', and V_{T_d}.

Similarly, a comparison of the channel resistances in *weak inversion*, before and after aging, yields[6]

$$\begin{aligned} G(V_G) &= \ln\left[\frac{R_a - R_b}{R_b} + \frac{\Delta L}{L}\right] = \ln\left(\frac{\mu_{nd}\Delta L}{\mu_d L}\right) + \\ &+ \beta V_G \left(\frac{1}{N_{nd}} - \frac{1}{N_d}\right) - \beta\left(\frac{V'_{T_{nd}}}{N_{nd}} - \frac{V'_{T_d}}{N_d}\right) \end{aligned} \tag{10.4}$$

where V'_T are the threshold voltages in weak inversion, $V'_T = V_T - 0.5\phi_F(1 + C_{it}/C_{ox})$, $N = (C_d + C_{ox} + C_{it})/C_{ox}$, $\beta = q/kT$, and $C_{it} = qD_{it}$.

The semilogarithmic plot of $\ln G(V_G)$ versus V_G is shown in Figure 10.5b. A straight line, corresponding to Equation (10.4), is obtained

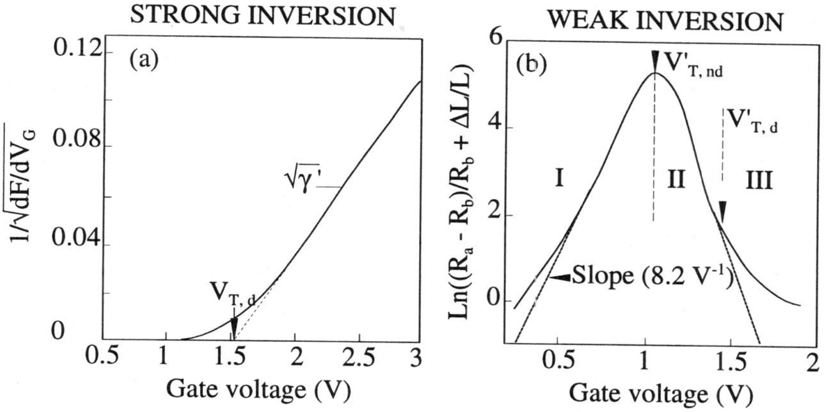

Figure 10.5 Illustration of the extraction of the damaged region parameters in (a) strong inversion and (b) weak inversion regions (after Haddara et al.[6]).

as long as the two channel regions are in weak inversion (zone I). The determination of the coefficient N_d and of the local density of interface traps $D_{it,d}$ proceeds directly from the measurement of the slope $\beta(1/N_{nd} - 1/N_d)$. The peak and inflection point, respectively, give $V'_{T_{nd,d}}$ and define zones II and III of the curve shown in Figure 10.5b. In region II, only the nondefective region has reached strong inversion, whereas the defective region is still in weak inversion and dominates the transistor operation. In region III, the damaged region becomes strongly inverted as well.

This extraction technique can be extended to p-channel MOSFETs.[7] First, the channel "shortening" is estimated from the transconductance increase after stress

$$\frac{\Delta L}{L} \simeq 1 - \frac{g_{m,b}}{g_{m,a}} \tag{10.5}$$

Then the value $\Delta L/L$, which is no longer negligible with respect to $(R_a - R_b)/R_b$, is inserted into Equation (10.4). The plot $\ln G(V_G)$ versus V_G again gives the amount of trapped electrons (from ΔV_T) and the generated interface traps (from the slope).

These methods have been validated by 2–D simulations. The experiment proves that they are efficient in characterizing *large* densities of localized defects—for instance, those currently formed at the back interface or after severe front-channel stress. Obviously, the accuracy is lowered if the damage is irrelevant ($N_d \simeq N_{nd}$) or situated too close to the drain.[2] In LDD devices, the strong-inversion method needs some adaptation in order to include the modification of the series resistances during aging.

The localized nature of generated defects is demonstrated by the comparison of $I_D(V_D)$ curves, measured in the normal and reverse (source and drain interchanged) modes of operation. Figure 10.1b shows that in the forward mode, as V_D increases, the drain current of the stressed transistor approaches that of the virgin device. This is due to the gradual expansion of the pinch-off region, which eventually covers the whole defective zone ΔL. Since the minority carriers now flow far from the interface, they do not experience the defect influence, nor can they get trapped. The defects become "inactive," and I_D recovers the prestress value.

In contrast, in the reverse mode, the defects are always located above the channel, near the source; they are fully active and make I_D smaller than it is in the virgin device. It is clear that the difference between normal and reverse curves is more prominent in weak inversion where the current exponentially depends on the surface potential and its defect-related perturbations.

A raw estimate of ΔL is obtained by drawing $I_D(V_D)$ in the forward mode. For a certain bias $V_D^\star$, the prestress values are recovered, which means that the depletion region has masked the whole defective zone. The extension of the surface depletion region, calculated for $V_D^\star$, gives ΔL. In principle, it is possible to resolve the lateral profile of defects by computing the depletion width as a function of V_D and by using differential analysis. Various other parameters (for example, $V_T(V_D)$, $g_m(V_D)$, $I_{sub}(V_D)$, and noise) present an asymmetrical behavior as well and can be used in a similar way.

The lateral shift of $I_G(V_G)$ curves (measured by the floating-gate technique[8]) is reminiscent of the defect-induced variation of the local vertical field: $\Delta Q_{ox} \simeq C_{ox} \Delta V_G$. Valuable information on the local defect density is found from charge pumping (Fig. 10.4a), as discussed in Section 9.2.

---

[2]In the case of gentle degradations, when the channel still looks "homogeneous," the shifts of the threshold voltage and transconductance give a first-order estimation of the amount of generated defects.

These various methods show that in 1 μm MOSFETs, ΔL depends on stressing biases and expands with time, typical values being 0.1–0.2 μm. The question is whether or not ΔL will shrink in mesoscopic transistors. Several arguments suggest that, in 0.1 μm MOSFETs, the degradation tends to cover the whole channel:[1] (i) parameter extraction with Equation (10.3) yields $\Delta L \simeq L$, which is consistent with field maps, (ii) the charge-pumping rectangle is just magnified after stress, not distorted as in Figure 10.4a, (iii) the amounts of interface traps extracted from the subthreshold slope, the weak-inversion two-piece model, or charge pumping are rather equivalent, and (iv) if V_D is increased, the forward I_D never reaches the prestress value (Fig. 10.1b). However, the defect distribution along the channel is not perfectly uniform, since an asymmetry still remains on $I_D(V_D)$ curves.

10.1.4 Aging Monitors and Device Lifetime

Stress-induced defects influence the drain current via three main parameters: inversion charge, carrier mobility, and series resistances. The mobility always decreases with increasing defect density, while the variations of the channel charge and series resistances depend on the nature and charging of the defects. This picture is complicated by the 2–D potential distribution in the damaged zone and the two-piece configuration of the transistor.

Frequently used aging monitors are V_T, g_m, and I_D. Several methods for extracting the threshold voltage have been discussed in Section 8.5.3: linear extrapolation from ohmic $I_D(V_G)$ characteristics or from $I_D/\sqrt{g_m}(V_G)$ curves, correspondence to a fixed current value, and peak of the second current derivative. For I_D to stand as a sensitive monitor, it must be measured in weak inversion, reverse-mode, and saturation region. The following are a few symptoms and a brief qualitative diagnoses for n–MOSFET aging:

- V_T *increases*—Electron trapping and/or creation of acceptor-like interface traps.

- V_T *decreases*—Hole trapping.

- g_m *increases*—Positive charges inducing a channel shortening or numerous negatively charged defects in a narrow region ΔL (transconductance overshoot).

- *Degraded subthreshold slope*—Interface-trap generation in a wide ΔL or a strong electron trapping in a narrow ΔL.

More insight into the nature of defects is obtained by correlating several parameters: V_T vs. g_m, g_m vs. I_D, etc. However, in most experimental situations, the monitoring of the above macroscopic parameters is not sufficient for the unambiguous identification of the defects. Since both the *substrate current* and the aging are direct consequences of carrier impact ionization, I_{sub} is often used as a monitor.

The change in I_{sub} and the poststress asymmetry of $I_{sub}(V_D)$ curves reflect the variation of the electric field in the channel due to generated oxide charges. In the normal mode, negatively charged defects near the drain cause the vertical field to decrease and the lateral field to increase; hence, I_{sub} increases. In the reverse mode, the defects are responsible for an additional potential drop near the "source," which in turn lowers the lateral field peak and I_{sub}.

The *gate current* is the most direct method with which to assess the injection rate of hot carriers into the gate oxide. Extremely small currents ($I_G \simeq 10^{-19}$ A) can be sensed by the *floating-gate* technique.[8] The transistor gate is initially biased at $V_G = V_{FG}(t = 0)$, and V_D is chosen in the saturation region. After equilibrium is reached, the gate is isolated by lifting up the gate probe. In this floating mode, the MOSFET operation is controlled by the gate charge $Q_G = C_{ox}V_{FG}(t)$. This charge is gradually modified by the injection current, $\Delta Q_G = I_G \Delta t$; hence, the floating-gate bias V_{FG} and I_D vary with time. The gate current is therefore deduced from the drain current transient:

$$I_G(t) = \frac{\Delta Q_G(t)}{\Delta t} = C_{ox}\frac{\Delta V_{FG}(t)}{\Delta t} = \frac{C_{ox}}{g_m}\frac{\Delta I_D(t)}{\Delta t} \tag{10.6}$$

For each sampling value $I_D(t)$, the corresponding gate voltage is known from the $I_D(V_G)$ curve. This treatment makes available a portion of the $I_G(V_G)$ curve, the whole curve being reconstructed by repeating the procedure for distinct values $V_{FG}(t = 0)$. Since carrier injection takes place near the drain, I_G is highly sensitive to aging. It can drastically decrease during stress, since the build-up of a negative charge blocks the injection of fresh electrons.

Gate-induced drain leakage (GIDL) has been proposed as an aging monitor. The GIDL current is measured at large drain bias ($V_D \geq 2$ V) by scanning the gate bias from depletion to strong accumulation. This off-region current is nothing but the gate-controlled reverse current of the drain junction. It is extremely sensitive to aging, since locally generated defects affect the surface potential very near the drain junction. Therefore, the poststress GIDL current is highly asymmetrical when measured in direct and reverse modes.[9]

Other sensitive (and time-consuming) techniques—charge pumping, noise, capacitance- or current–DLTS, dynamic transconductance, gated-diode leakage—have been discussed in previous chapters. Optical measurements must also be mentioned, in particular the *photodepopulation* technique. This method consists in monitoring the variation of the MOS capacitance in deep depletion after switching on a monochromatic light. The carrier photoemission from the interface traps induces a transient response of the photocapacitance, which is used to determine the energy profile and capture cross section of the traps.[10]

Emission microscopy is based on the detection of the light emission near the drain, which results from the recombination of electron-hole pairs generated by impact ionization. The electron temperature is estimated from the emission spectra.

In the industrial context, the merits of a technology are qualified in terms of *device lifetime.* The idea of this empirical approach is to extrapolate, from *accelerated* aging, the transistor lifetime under standard conditions of operation (5 V or 3 V). The degradation rate of a given monitor X as a function of stress time normally follows a power law[11]

$$\frac{\Delta X}{X} = At^n, \qquad \text{with} \qquad A = A_1 \exp\left(\frac{-\alpha}{V_D}\right) \tag{10.7}$$

where the coefficients n, α, and A_1 depend on technology parameters and gate biasing.

Linear extrapolation in double logarithmic scale yields the "accelerated" lifetime τ. It is defined as the stress time that would be required to reach a certain amount of tolerated degradation: 10–100 mV shift of ΔV_T, 1%–10% variation of ΔI_D or Δg_m. From Equation (10.7), it follows that τ exponentially depends on $1/V_D$, exactly as the substrate current does (see also Eq. (8.2)). This is not surprising if one recalls that both the substrate current and the degradation mechanisms are related to the peak of the lateral field (or, in deep submicron MOSFET, to the carrier temperature). A linear dependence is therefore obtained between $\ln \tau$ and $\ln I_{sub}$ or between $\ln \tau$ and $1/V_D$. The latter solution is more convenient in SOI where, in general, I_{sub} is not measurable because no body contact is available.[3]

The determination of the "actual" device lifetime $\tau^\star$ requires a two-step procedure. First, several devices are stressed with different high drain biases (at constant V_G), and the accelerated lifetimes $\tau(V_D)$ are extrapolated from $\ln \Delta X/X$ vs. $\ln t$ curves. Then, $\ln \tau$ is drawn versus $1/V_D$ (stressing voltage). A second extrapolation gives the predicted lifetime $\tau^\star$ for nominal drain bias ($V_D = 5\,\mathrm{V}$). A detailed analysis shows that more accurate evaluations can be obtained by drawing $\ln \tau I_D/W$ versus $1/(V_D - V_{D,sat})$.[11]

The practical interest for such predictions is obvious, but we have to keep in mind several basic limitations: (1) the discrepancy between the duration of the stress (minutes or hours) and the extrapolated lifetime (10–100 years!), (2) defect saturation effects (i.e., sublinear variations of Δg_m and ΔV_T with time), and (3) the expansion of ΔL due to the variation of the local field induced by the charge of created defects. In general, the parameters capable to account for defect localization are more suitable for lifetime predictions than averaged parameters such as $V_{T,a}$ and $g_{m,a}$.

10.1.5 Hot-Carrier Degradation of SIMOX Buried Oxides

Ultrathin film SOI technologies aggressively compete for sub-quarter-micron CMOS circuits. In this respect, the transistor immunity to aging effects is regarded as a key argument for promoting a technological approach. Most stressing tests have been performed on SIMOX. It was found that, in n–MOSFETs, defects are exclusively created at the interface where carriers flow.[12] However, the critical aging mechanism in SIMOX p–MOSFETs is quite different: during the front-channel stress, electrons generated by impact ionization are massively trapped into the buried oxide and may cause a significant degradation of the front-channel properties.[13]

The analysis of the buried-oxide degradation is therefore important for at least two reasons: (1) the analysis will help identify the nature of the defects, understand the microstructure of the BOX, and optimize its properties, (2) the BOX is more subject to degradation than the gate oxide, and its defects may jeopardize, via coupling effects, the performance

---

[3]Alternatively, I_{sub} can be replaced by the gate current. The plot of $\ln I_{sub}$ vs. $\ln I_G$ is indeed a straight line.

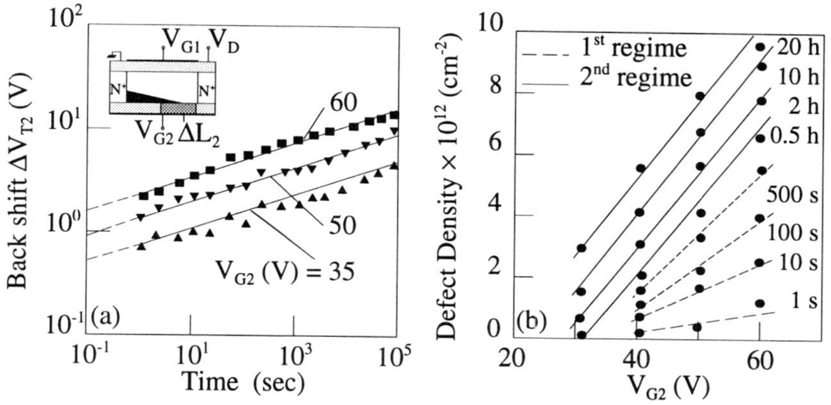

Figure 10.6 (a) Back-channel threshold-voltage shift versus stress time and (b) corresponding density of localized defects versus stressing gate bias for back-channel SIMOX n–MOSFETs ($V_D = 6\,V$, $L = 1.4\,\mu m$; after Ouisse et al.[5]).

of CMOS circuits.

The investigation is conducted simply by stressing the back channel. Typical threshold-voltage variations are shown in Figure 10.6a. The shift follows a power law, $\Delta V_{T_2} \sim t^n$, and exceeds 10 V after only a few hours of stress. The subthreshold slope is also degraded.

The transconductance exhibits a clear overshoot (Fig. 10.3b) as predicted by Equation (10.2) and Figure 10.3a. The thickness of the buried oxide makes the field-effect mobility far less sensitive to the gate bias (smaller θ coefficient in Equation (8.41)) than at the front channel, leading to overshoots larger than 20%–40%.[5] Using the two-piece model of Equations (10.3) and (10.4), the parameters of the defective region can easily be extracted. The high density of locally generated defects ($\sim 10^{12}\,cm^{-2}$) as well as the large extension of the damage (increasing with V_D to reach $\Delta L \simeq 0.5\,\mu m$) indicate the vulnerability of the BOX.

The variation of the local density of defects as a function of time and gate bias (Fig. 10.6b) suggests a two-step process. First, the generation rate increases with gate voltage, which is unusual in bulk transistors. This is due to the filling of preexisting electron traps that saturates after 10 minutes of stress. Then, the defect generation rate becomes independent of V_{G_2} being dominated by the formation of new defects

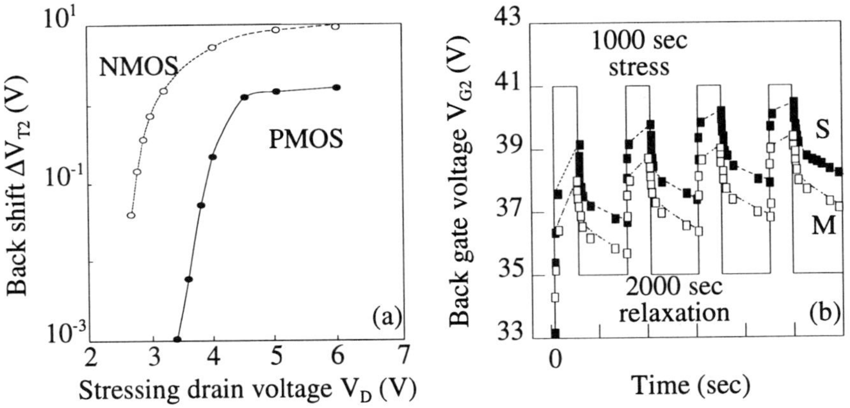

Figure 10.7 Back-channel threshold-voltage shifts in SIMOX MOS transistors: (a) low drain-voltage stress of back n-channels and front p-channels,[16] (b) sequence of degradation ($V_{G_2} = 50\,\mathrm{V}$, $V_D = 6\,\mathrm{V}$) and relaxation ($V_{G_2} = V_D = 0\,\mathrm{V}$) in single-implanted (S) and multiple-implanted (M) SIMOX.[17]

(interface states and electron traps). This second mechanism prevails for low gate voltages. The electron trapping is reversible and can be accelerated under negative bias. The relation between relaxation time and V_{G_2} supports the scenario of electron detrapping by tunneling.[5]

The high density of electron traps is an intrinsic feature of SIMOX oxides. Their presence and filling has been confirmed by homogeneous Fowler–Nordheim and avalanche injection[14] or photoinjection experiments.[15] The question is whether electron traps will continue to limit SIMOX properties even for transistors operated at lower drain bias. Fortunately, the experiment of Figure 10.7a shows an abrupt decrease of ΔV_{T_2} with V_D. The aging kinetics is very well reproduced by Equation (10.7), where the parameters A_1, α, and n depend only on channel length and stressing gate bias.[16] For the region of practical interest, $1.5 < V_D < 3.3\,\mathrm{V}$, the degradation decreases by increasing V_{G_2}, which contrasts with the case of accelerated stress (Fig. 10.6a). By choosing a very strict criterion of tolerated degradation, $\Delta V_{T_2} = 100\,\mathrm{mV}$, it is found that the expected lifetime can exceed 10 years for an operating drain bias of 2 V.

Triple- and single-implanted SIMOX oxides have been compared by monitoring ΔV_{T_2} during repeated cycles of back-channel stress and subsequent relaxation.[17] In Figure 10.7b, ΔV_{T_2} shows a rapid increase at the beginning of each stress period (fast filling of preexisting electron traps), followed by a slower increase (formation of irreversible damage). During each relaxation step, ΔV_{T_2} decreases by about the same amount, which indicates a (partial) reemission of electrons from traps with constant density and short enough time constants.

Triple SIMOX exhibits a reduced voltage shift and same amount of relaxation. It is concluded that while the trapping rate is similar, the irreversible damage is less severe than in single-implanted SIMOX. In order to avoid underestimations of the device lifetime, it is important to account for irreversible damage only, by allowing the devices to relax after the stress.

10.1.6 Gate Oxide Degradation

Early measurements have shown that the aging of thick-film (partially depleted) SOI n–MOSFETs is comparable with that of bulk-silicon transistors and more accentuated than in thin-film (fully depleted) SOI.[18] This advantage of thin SOI devices comes from an attenuation of the lateral electric field peak due to the implicit reduction of the junction depth.

Several investigators have observed more or less degradation in fully depleted transistors, according to the degree of optimization of the LDD region.[19, 20] However, the damage occurring in the gate oxide and interface does not differ qualitatively from that encountered in bulk-Si MOSFETs.[21] Neither is it highly sensitive to the type of SIMOX substrate (single- or multiple-implantation, variable anneal temperature and duration, etc.).

The aging kinetics of the front channel obeys the usual power law, the degradation being accentuated for lower gate bias (Fig. 10.8a).[22, 23] It may happen that the defects are generated exclusively above the LDD region, leading to the aging signature of Figure 10.8b: the transconductance is dramatically reduced (by increased series resistances), whereas the threshold voltage does not change.

Numerical simulations currently demonstrate that the lateral field in fully depleted SOI transistors may be smaller than in bulk Si (with thin junctions) but increases as the film thickness is reduced. Nevertheless, no accelerated aging was detected in ultrathin transistors.[23] In fact,

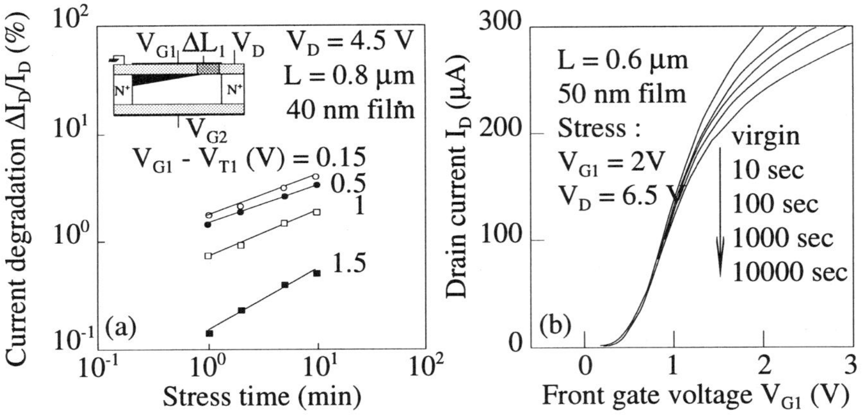

Figure 10.8 Front-channel degradations of ultrathin SIMOX n–MOSFETs: (a) inversion mode, (b) accumulation mode.[22, 23]

the comparison of 50 nm and 70 nm films tends to prove that thinner films are more resistant to hot carriers. But, once the difference in series resistances is compensated by adding an external resistor to the thicker transistor, the results become roughly identical. Indeed, for such a comparison to be fair, not only the stressing voltages but also the current value should be the same.

Another possibility is to analyze the maximum impact ionization rate, $I_{sub,max}/I_D$, measured at fixed V_D by varying V_{G_1}. According to Ohmura and Izumi,[24] this ratio can be reduced if (i) the Si layer is thinner (since impact ionization becomes surface controlled, thus less efficient), (ii) the buried oxide is thicker, and (iii) the channel length is shorter (for $L < 0.2\,\mu$m, nonstationary transport occurs). As a matter of fact, in 0.1 μm SIMOX transistors, fabricated on 30-nm-thick film, the lifetime may exceed 10 years for $V_D \simeq 1.5$–2 V. Similar trends were observed for the normalized gate current I_G/I_D.[22]

The aging kinetics of the front channel obeys the usual power law, the degradation being accentuated for lower gate bias The stress of accumulation-mode transistors does not reveal any particular immunity problem or extravagant mechanism of failure.[23] The degradation laws and the response to biasing or technological conditions mimic the case of inversion-mode MOSFETs.

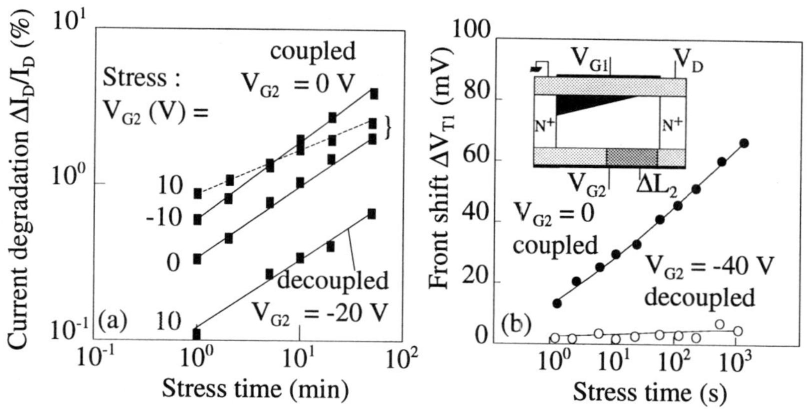

Figure 10.9 Influence of the back-gate bias during (a) the stress ($V_{G_1} - V_{T_1} = 0.15\,\text{V}$, $V_D = 3.8\,\text{V}$) and (b) the test of front-channel 40-nm-thick SIMOX n–MOSFETs.[22, 23]

A relevant feature of fully depleted SOI is associated with the role of the back-gate bias during the front-channel stress. The parallel curves obtained in Figure 10.9a show that the time dependence of the aging does not depend on V_{G_2}. In contrast, the amount of the degradation does: it is much lower for inversion than for accumulation at the back interface, in agreement with the relationship between the lateral field and V_{G_2} (in accumulation, the front threshold voltage is larger, the saturation voltage is reduced and, therefore, $V_D - V_{D_{sat}}$ increases). This argument is promoting Gate-All-Around transistors or dual-gate MOSFETs with volume inversion, where very limited degradation was indeed observed.[25]

For the purpose of characterization, it is important to stress all devices in similar conditions. Otherwise, a mediocre transistor, stressed with the back interface inverted or depleted, may look more immune than an advanced device stressed with the back interface in accumulation. A constant back-surface potential is not necessarily guaranteed by $V_{G_2} = 0$. The simplest way is to compare only devices stressed with the back channel accumulated. However, this leads to a very unpopular underestimation of the device lifetime (Fig. 10.9a). Other interface coupling effects, including the aging of p-channel SIMOX transistors, will be described next.

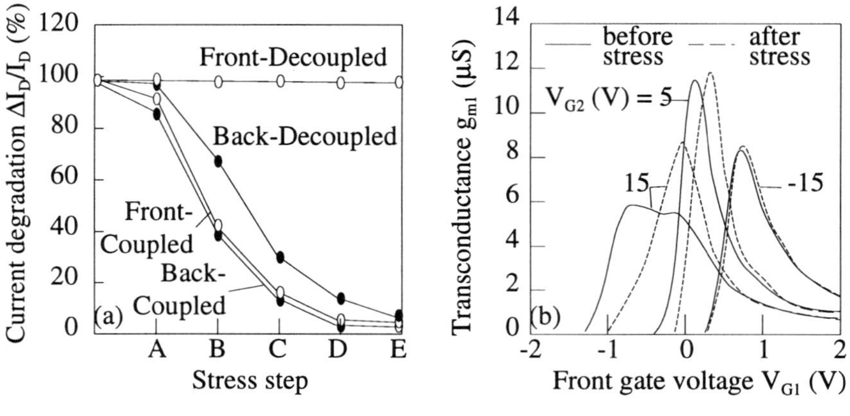

Figure 10.10 Normalized variations of saturation currents in a fully depleted SIMOX MOSFET after several stresses of the back channel; the front- and back-channel currents are measured with the opposite gate grounded (coupled mode) or accumulated (uncoupled mode). (b) Influence of back-gate bias on the test of front-channel transconductance (after Cristoloveanu et al.[12]).

10.1.7 Coupling Degradation Mechanisms

The main coupling mechanism in fully depleted SOI MOSFETs involves the influence of localized defects, generated at one interface, on the properties of the opposite channel.

In the case shown in Figure 10.9b, the back channel was stressed, and this resulted in a large degradation of the back interface. The front-channel threshold voltage, monitored for $V_{G_2} = 0$, is shifted as well which would imply that during back-channel stress, many defects have also been generated at the front interface. Such a confusing conclusion is totally negated by monitoring ΔV_{T_1} with $V_{G_2} = -40\,\text{V}$. Since the back interface is now accumulated, the influence of buried-oxide defects is masked and no shift of front threshold voltage remains (Fig. 10.9b). In other words, the presumed front-interface defects were just *apparent*, the illusion being due to the coupling between the front channel and the back-interface defects.

A second example is shown in Figure 10.10a, where again the back channel was stressed.[12] The drain-current value in weak inversion, saturation region, and reverse mode was selected as a monitor. The front-channel current, measured with the back interface depleted (coupled mode), decreases rapidly during stress, insinuating the formation of localized front interface defects. However, when the buried-interface defects are screened by an accumulation layer, the front channel current, which now would be affected only by front-interface defects, remains constant during stress.

The correct scenario is that negatively charged defects were generated exclusively at the back interface, increasing the threshold voltage V_{T_2}. This is equivalent to a decrease of the effective back-gate voltage. Since V_{T_1} increases with decreasing V_{G_2}, it follows that the front current reduces accordingly. In a similar way, the front-channel transconductance measured with $V_{G_2} = -15\,\mathrm{V}$ (accumulation) does not change during stress (Fig. 10.10b), but looks unexpectedly increased if measured with $V_{G_2} = +15\,\mathrm{V}$ (inversion).

The back current gradually decreases during stress. However, a very subtle coupling effect distinguishes the variations measured with the front interface depleted or accumulated (Fig. 10.10a). This means that the efficiency of back-interface defects is controlled by the front-interface potential, and the apparent density of defects varies according to V_{G_1}. Symmetrically, the choice of V_{G_2} is not innocent for the evaluation of the localized damage formed at the front interface. We now are less surprised by the special slope of the dotted curve in Figure 10.9a, which is just an artifact brought about by interface coupling.

These coupling mechanisms can easily be reproduced by two-dimensional simulations. In Figure 10.11, the front-channel transconductance is computed, for depletion and inversion at the back interface, before stress and after defects were localized at either the front or the back interface.[26]

The understanding of interface coupling mechanisms makes easier the aging analysis of p-channel SIMOX transistors. During front-channel stress, the critical role is played by electrons, which are generated by impact ionization and get trapped into the buried oxide.[13] Evidence for this is given in Figure 10.12a, where the back-channel threshold voltage (monitored with the front interface in accumulation) is dramatically shifted.

In spite of the fact that a number of defects were simultaneously generated at the front interface, the electron trapping into the BOX prevails. It is equivalent to $V_{G_2} < 0$ and activates the interface coupling.

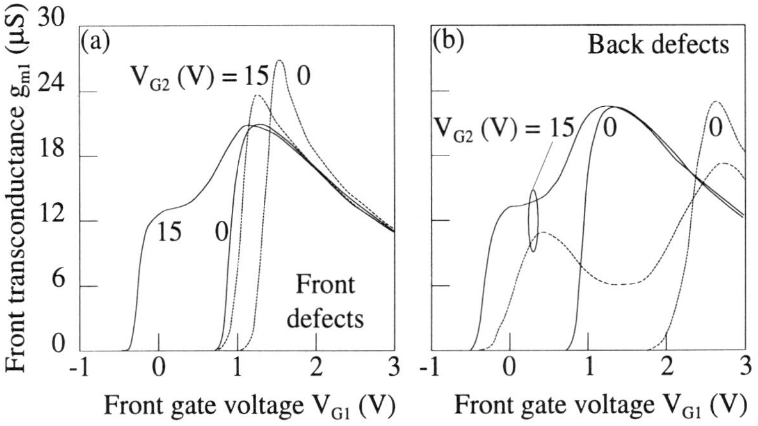

Figure 10.11 Two-dimensional simulations of the front-channel transconductance before (——) and after (- - -) the formation of localized traps ($D_{it_{1,2}} = 2 \times 10^{12}\,\mathrm{cm}^{-2}$, $\Delta L = 0.2\,\mu\mathrm{m}$) at either (a) the front or (b) the back interface ($L = 1\,\mu\mathrm{m}$, $t_{si} = 50\,\mathrm{nm}$). The curves are given for inversion and depletion at the back interface.[26]

This contributes to a substantial overestimation of the "virtual" aging of the front interface. The actual amount of "physical" defects is measured by accumulating the back interface. The true aging damage turns out to be far more reasonable and is comparable with that observed in partially depleted SIMOX transistors that are free from coupling problems (see Figure 10.12b).

It is worth remembering that the back-channel threshold voltage shifts, obtained by stressing either the back channel of n–MOSFETs or the front channel of p–MOSFETs, are strongly correlated (Fig. 10.7a), since both are governed by electron trapping into the BOX. This implies that the aging of p-channels may be predicted from back-channel stress tests in n–MOSFETs.[16]

The coupling mechanisms can be taken advantage of for achieving new device functions. The experiment proposed by Ioannou's group[27] consists of a sequence of front-channel and back-channel stresses. The drain bias is selected in the avalanche region, where the bipolar transistor is very effective. In this special situation, it is known that many carriers flow in the volume of the film and can get trapped at *both* interfaces.[28]

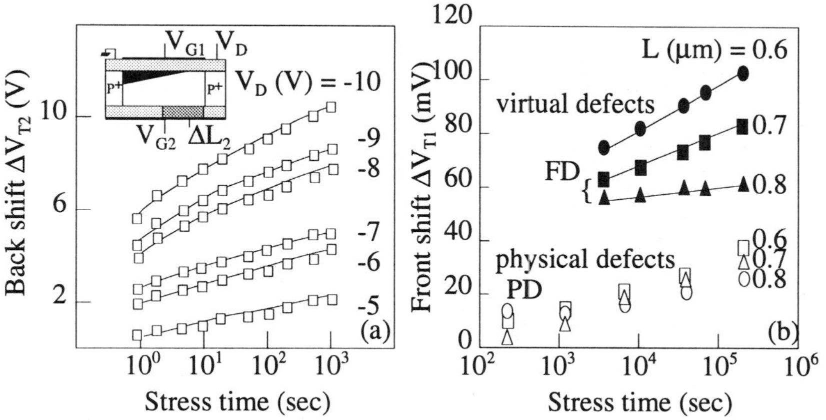

Figure 10.12 (a) Back- and (b) front-threshold voltage shifts during the stress ($V_{G_1} = -1.5\,\text{V}$, $V_{G_2} = 0$) of the front channels in partially and fully depleted SIMOX p–MOSFETs.[13]

The opposite gate is biased in accumulation with the aim of helping the injection of majority carriers.

Perfect symmetrical variations of the front and back threshold voltages are monitored throughout the stress (Fig. 10.13). This indicates that during front-channel stress electrons get trapped into the gate oxide and holes into the BOX, leading to a positive shift of the front threshold voltage and to a negative shift of the back threshold voltage.

Reciprocal trapping occurs during back-channel stress periods. The exciting point is that the traps charged during front-channel stress are totally discharged following the back-channel stress. The prestress values of the threshold voltages are recovered, and the transistor appears virginal again. Only a minor amount of permanent damage (interface states) is visible at the front channel, where the bottom level of V_{T_1} increases slightly with time.

This principle features a new type of flash memory. The front-channel stress represents the writing stage. The novelty comes from the erasing scheme, achieved by hot hole injection into the front gate via back-channel stress. In actual memories, the writing/erasing cycles are expected to be short enough, since carriers will be directly injected into the floating gate instead of getting trapped into the oxide.

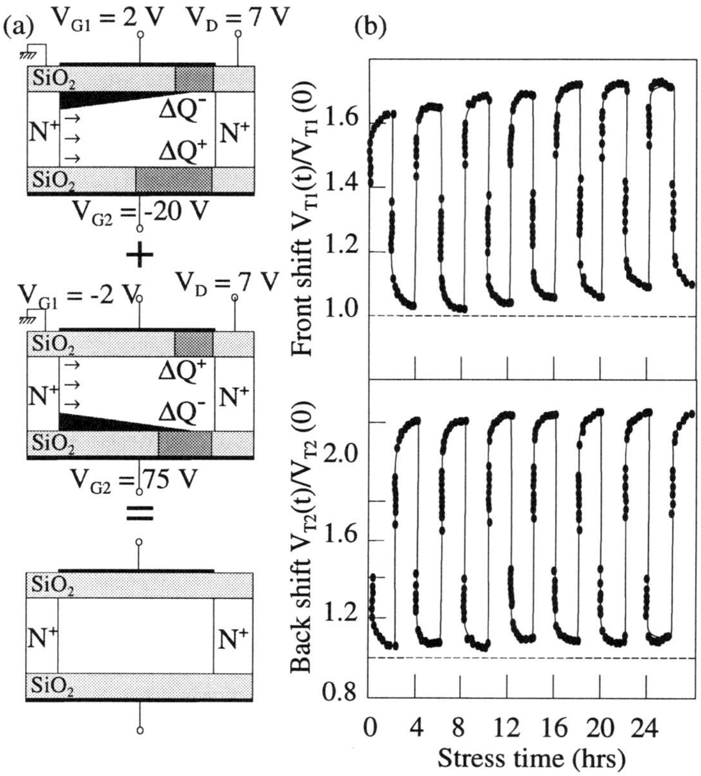

Figure 10.13 (a) Stress cycles showing (b) a memory effect on the front and back threshold-voltage variations (after Zaleski et al.[27]).

10.2 Radiation-Induced Defects

It is the thinness of the silicon film and its perfect isolation that allow SOI devices to resist *transient* radiation effects far better than bulk-Si counterparts. The leakage currents and the soft errors (bit upsets) are reduced by orders of magnitude.

A still-unsolved problem is the *permanent* radiation damage related to cumulative dose effects. Since most of the defects are accumulated into the buried oxide (very thick as compared to the gate oxide), their evaluation has to be given priority.

10.2.1 Buried-Oxide Degradation

We first review the methods used for *wafer-level* inspection.

MOS capacitors can easily be formed by removing the Si overlay and depositing a metal gate on the bared buried oxide (Fig. 10.14a).[4] The buried oxide acts as the capacitor insulator, and simple one-interface $C(V)$ characteristics, as in bulk Si, are obtained (Section 6.5). The radiation-induced shift and stretching of $C(V)$ curves (Fig. 10.14a') is indicative of the nature and density of the defects generated within the buried oxide and at the interface with the silicon substrate. No information is available on the damage occurring at the upper oxide interface or within the Si film. This method is actually useful for monitoring the oxide degradation only. Indeed, the damage affecting the bottom oxide interface is of little relevance for the reliability of SOI circuits. It is pure illusion to attempt extrapolations of the defects that might have been generated at the upper oxide interface from those measured at the bottom interface.

SIS and MOSIS capacitors. Silicon–Insulator–Silicon (SIS) and Metal–Oxide–Silicon–Insulator–Silicon (MOSIS) capacitors, shown in Figure 10.14b, are formed by maintaining the Si film undoped. In this way, the characteristics of these capacitors contain all valuable information on film, oxide, and interface damage.[30] We have noted in Section 6.1 that the $C(V)$ curves are complex due to the superposed effects of two or three depletion regions (Fig. 10.14b'). The pre- and postradiation curves can be reproduced by simulation. Nevertheless, the experimental parameter extraction is rather difficult, since up to three interfaces and two oxides are simultaneously affected by the radiation.

The Ψ–MOSFET technique, described in Section 4.6, is a very convenient radiation-monitoring method. Pseudo–MOS transistors, with either accumulation or inversion channels, can be activated by choosing an appropriate back-gate bias (Fig. 10.14c). Source and drain probes are used to plot $I_D(V_G)$ characteristics (Fig. 10.14c') and related transconductance $g_m(V_G)$ curves. The shifts of the threshold and flat-band voltages combined with the lowering of

[4] Alternatively, the film may be heavily doped to behave as a gate.[29]

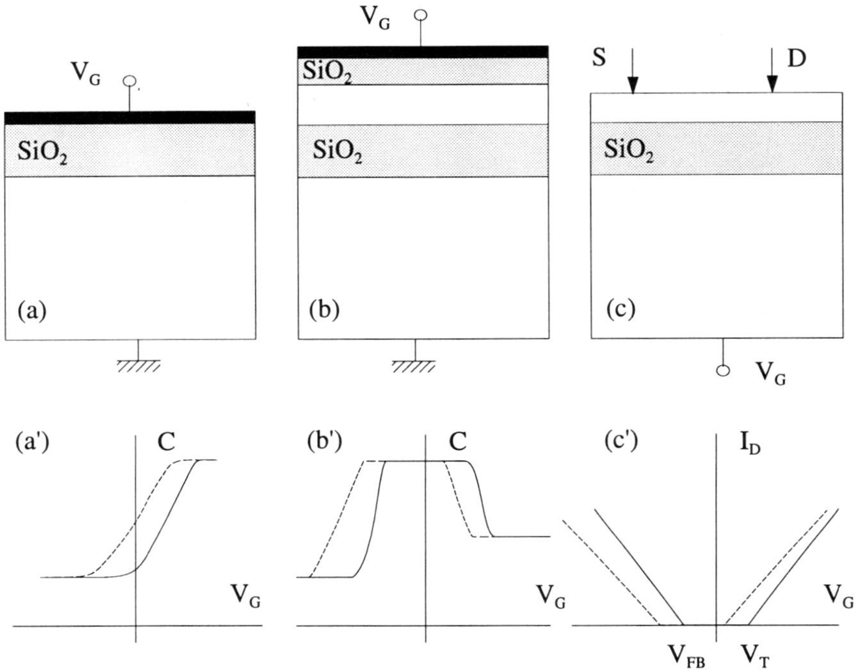

Figure 10.14 Schematic experimental setup and typical characteristics, respectively, of (a,a') buried-oxide capacitance, (b,b') MOSIS capacitance, and (c,c') pseudo–MOS transistor (Ψ–MOSFET) measurements in SOI wafers before (——) and after (- - -) exposure to radiation.

the subthreshold slope and transconductance yield substantial information on the degradation of the buried oxide and its upper interface. The measurement may need adjustment of the probe pressure and drain bias to limit series-resistance effects, as well as a *four-point probe* calibration for determining "absolute" mobility values. An example of a characteristics shift is shown in Figure 10.14c'.

We now discuss in more detail the back-channel operation of *fully-processed* MOS transistors, which obviously stand as the most accurate test vehicle. Figure 10.15a illustrates the degradation of the buried

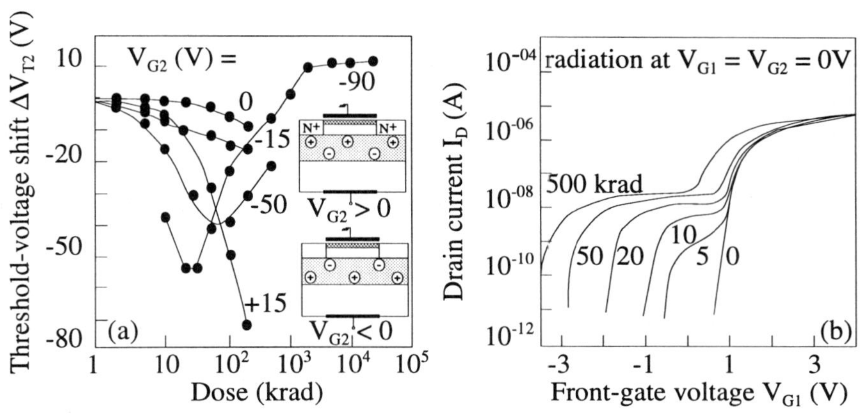

Figure 10.15 Radiation-induced degradation of (a) the back-channel threshold voltage[31] and (b) the sidewall current in LOCOS-isolated SIMOX MOSFETs.

SIMOX oxide. It is noteworthy that the threshold-voltage shift[5] depends on the bias applied on the back gate during exposure, thus revealing several distinct mechanisms.[31] For positive bias, radiation-induced electron-hole pairs are separated by the oxide field, the holes being pushed towards the upper oxide interface. A *net* positive charge effect is obtained that is substantiated by a very large decrease in V_{T_2}. In contrast, for zero bias, most of the electron-hole pairs immediately recombine, and the threshold voltage shift is much limited.

A special SIMOX effect occurs for large negative biases, where again the pairs are separated, but the holes are now attracted to the bottom oxide interface. The positive-charge effect still dominates for low radiation doses. Above 100 krad, however, a rebound is observed, leading to a large increase in V_{T_2}, which eventually recovers the initial value. This overall negative-charge effect accounts for either electron trapping or interface-trap formation.

Here, additional back-channel measurements are needed to identify the main degradation mechanism. Modifications of subthreshold slope,

[5]Another useful parameter is the midgap voltage, which corresponds to a band bending of $\psi_s = \phi_F$. It is obtained by extrapolating the ohmic region subthreshold slope $\ln I_D(V_G)$ to a very low current level ($\simeq 10^{-14}$ A), namely, the drain current at threshold divided by N_A/n_i.

transconductance, and charge pumping show that interface traps do form during radiation, but their density is far too small to explain the large shift (+60 V) of the threshold voltage. This guarantees that electron trapping is the major and very typical source of degradation of SIMOX oxides. Since SIMOX is an ultradry substoechiometric oxide, the traps are presumably related to the high density of oxygen vacancies. Etch-back experiments reveal that the density of traps decreases linearly from the top interface to the bottom interface of the buried oxide.[29]

10.2.2 Front-Interface, Edge, and Film Damage

The degradation of the front interface and gate oxide is monitored by operating the front-channel transistor and using conventional methods (static characteristics, charge pumping, low-frequency noise, etc.). Neither the techniques nor the evaluated damage differ substantially from the case of bulk-Si MOSFETs. Positive gate bias is known to enhance interface-trap generation.

In fact, most defects are created on the LOCOS edges. Figure 10.15b shows that during radiation a hump may appear in the subthreshold characteristics, highlighting the earlier activation of the parasitic side-wall transistor. This is because hole trapping induces a more rapid negative shift of the threshold voltage on the edges than at the front channel. In wide MOSFETs, the current level on the sidewalls is much smaller than at the front channel, so the two threshold voltages can be independently monitored during radiation. In narrow transistors, however, the two currents may be on the same order of magnitude, leading to rather confusing transconductance and subthreshold characteristics. Care is needed to discriminate the kind of radiation-induced coupling effect: between front channel and sidewalls or between front and back interfaces.

Transient current measurements, described in Section 9.5, confirm that the generation velocity on the LOCOS edge interface may increase considerably with radiation dose. Also, a decrease in the generation lifetime was observed.[32] It is not clear yet whether new defects were created within the film or whether preexisting generation centers were just being activated by the radiation.

By comparison, mesa-isolated SOI transistors are very resistant to radiation and behave almost like edgeless MOSFETs with enclosed geometry. The solutions for suppressing edge effects consist in overdoping

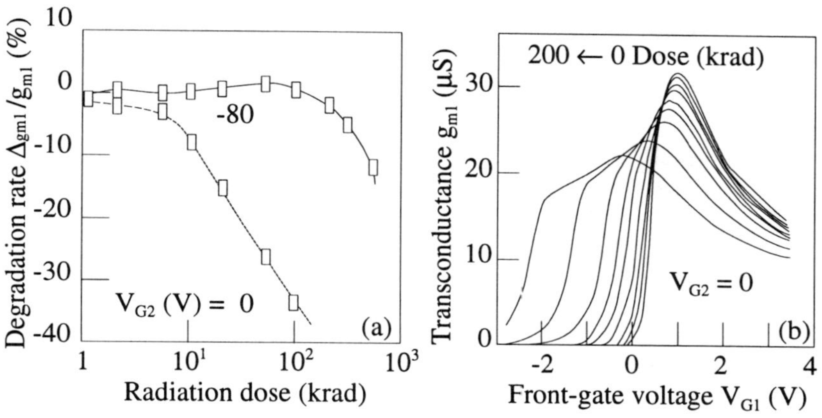

Figure 10.16 (a) Radiation-induced degradation of the front-channel transconductance peak in coupled ($V_{G_2} = 0$) and uncoupled ($V_{G_2} = -80\,V$) modes; (b) distortion of the front-channel transconductance during radiation (after Ouisse et al.[33]).

the sidewalls and hardening the LOCOS oxide.

10.2.3 Special Radiation Effects

(a) Radiation-induced interface coupling. In fully depleted transistors, the severe degradation of the buried oxide induces an overestimation of the front-interface damage. In Figure 10.16a, a large decrease in transconductance is measured for $V_{G_2} = 0$, which may be misattributed to the generation of many front-interface defects. When the measurement is repeated, with the back channel being this time accumulated in order to screen buried-oxide defects, it comes out that the mobility is rather unaffected for doses up to 100 krad.

A closer look at transconductance curves (Fig. 10.16b) shows that a distortion occurs during radiation, exactly as if the back-gate voltage has been increased (see also Figure 8.22a in Chapter 8). Indeed, the hole trapping within the BOX changes the back surface potential and looks like an apparently larger V_{G_2}.

It follows that the variation of the front-channel mobility, illustrated by the dotted line in Figure 10.16a, does not necessarily imply physical damage.[33] This is just a consequence of the modification of the transis-

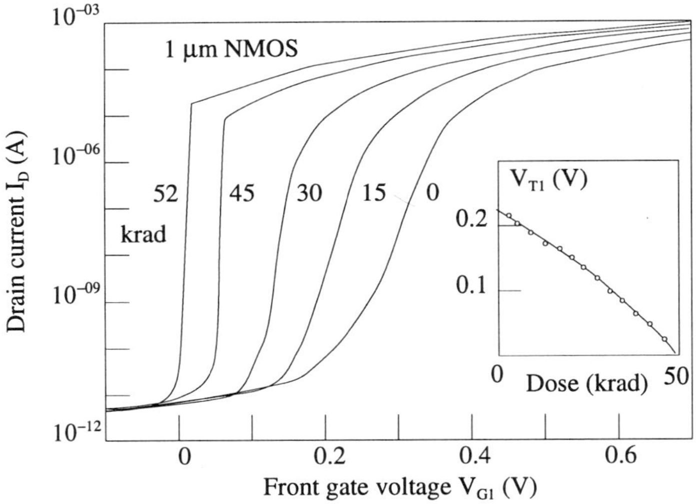

Figure 10.17 Shift of the subthreshold characteristics (measured with $V_D = 5\,\text{V}$) and threshold voltage (insert, $V_D = 50\,\text{mV}$) during radiation. For doses above 45 krad, the floating-body effect is amplified, leading to current jumps and hysteresis (after Ouisse et al.[34]).

tor operating region, from depletion to inversion at the back channel. As a matter of fact, the main mechanism that physically degrades the front interface, for moderate doses, is the trapping of holes in the LDD region. This leads to a lowering of the series resistances and a slight increase in the transconductance (plain line).

The suppression of interface coupling by accumulation of one interface is a necessary condition for the accurate test of the opposite interface. Most of the radiation-hard circuits are actually partially depleted. The bottom of the film receives a much higher doping that screens the influence of buried-oxide defects. This erases the (apparent) excess radiation damage at the front interface brought about by the coupling.

(b) Radiation-induced floating-body effects. During exposure, the subthreshold characteristics of partially depleted MOSFETs are gradually shifted until an abrupt transition between off and on states is observed (Fig. 10.17). We already have seen in Section 8.2 that this corresponds to a hysteresis induced by the positive feedback of impact ionization and floating-body effects.

One may wonder how the radiation can influence floating-body mechanisms. The explanation shown in Figure 10.17 is again based on the argument that, for gentle doses, series resistances are reduced during exposure. Thus, more impact ionization occurs and eventually triggers the hysteresis.[34] This example calls attention to the importance of the LDD doping trade-off: a low doping enhances the sensitivity to radiations, whereas a high doping is no longer efficient in reducing the lateral electric field.

(c) Asymmetrical radiation effects. A more complex situation is reached when the radiation is conducted while biasing the drain and keeping the source and the two gates grounded. In this case, the vertical oxide field is inhomogeneous along the channels, increasing from zero at the source to a maximum near the drain. The damage generation rate follows the field profile, being much larger on the drain side where the pairs are more efficiently separated and carrier trapping occurs.

The existence of an asymmetrical defect profile can be visualized by comparing $I_D(V_D)$ curves measured in normal and reverse modes. In the reverse mode, most defects are located besides the new "source" and have full action on the channel, whatever V_D may be. In the normal mode, some of them are situated above/below the drain depletion region. As the pinch-off point moves away from the drain, more defects become inactive and the current is less affected. As a consequence, the effective density of defects here is drain-voltage controlled.

Figure 10.18a shows that, at the back interface, the current is larger in the reverse mode than in the direct mode, which is typical for positively charged defects. The dominating mechanism is the trapping of holes into the SIMOX oxide that naturally happens for low oxide fields (see also Figure 10.15a). The front-channel characteristics are still asymmetrical but show an opposite behavior (Fig. 10.18b). The normal-mode current now is larger and indicates a predominance of negatively charged defects, probably interface traps. This contrast is not surprising if one keeps in mind that the electric field, which is 20 times larger in the gate oxide than in the BOX, repels the holes from the interface.[35]

We have tried, in this chapter, to illustrate the complexity of the degradation mechanisms in SOI transistors. There is a difference in the nature and amount of defects that may simultaneously be generated in the gate oxide, in the buried oxide, and at the two interfaces. We must also keep in mind that the damage formed at one interface may affect

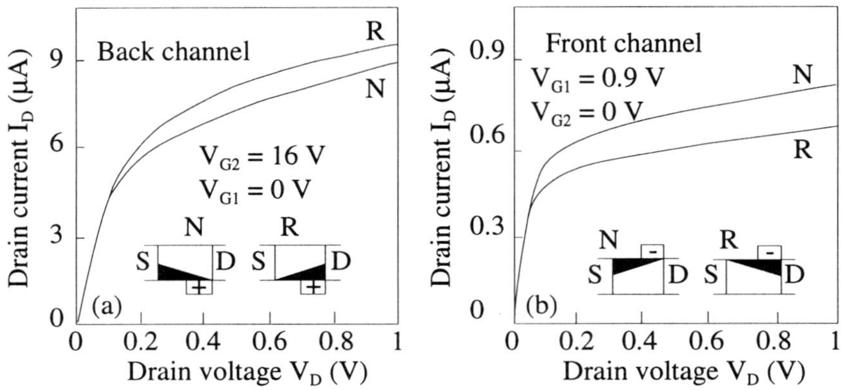

Figure 10.18 Subthreshold currents versus drain voltage measured, in normal (N) and reverse (R) modes of operation, at (a) the back and (b) the front channels of a SIMOX MOSFET irradiated at $V_D = 5\,\mathrm{V}$ and $V_{G_{1,2}} = 0$ (after Cristoloveanu et al.[35]).

the real properties of the opposite channel.

For characterization purposes, the biases used during degradation as well as the test conditions should be carefully checked to avoid misleading conclusions due to interface coupling. The characterization must be aimed at a segregation of the defects belonging to each interface. After front-channel degradation, it is essential to evaluate the physical damage by accumulating the back interface. Subsequently, the front interface must be accumulated for the evaluation of the back channel to be free from the influence of front oxide damage.

The analysis is even more complicated, since three different types of coupling may be superposed: longitudinal coupling between the defective and nondefective regions, vertical coupling between the two interfaces, and transversal coupling between the main channel and the sidewalls.

In SIMOX, front- and back-channel degradations conducted independently have demonstrated that the gate oxide behaves normally, whereas the buried oxide is extremely vulnerable to electron trapping. Technological solutions exist to ignore the BOX damage induced by radiations or hot carrier injection.

Distinct signatures were found for n- and p-channel MOSFETs subsequent to hot-carrier stress. Although in n–MOSFETs, defects are essentially created at the interface where carriers flow, the aging of p-channel SIMOX MOSFETs may be governed by the BOX degradation. However, the hot-carrier reliability is not seen as a critical limit for the operation of fully or partially depleted SOI circuits.

Bibliography

[1] S. Cristoloveanu, "Physical mechanisms of hot-carrier-induced degradation in deep-submicron MOSFETs," *ESSDERC'93 Conf. Proc.*, J. Borel, P. Gentil, J.P. Noblanc, A. Nouailhat, and M. Verdone (eds.), Ed. Frontières, Gif-sur-Yvette, France, p. 797, 1993.

[2] P. Heremans, R. Bellens, G. Groeseneken, A.. v. Schwerin, W. Weber, M. Brox, and H.E. Maes, "The mechanisms of hot carrier degradation," in *Hot Carrier Design Considerations for MOS Devices and Circuits*, C.T. Wang (ed.), Van Nostrand Reinhold, New York, 1992.

[3] S. Cristoloveanu, H. Haddara, and N. Revil, "Defect localization induced by hot carrier injection in short-channel MOSFETs: concept, modeling and characterization," *Microelectron. Reliability*, vol. 33, p. 1365, 1993.

[4] H. Haddara and S. Cristoloveanu, "Two-dimensional modeling of locally damaged short-channel MOSFETs operating in the linear region," *IEEE Trans. Electron Devices*, vol. 34, p. 378, 1987.

[5] T. Ouisse, S. Cristoloveanu, and G. Borel, "Hot carrier-induced degradation of the back interface in short-channel silicon-on-insulator MOSFETs," *IEEE Electron Device Lett.*, vol. 12, p. 290, 1991.

[6] H. Haddara and S. Cristoloveanu, "Parameter extraction method for inhomogeneous MOSFETs, locally damaged by hot carrier injection," *Solid-State Electron.*, vol. 31, p. 1573, 1988.

[7] A. Hassein-Bey and S. Cristoloveanu, "Modelling and characterization of submicron p-channel MOSFETs locally degraded by hot carrier injection," *INFOS'91 Conf. Proc.*, Adam Hilger, Bristol, UK, p. 251, 1991.

[8] N.S. Saks, P.L. Heremans, L. Van Den Hove, H.E. Maes, R.F. de Keersmaecker, and G.J. Declerck, "Observation of hot-hole injection in NMOS transistors using a modified floating-gate technique," *IEEE Trans. Electron Devices*, vol. 33, p. 1529, 1986.

[9] B. Zhang, A. Balasinsky, and T.P. Ma, "Hot-carrier effects on gate-induced-drain-leakage (GIDL) current in thin-film SOI/NMOSFETs," *IEEE Electron Device Lett.*, vol. 15, p. 169, 1994.

[10] J.R. Morante, A. Herms, J. Samitier, A. Cornet, and E. Lora-Tamayo, "Optical and thermal cross-sections of interface states from photocapacitance measurements," *Solid-State Electron.*, vol. 293, p. 1195, 1986.

[11] C. Hu, S.C. Tam, F.C. Hsu, P.K. Ko, T.Y. Chan, and K.W. Terrill, "Hot-electron-induced MOSFET degradation—model, monitor, and improvement," *IEEE Trans. Electron Devices*, vol. 32, p. 375, 1985.

[12] S. Cristoloveanu, S.M. Gulwadi, D.E. Ioannou, G.J. Campisi, and H.L. Hughes, "Hot-electron-induced degradation of front and back channels in partially and fully depleted SIMOX MOSFETs," *IEEE Electron Device Lett.*, vol. 13, p. 603, 1992.

[13] S. Cristoloveanu and T. Ouisse, "Properties of SIMOX and related systems," in *The Physics and Chemistry of SiO_2 and the Si–SiO_2 Interface*, C.R. Helms and B.E. Deal (eds.), Plenum Press, New York, p. 309, 1993.

[14] P.K. Hurley, S. Hall, and W. Eccleston, "Electron trapping in SiO_2 formed by oxygen implantation," *IEEE Electron Device Lett.*, vol. 13, p. 238, 1992.

[15] A.G. Revesz, V.V. Afanas'ev, and H.L. Hughes, "Photo-injection studies of buried oxide layers in SIMOX and BESOI structures," *1993 IEEE Int. SOI Conf. Proc.*, p. 24, 1993.

[16] E. Guichard, S. Cristoloveanu, G. Reimbold, and G. Borel, "Detailed analysis of buried oxide degradation induced by hot-carrier injection," *ESSDERC'93 Conf. Proc.*, J. Borel, P. Gentil, J.P. Noblanc, A. Nouailhat, and M. Verdone (eds.), Ed. Frontières, Gif-sur-Yvette, France, p. 825, 1993.

[17] A. Karmann, G. Reimbold, and S. Cristoloveanu, "Degradation and relaxation of the buried oxide in SIMOX MOSFETs," in *Silicon-On-Insulator Technology and Devices*, S. Cristoloveanu (ed.), The Electrochemical Society, Pennington, NJ, p. 351 , 1994.

[18] J.P. Colinge, "Hot-electron effects in silicon-on-insulator n-channel MOSFETs," *IEEE Trans. Electron Devices*, vol. 34, p. 2173, 1987.

[19] G. Reimbold and A.J. Auberton-Hervé, "Aging analysis of nMOS of a 1.3-μm partially depleted SIMOX technology; comparison with a 1.3-μm bulk technology," *IEEE Trans. Electron Devices*, vol. 40, p. 364, 1993.

[20] J.Y. Choi, R. Sundaresan, and J.G. Fossum, "Monitoring hot-electron-induced degradation in floating-body SOI MOSFETs," *IEEE Electron Device Lett.*, vol. 11, p. 156, 1990.

[21] E. Guichard, S. Cristoloveanu, G. Reimbold, and G. Borel, "Full analysis of alternating injection in SOI transistors: comparison to bulk transistors," *IEDM'94 Techn. Digest*, p. 315, 1994.

[22] L.T. Su, H. Fang, J.E. Chung, and D.A. Antoniadis, "Hot-carrier effects in fully-depleted SOI nMOSFETs," *IEDM'92 Techn. Digest*, p. 349, 1992.

[23] O. Faynot, S. Cristoloveanu, A-J. Auberton-Hervé, and G. Reimbold, "Hot-carrier degradation in ultra-thin fully-depleted accumulation-mode SIMOX n-MOSFETs," *Microelectron. Eng.*, vol. 22, p. 407, 1993.

[24] Y. Ohmura and K. Izumi, "Hot-carrier immunity of a 0.1–μm–gate ultrathin-film MOSFET/SIMOX," *SSDM'92 Conf. Proc.*, p. 496, 1992.

[25] D. Flandre, P. Francis, J.P. Colinge, and S. Cristoloveanu, "Comparison of hot-carier effects in thin-film SOI and gate-all-around accumulation-mode p-MOSFETs," *1993 IEEE Int. SOI Conf. Proc.*, p. 160, 1993.

[26] A. Hassein-Bey and S. Cristoloveanu, "Simulation of interface coupling effects in ultra-thin silicon on insulator MOSFETs," *COMPEL*, vol. 11, p. 513, 1992.

[27] A. Zaleski, D.E. Ioannou, G.J. Campisi, and H.L. Hughes, "Successive charging/discharging of gate oxides in SOI MOSFETs by sequential hot-electron stressing of front/back channel," *IEEE Electron Device Lett.*, vol. 14, p. 435, 1993.

[28] A. Yoshino, T.P. Ma, and K. Okumura, "Hot-carrier effects in fully depleted submicrometer NMOS/SIMOX as influenced by back interface degradation," *IEEE Electron Device Lett.*, vol. 13, p. 522, 1992.

[29] D. Hervé, J.L. Leray, and R. Devine, "Comparative study of radiation induced electrical and spin active defects in buried SiO_2 layers," *J. Appl. Phys.*, vol. 72, p. 3634, 1992.

[30] F.T. Brady and S.S. Li, "Radiation-induced charge trapping in implanted buried oxides," *J. Appl. Phys.*, vol. 68, p. 6143, 1990.

[31] T. Ouisse, S. Cristoloveanu, and G. Borel, "Electron trapping in irradiated SIMOX buried oxides," *IEEE Electron Device Lett.*, vol. 12, p. 312, 1991.

[32] A.M. Ionescu, S. Cristoloveanu, A. Chovet, P. Jarron, E. Heijne, F. Faccio, and G. Rossi, "A systematic investigation of radiation effects in MOS/SIMOX structures," *Microelectron. Eng.*, vol. 22, p. 391, 1993.

[33] T. Ouisse, S. Cristoloveanu, and G. Borel, "Influence of series resistances and interface coupling on the transconductance of fully-depleted silicon-on-insulator MOSFETs," *Solid-State Electron.*, vol. 35, p. 141, 1992.

[34] T. Ouisse, G. Ghibaudo, J. Brini, S. Cristoloveanu, and G. Borel, "Investigation of floating body effects in silicon-on-insulator metal-oxide-semiconductor field-effect transistors," *J. Appl. Phys.*, vol. 70, p. 3912, 1991.

[35] S. Cristoloveanu, D.E. Ioannou, R.K. Lawrence, G.J. Campisi, and H.L. Hughes, "Asymmetrical irradiation effects in SIMOX MOSFETs," *RADECS'93 Conf. Proc.*, IEEE, p. 373, 1994.

List of Symbols

The following table illustrates the specific meanings for subscripts or superscripts used in the generic symbols.

Subscript or Superscript	Description
1	Front gate/interface
2	Back gate/interface
3	BOX–substrate interface
acc	Accumulation region
d	Depletion
dep	Depletion region
eff	Effective value
inv	Inversion region
max	Maximum value
min	Minimum value
n	Electron
o	Equilibrium
ox	Oxide
p	Hole
plat	Plateau value
si	Silicon film
A, D	Acceptor/Donnor
C, V	Conduction/Valence band
D	Drain
G	Gate
S	Source
HF	High-frequency
LF	Low-frequency

Selected Symbols

Symbol	Description	Symbol	Description
α	Absorption coefficient	C_{LF}	LF capacitance
$\alpha_{1,2}$	Coupling factors	C_T	Total capacitance
β	Current gain	C_d	Depletion capacitance
γ_H	Hall scattering factor	C_{inv}	Inversion-layer capacitance
ϵ_{ox}	Permittivity of oxide	C_{it}	Interface-state capacitance
ϵ_o	Permittivity in vacuum	C_{ox}	Oxide capacitance
ϵ_{si}	Permittivity of silicon	$C_{1,2}$	Proportionality factors
η	Quantum efficiency	D	Diffusion constant
λ	Wavelength	D_{it}	Interface-trap density
λ_{ox}	Tunneling length	$D_{n,p}$	Diffusion coefficients
μ	Carrier mobility	d	Thickness, Distance
μ_H	Hall mobility	E	Energy, Electric field
μ_{eff}	Effective mobility	E_C	Conduction-band energy
μ_{fe}	Field-effect mobility	E_D	Donor activation energy
ρ	Resistivity	E_F	Fermi energy
σ	Carrier cross section	E_G	Energy bandgap
σ_{pc}	Photoconductivity	E_V	Valence-band energy
τ	Carrier lifetime, Scattering constant	E_i	Intrinsic energy level
τ_g	Generation lifetime	E_{max}	Maximum electric field
τ_{pc}	PC lifetime	E_{ox}	Oxide electric field
ϕ	Photon flux density	E_s	Vertical field at surface
ϕ_F	Fermi potential	E_t	Trap energy
ϕ_{ms}	Work function	e	Emission rate
ψ_{fb}	Flat-band potential	$f(E)$	Distribution function
ψ_{ox}	Oxide potential drop	f	Frequency
ψ_s	Surface potential	f_T	Unity gain frequency
ω	Angular frequency	G	Conductance
ω_d	Depletion width	G_p	Parallel conductance
A_i, B_i	Ionization coefficients	g_D	Dynamic conductance
B	Magnetic field	g_m	Transconductance
C	Capacitance	h	Planck constant
C_{CB}	Collector capacitance	I	Current
C_{EB}	Emitter capacitance	I_B	Body current
C_{HF}	HF capacitance	I_{CP}	Charge-pumping current

Symbol	Description	Symbol	Description
I_D	Drain current	R_L	Load resistance
$I_{D,sat}$	Saturation drain current	R_{SR}	Spreading resistance
I_R	Reverse current	$R_\square$	Sheet resistance
I_{ch}	Channel current	R_s	Reflection coefficient
I_{dif}	Diffusion current	RHS	Right-hand side
I_{fe}	Field-enhanced current	S	Subthreshold swing Noise, G–R velocity
I^b_{gen}	Bulk-generation current	S_I	Current noise
I_o	Light intensity	S_V	Voltage noise
K_f	Noise factor	T	Temperature
k	Boltzmann constant	t	Time
L	Diffusion length, Gate length	t_f	Fall time
L_D	Debye length	t_{ox}	Oxide thickness
L_{eff}	Effective channel length	t_r	Rise time
L_s	Critical length	t_{si}	Silicon-film thickness
LHS	Left-hand side	V	Voltage
M	Multiplication factor	V_B	Body potential
N_A	Acceptor impurity density	V_D	Drain voltage
N_C	Conduction-band states	V_{Dsat}	Saturation voltage
N_D	Donor impurity density	V_F	Forward-bias voltage
N_V	Valence-band state density	V_{FB}	Flat-band voltage
n	Electron concentration, Ideality factor	V_G	Gate voltage
n_i	Intrinsic concentration	V_H	Hall voltage
n_{ox}	Refractive index of oxide	V_R	Reverse-bias voltage
n_{si}	Refractive index of silicon	V_T	Threshold voltage
$\overline{\Delta N^2}$	Variance of carrier number	V_{Te}	Sidewall threshold
Δn	Excess concentration	V_{in}	Input voltage
p	Hole concentration	V_{out}	Output voltage
Δp	Excess-hole concentration	V_{ox}	Oxide potential drop
Q_{acc}	Accumulation charge	V_{rt}	Reach-through voltage
Q_d	Depletion charge	BV	Breakdown voltage
Q_{ox}	Oxide charge	v_{th}	Thermal velocity
Q_s	Surface charge	W	Gate width
q	Electron charge	W_d	Depletion width
R	Resistance	Y_{it}	Interface admittance
R_H	Hall coefficient	Z	Impedance

Index